STUDENT SOLUTIONS MANUAL

to accompany

Physics Fifth Edition

John D. Cutnell
Kenneth W. Johnson

Southern Illinois University at Carbondale

John Wiley & Sons, Inc.

New York • Chichester • Weinheim • Brisbane • Singapore • Toronto

Cover Photo: ©Karl Weatherly / Tony Stone Images, New York.

To order books or for customer service call 1-800-CALL-WILEY (225-5945).

ISBN 0-471-35583-6

Printed in the United States of America

10 9 8 7 6 5 4 3

Printed and bound by Courier Stoughton, Inc.

PREFACE

This volume contains the complete solutions to those problems in the text that are marked with an **SSM** icon. There are about 600 such problems, which represent 25% of the total number of problems in the text.

The solutions are worked out with great care, and all the steps are included. Most of the solutions are comprised of two parts, a REASONING part followed by a SOLUTION part. In the REASONING part we explain what motivates our procedure for solving the problem, before any algebraic or numerical work is done. During the SOLUTION part, numerical calculations are performed, and the answer to the problem is obtained.

For ease of use and consistency, the notation, terminology, laws, theorems, and definitions employed in this manual are the same as those in the text.

Some of the solutions are preceded with a **WWW** icon. These solutions are also available on the World Wide Web at http://www.wiley.com/college/cutnell.

CONTENTS

CHAPTER 1 INTRODUCTION AND MATHEMATICAL CONCEPTS

PROBLEMS

---

1. ***REASONING AND SOLUTION*** We use the fact that 0.200 g = 1 carat and that, under the conditions stated, 1000 g has a weight of 2.205 lb to construct the two conversion factors: (0.200 g)/(1 carat) = 1 and (2.205 lb)/(1000 g) = 1. Then,

$$(3106 \text{ carats}) \left(\frac{0.200 \text{ g}}{1 \text{ carat}} \right) \left(\frac{2.205 \text{ lb}}{1000 \text{ g}} \right) = \boxed{1.37 \text{ lb}}$$

---

5. ***REASONING*** When converting between units, we write down the units explicitly in the calculations and treat them like any algebraic quantity. We construct the appropriate conversion factor (equal to unity) so that the final result has the desired units.

SOLUTION
a. Since 1.0×10^3 grams = 1.0 kilogram, it follows that the appropriate conversion factor is $(1.0 \times 10^3 \text{ g})/(1.0 \text{ kg}) = 1$. Therefore,

$$\left(5 \times 10^{-6} \text{ kg}\right) \left(\frac{1.0 \times 10^3 \text{ g}}{1.0 \text{ kg}} \right) = \boxed{5 \times 10^{-3} \text{ g}}$$

b. Since 1.0×10^3 milligrams = 1.0 gram,

$$\left(5 \times 10^{-3} \text{ g}\right) \left(\frac{1.0 \times 10^3 \text{ mg}}{1.0 \text{ g}} \right) = \boxed{5 \text{ mg}}$$

c. Since 1.0×10^6 micrograms = 1.0 gram,

$$\left(5 \times 10^{-3} \text{ g}\right) \left(\frac{1.0 \times 10^6 \text{ } \mu\text{g}}{1.0 \text{ g}} \right) = \boxed{5 \times 10^3 \text{ } \mu\text{g}}$$

---

9. ***REASONING*** The volume of water at a depth d beneath the rectangle is equal to the area of the rectangle multiplied by d. The area of the rectangle = (1.20 nautical miles) × (2.60 nautical miles) = 3.12 (nautical miles)^2. Since 6076 ft = 1 nautical mile and 0.3048 m = 1 ft, the conversion factor between nautical miles and meters is

$$\left(\frac{6076 \text{ ft}}{1 \text{ nautical mile}}\right)\left(\frac{0.3048 \text{ m}}{1 \text{ ft}}\right) = \frac{1.852 \times 10^3 \text{ m}}{1 \text{ nautical mile}}$$

SOLUTION The area of the rectangle of water in m² is, therefore,

$$\left[3.12 \text{ (nautical miles)}^2\right]\left(\frac{1.852 \times 10^3 \text{ m}}{1 \text{ nautical mile}}\right)^2 = 1.07 \times 10^7 \text{ m}^2$$

Since 1 fathom = 6 ft, and 1 ft = 0.3048 m, the depth d in meters is

$$(16.0 \text{ fathoms})\left(\frac{6 \text{ ft}}{1 \text{ fathom}}\right)\left(\frac{0.3048 \text{ m}}{1 \text{ ft}}\right) = 2.93 \times 10^1 \text{ m}$$

The volume of water beneath the rectangle is

$$(1.07 \times 10^7 \text{ m}^2)(2.93 \times 10^1 \text{ m}) = \boxed{3.13 \times 10^8 \text{ m}^3}$$

---

11. $\boxed{\text{WWW}}$ **REASONING**   The shortest distance between the two towns is along the line that joins them. This distance, h, is the hypotenuse of a right triangle whose other sides are $h_o = 35.0$ km and $h_a = 72.0$ km, as shown in the figure below.

SOLUTION The angle θ is given by $\tan \theta = h_o / h_a$
so that

$$\theta = \tan^{-1}\left(\frac{35.0 \text{ km}}{72.0 \text{ km}}\right) = \boxed{25.9° \text{ S of W}}$$

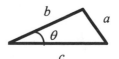

We can then use the Pythagorean theorem to find h.

$$h = \sqrt{h_o^2 + h_a^2} = \sqrt{(35.0 \text{ km})^2 + (72.0 \text{ km})^2} = \boxed{80.1 \text{ km}}$$

---

19.   **REASONING**   The law of cosines is
$$a^2 = b^2 + c^2 - 2bc \cos\theta$$

where a is the side opposite angle θ, and b and c are the other two sides.
Solving for θ, we have

$$\theta = \cos^{-1}\left[(b^2 + c^2 - a^2)/(2bc)\right]$$

SOLUTION For $a = 190$ cm, $b = 95$ cm, $c = 150$ cm,

$$\theta = \cos^{-1}\left[\frac{(95 \text{ cm})^2 + (150 \text{ cm})^2 - (190 \text{ cm})^2}{2(95 \text{ cm})(150 \text{ cm})}\right] = 99°$$

Thus, the angle opposite the side of length 190 cm is $\boxed{99 \text{ degrees}}$.

Similarly, when $a = 95$ cm, $b = 190$ cm, $c = 150$ cm, we find that the angle opposite the side of length 95 cm is $\boxed{\theta = 3.0 \times 10^1 \text{ degrees}}$.

Finally, when $a = 150$ cm, $b = 190$ cm, $c = 95$ cm, we find that the angle opposite the side of length 150 cm is $\boxed{\theta = 51 \text{ degrees}}$.

---

23. $\boxed{\text{WWW}}$ **REASONING** When a vector is multiplied by -1, the magnitude of the vector remains the same, but the direction is reversed. Vector subtraction is carried out in the same manner as vector addition except that one of the vectors has been multiplied by -1.

SOLUTION Since both vectors point north, they are colinear. Therefore their magnitudes may be added by the rules of ordinary algebra.
a. Taking north as the positive direction, we have

$$\mathbf{A} - \mathbf{B} = \mathbf{A} + (-\mathbf{B}) = +2.43 \text{ km} + (-7.74 \text{ km}) = -5.31 \text{ km}$$

The minus sign in the answer for $\mathbf{A} - \mathbf{B}$ indicates that the direction is south so that

$$\boxed{\mathbf{A} - \mathbf{B} = 5.31 \text{ km, south}}$$

b. Similarly,

$$\mathbf{B} - \mathbf{A} = \mathbf{B} + (-\mathbf{A}) = +7.74 \text{ km} + (-2.43 \text{ km}) = +5.31 \text{ km}$$

The plus sign in the answer for $\mathbf{B} - \mathbf{A}$ indicates that the direction is north, so that

$$\boxed{\mathbf{B} - \mathbf{A} = 5.31 \text{ km, north}}$$

---

25. **REASONING** Using the component method, we find the components of the resultant $\mathbf{R}$ that are due east and due north. The magnitude and direction of the resultant $\mathbf{R}$ can be determined from its components, the Pythagorean theorem, and the tangent function.

SOLUTION The first four rows of the table below give the components of the vectors $\mathbf{A}$, $\mathbf{B}$, $\mathbf{C}$, and $\mathbf{D}$. Note that east and north have been taken as the positive directions; hence vectors pointing due west and due south will appear with a negative sign.

Vector	East/West Component	North/South Component
A	+ 2.00 km	0
B	0	+ 3.75 km
C	– 2.50 km	0
D	0	–3.00 km
R = A + B + C + D	– 0.50 km	+ 0.75 km

The fifth row in the table gives the components of **R**. The magnitude of **R** is given by the Pythagorean theorem as

$$R = \sqrt{(-0.50\ \text{km})^2 + (+0.75\ \text{km})^2} = \boxed{0.90\ \text{km}}$$

The angle θ that **R** makes with the direction due west is

$$\theta = \tan^{-1}\left(\frac{0.75\ \text{km}}{0.50\ \text{km}}\right) = \boxed{56°\ \text{north of west}}$$

27. **WWW** **REASONING AND SOLUTION** The single force needed to produce the same effect is equal to the resultant of the forces provided by the two ropes. The figure below shows the force vectors drawn to scale and arranged tail to head. The magnitude and direction of the resultant can be found by direct measurement using the scale factor shown in the figure.

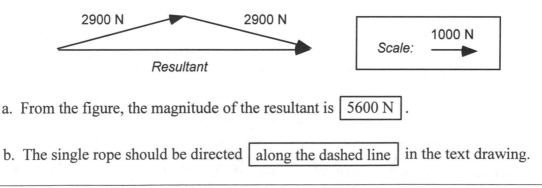

a. From the figure, the magnitude of the resultant is $\boxed{5600\ \text{N}}$.

b. The single rope should be directed $\boxed{\text{along the dashed line}}$ in the text drawing.

31. **REASONING AND SOLUTION** In order to determine which vector has the largest x and y components, we calculate the magnitude of the x and y components explicitly and compare them. In the calculations, the symbol u denotes the units of the vectors.

$A_x = (100.0 \text{ u}) \cos 90.0° = 0.00 \text{ u}$ $\qquad A_y = (100.0 \text{ u}) \sin 90.0° = 1.00 \times 10^2 \text{ u}$

$B_x = (200.0 \text{ u}) \cos 60.0° = 1.00 \times 10^2 \text{ u}$ $\qquad B_y = (200.0 \text{ u}) \sin 60.0° = 173 \text{ u}$

$C_x = (150.0 \text{ u}) \cos 0.00° = 150.0 \text{ u}$ $\qquad C_y = (150.0 \text{ u}) \sin 0.00° = 0.00 \text{ u}$

a. $\boxed{\textbf{C} \text{ has the largest } x \text{ component}}$.

b. $\boxed{\textbf{B} \text{ has the largest } y \text{ component}}$.

---

35. **REASONING**   The x and y components of **r** are mutually perpendicular; therefore, the magnitude of **r** can be found using the Pythagorean theorem. The direction of **r** can be found using the definition of the tangent function.

SOLUTION According to the Pythagorean theorem, we have

$$r = \sqrt{x^2 + y^2} = \sqrt{(-125 \text{ m})^2 + (-184 \text{ m})^2} = \boxed{222 \text{ m}}$$

The angle θ is

$$\theta = \tan^{-1}\left(\frac{184 \text{ m}}{125 \text{ m}}\right) = \boxed{55.8°}$$

---

39. $\boxed{\text{WWW}}$ **REASONING AND SOLUTION**   The force **F** can be first resolved into two components; the z component F_z and the projection onto the x-y plane, F_p as shown in the following figure on the left. According to that figure,

$$F_p = F \sin 54.0° = (475 \text{ N}) \sin 54.0° = 384 \text{ N}.$$

The projection onto the x-y plane, F_p, can then be resolved into x and y components.

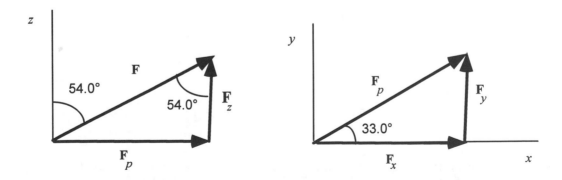

a. From the figure on the right,

$$F_x = F_p \cos 33.0° = (384 \text{ N}) \cos 33.0° = \boxed{322 \text{ N}}$$

b. Also from the figure on the right,

$$F_y = F_p \sin 33.0° = (384 \text{ N}) \sin 33.0° = \boxed{209 \text{ N}}$$

c. From the figure on the left,

$$F_z = F \cos 54.0° = (475 \text{ N}) \cos 54.0° = \boxed{279 \text{ N}}$$

41. **REASONING**   The individual displacements of the golf ball, **A**, **B**, and **C** are shown in the figure. Their resultant, **R**, is the displacement that would have been needed to "hole the ball" on the very first putt. We will use the component method to find **R**.

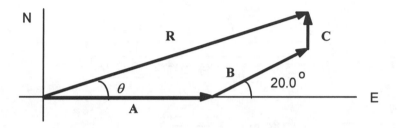

SOLUTION   The components of each displacement vector are given in the table below.

Vector	x Components	y Components
A	(5.0 m) cos 0° = 5.0 m	(5.0 m) sin 0° = 0
B	(2.1 m) cos 20.0° = 2.0 m	(2.1 m) sin 20.0° = 0.72 m
C	(0.50 m) cos 90.0° = 0	(0.50 m) sin 90.0° = 0.50 m
R = A + B + C	7.0 m	1.22 m

The resultant vector **R** has magnitude

$$R = \sqrt{(7.0 \text{ m})^2 + (1.22 \text{ m})^2} = \boxed{7.1 \text{ m}}$$

and the angle θ is

$$\theta = \tan^{-1}\left(\frac{1.22 \text{ m}}{7.0 \text{ m}}\right) = 9.9°$$

Thus, the required direction is $\boxed{9.9° \text{ north of east}}$.

47. ***REASONING AND SOLUTION***   We take due north to be the direction of the $+y$ axis. Vectors **A** and **B** are the components of the resultant, **C**. The angle that **C** makes with the x axis is then $\theta = \tan^{-1}(B/A)$.   The symbol u denotes the units of the vectors.

    a. Solving for B gives

$$B = A \tan \theta = (6.00 \text{ u}) \tan 60.0° = \boxed{10.4 \text{ u}}$$

    b. The magnitude of **C** is

$$C = \sqrt{A^2 + B^2} = \sqrt{(6.00 \text{ u})^2 + (10.4 \text{ u})^2} = \boxed{12.0 \text{ u}}$$

---

49. ***REASONING***   Since the finish line is coincident with the starting line, the net displacement of the sailboat is zero. Hence the sum of the components of the displacement vectors of the individual legs must be zero. In the drawing in the text, the directions to the right and upward are taken as positive.

    SOLUTION In the horizontal direction $R_h = A_h + B_h + C_h + D_h = 0$

$$R_h = (3.20 \text{ km}) \cos 40.0° - (5.10 \text{ km}) \cos 35.0° - (4.80 \text{ km}) \cos 23.0° + D \cos \theta = 0$$

$$D \cos \theta = 6.14 \text{ km.} \qquad (1)$$

In the vertical direction $R_v = A_v + B_v + C_v + D_v = 0$.

$$R_v = (3.20 \text{ km}) \sin 40.0° + (5.10 \text{ km}) \sin 35.0° - (4.80 \text{ km}) \sin 23.0° - D \sin \theta = 0.$$

$$D \sin \theta = 3.11 \text{ km} \qquad (2)$$

Dividing (2) by (1) gives

$$\tan \theta = (3.11 \text{ km})/(6.14 \text{ km}) \quad \text{or} \quad \theta = \boxed{26.9°}$$

Solving (1) gives

$$D = (6.14 \text{ km})/\cos 26.9° = \boxed{6.88 \text{ km}}$$

---

53. **REASONING** The ostrich's velocity vector **v** and the desired components are shown in the figure at the right. The components of the velocity in the directions due west and due north are v_W and v_N, respectively. The sine and cosine functions can be used to find the components.

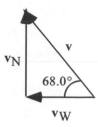

SOLUTION
a. According to the definition of the sine function, we have for the vectors in the figure

$$\sin\,\theta = \frac{v_N}{v} \quad\text{or}\quad v_N = v\sin\,\theta = (17.0\ \text{m/s})\sin 68° = \boxed{15.8\ \text{m/s}}$$

b. Similarly,

$$\cos\,\theta = \frac{v_W}{v} \quad\text{or}\quad v_W = v\cos\,\theta = (17.0\ \text{m/s})\cos 68.0° = \boxed{6.37\ \text{m/s}}$$

---

57. **REASONING AND SOLUTION** A single rope must supply the resultant of the two forces. Since the forces are perpendicular, the magnitude of the resultant can be found from the Pythagorean theorem.

a. Applying the Pythagorean theorem,

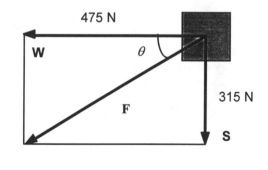

$$F = \sqrt{(475\ \text{N})^2 + (315\ \text{N})^2} = \boxed{5.70\times 10^2\ \text{N}}$$

b. The angle θ that the resultant makes with the westward direction is

$$\theta = \tan^{-1}\left(\frac{315\ \text{N}}{475\ \text{N}}\right) = 33.6°$$

Thus, the rope must make an angle of $\boxed{33.6°\ \text{south of west}}$.

CHAPTER 2 | *KINEMATICS IN ONE DIMENSION*

1. ***REASONING AND SOLUTION***
   a. The total distance traveled is found by adding the distances traveled during each segment of the trip:

   $$6.9 \text{ km} + 1.8 \text{ km} + 3.7 \text{ km} = \boxed{12.4 \text{ km}}$$

   b. All three segments of the trip lie along the east-west line. Taking east as the positive direction, the individual displacements can then be added to yield the resultant displacement.

   $$6.9 \text{ km} + (-1.8 \text{ km}) + 3.7 \text{ km} = +8.8 \text{ km}$$

   The displacement is positive, indicating that it points due east. Therefore,

   $$\text{Displacement of the whale} = \boxed{8.8 \text{ km, due east}}$$

3. ***REASONING AND SOLUTION*** The moving plane can be seen through the window of the stationary plane for the time that it takes the moving plane to travel its own length. To find the time, we write Equation 2.1 as

   $$\text{Elapsed time} = \frac{\text{Distance}}{\text{Average speed}} = \frac{36 \text{ m}}{45 \text{ m/s}} = \boxed{0.80 \text{ s}}$$

9. $\boxed{\text{WWW}}$ ***REASONING*** Since the woman runs for a known distance at a known constant speed, we can find the time it takes for her to reach the water from Equation 2.1. We can then use Equation 2.1 to determine the total distance traveled by the dog in this time.

   SOLUTION The time required for the woman to reach the water is

   $$\text{Elapsed time} = \frac{d_{\text{woman}}}{v_{\text{woman}}} = \left(\frac{4.0 \text{ km}}{2.5 \text{ m/s}} \right) \left(\frac{1000 \text{ m}}{1.0 \text{ km}} \right) = 1600 \text{ s}$$

   In 1600 s, the dog travels a total distance of

   $$d_{\text{dog}} = v_{\text{dog}} t = (4.5 \text{ m/s})(1600 \text{ s}) = \boxed{7.2 \times 10^3 \text{ m}}$$

13.   **WWW**   *REASONING AND SOLUTION*
a. The magnitude of the acceleration can be found from Equation 2.4 ($v = v_0 + at$) as

$$a = \frac{v - v_0}{t} = \frac{3.0 \text{ m/s} - 0 \text{ m/s}}{2.0 \text{ s}} = \boxed{1.5 \text{ m/s}^2}$$

b. Similarly the magnitude of the acceleration of the car is

$$a = \frac{v - v_0}{t} = \frac{41.0 \text{ m/s} - 38.0 \text{ m/s}}{2.0 \text{ s}} = \boxed{1.5 \text{ m/s}^2}$$

c. Assuming that the acceleration is constant, the displacement covered by the car can be found from Equation 2.9 ($v^2 = v_0^2 + 2ax$):

$$x = \frac{v^2 - v_0^2}{2a} = \frac{(41.0 \text{ m/s})^2 - (38.0 \text{ m/s})^2}{2(1.5 \text{ m/s}^2)} = 79 \text{ m}$$

Similarly, the displacement traveled by the jogger is

$$x = \frac{v^2 - v_0^2}{2a} = \frac{(3.0 \text{ m/s})^2 - (0 \text{ m/s})^2}{2(1.5 \text{ m/s}^2)} = 3.0 \text{ m}$$

Therefore, the car travels 79 m – 3.0 m = $\boxed{76 \text{ m}}$ further than the jogger.

---

15.   *REASONING AND SOLUTION*   The initial velocity of the runner can be found by solving Equation 2.4 ($v = v_0 + at$) for v_0. Taking west as the positive direction, we have

$$v_0 = v - at = (+5.36 \text{ m/s}) - (+0.640 \text{ m/s}^2)(3.00 \text{ s}) = +3.44 \text{ m/s}$$

Therefore, the initial velocity of the runner is $\boxed{3.44 \text{ m/s, due west}}$.

---

19. *REASONING*   The average acceleration is defined by Equation 2.4 as the change in velocity divided by the elapsed time. We can find the elapsed time from this relation because the acceleration and the change in velocity are given.

SOLUTION
a. The time Δt that it takes for the VW Beetle to change its velocity by an amount $\Delta v = v - v_0$ is (and noting that 0.4470 m/s = 1 mi/h)

$$\Delta t = \frac{v - v_0}{a} = \frac{(60.0 \text{ mi/h})\left(\dfrac{0.4470 \text{ m/s}}{1 \text{ mi/h}}\right) - 0 \text{ m/s}}{2.35 \text{ m/s}^2} = \boxed{11.4 \text{ s}}$$

b. From Equation 2.4, the acceleration (in m/s^2) of the dragster is

$$a = \frac{v - v_0}{t - t_0} = \frac{(60.0 \text{ mi/h})\left(\dfrac{0.4470 \text{ m/s}}{1 \text{ mi/h}}\right) - 0 \text{ m/s}}{0.600 \text{ s} - 0 \text{ s}} = \boxed{44.7 \text{ m/s}^2}$$

---

23. ***REASONING AND SOLUTION***   The average acceleration of the plane can be found by solving Equation 2.9 $\left(v^2 = v_0^2 + 2ax\right)$ for a. Taking the direction of motion as positive, we have

$$a = \frac{v^2 - v_0^2}{2x} = \frac{(+6.1 \text{ m/s})^2 - (+69 \text{ m/s})^2}{2(+750 \text{ m})} = \boxed{-3.1 \text{ m/s}^2}$$

The minus sign indicates that the direction of the acceleration is opposite to the direction of motion, and the plane is slowing down.

---

27. ⬛WWW⬛ ***REASONING***   Since the belt is moving with constant velocity, the displacement ($x_0 = 0$ m) covered by the belt in a time t_{belt} is giving by Equation 2.2 (with x_0 assumed to be zero) as

$$x = v_{\text{belt}} t_{\text{belt}} \tag{1}$$

Since Clifford moves with constant acceleration, the displacement covered by Clifford in a time t_{Cliff} is, from Equation 2.8,

$$x = v_0 t_{\text{Cliff}} + \tfrac{1}{2} a t_{\text{Cliff}}^2 = \tfrac{1}{2} a t_{\text{Cliff}}^2 \tag{2}$$

The speed v_{belt} with which the belt of the ramp is moving can be found by eliminating x between Equations (1) and (2).

SOLUTION   Equating the right hand sides of Equations (1) and (2), and noting that $t_{\text{Cliff}} = \frac{1}{4} t_{\text{belt}}$, we have

$$v_{\text{belt}} \, t_{\text{belt}} = \frac{1}{2} a \left(\frac{1}{4} t_{\text{belt}} \right)^2$$

$$v_{\text{belt}} = \frac{1}{32} a \, t_{\text{belt}} = \frac{1}{32} (0.37 \text{ m/s}^2)(64 \text{ s}) = \boxed{0.74 \text{ m/s}}$$

---

35. **REASONING**   As the train passes through the crossing, its motion is described by Equations 2.4 ($v = v_0 + at$) and 2.7 $\left[x = \frac{1}{2}(v + v_0)t \right]$, which can be rearranged to give

$$v - v_0 = at \quad \text{and} \quad v + v_0 = \frac{2x}{t}$$

These can be solved simultaneously to obtain the speed v when the train reaches the end of the crossing. Once v is known, Equation 2.4 can be used to find the time required for the train to reach a speed of 32 m/s.

SOLUTION   Adding the above equations and solving for v, we obtain

$$v = \frac{1}{2} \left(at + \frac{2x}{t} \right) = \frac{1}{2} \left[(1.6 \text{ m/s}^2)(2.4 \text{ s}) + \frac{2(20.0 \text{ m})}{2.4 \text{ s}} \right] = 1.0 \times 10^1 \text{ m/s}$$

The motion from the end of the crossing until the locomotive reaches a speed of 32 m/s requires a time

$$t = \frac{v - v_0}{a} = \frac{32 \text{ m/s} - 1.0 \times 10^1 \text{ m/s}}{1.6 \text{ m/s}^2} = \boxed{14 \text{ s}}$$

---

37. **REASONING AND SOLUTION**   When air resistance is neglected, free fall conditions are applicable. The final speed can be found from Equation 2.9;

$$v^2 = v_0^2 + 2ay$$

where v_0 is zero since the stunt man falls from rest. If the origin is chosen at the top of the hotel and the upward direction is positive, then the displacement is $y = -99.4$ m. Solving for v, we have

$$v = -\sqrt{2ay} = -\sqrt{2(-9.80 \text{ m/s}^2)(-99.4 \text{ m})} = -44.1 \text{ m/s}$$

The speed at impact is the magnitude of this result or $\boxed{44.1 \text{ m/s}}$.

---

41. **REASONING AND SOLUTION** Since the balloon is released from rest, its initial velocity is zero. The time required to fall through a vertical displacement y can be found from Equation 2.8 $\left(y = v_0 t + \frac{1}{2}at^2\right)$ with $v_0 = 0$ m/s. Assuming upward to be the positive direction, we find

$$t = \sqrt{\frac{2y}{a}} = \sqrt{\frac{2(-6.0 \text{ m})}{-9.80 \text{ m/s}^2}} = \boxed{1.1 \text{ s}}$$

45. **REASONING AND SOLUTION** Equation 2.8 can be used to determine the displacement that the ball covers as it falls halfway to the ground. Since the ball falls from rest, its initial velocity is zero. Taking down to be the negative direction, we have

$$y = v_0 t + \frac{1}{2}at^2 = \frac{1}{2}at^2 = \frac{1}{2}(-9.80 \text{ m/s}^2)(1.2 \text{ s}) = -7.1 \text{ m}$$

In falling all the way to the ground, the ball has a displacement of $y = -14.2$ m. Solving Equation 2.8 with this displacement then yields the time

$$t = \sqrt{\frac{2y}{a}} = \sqrt{\frac{2(-14.2 \text{ m})}{-9.80 \text{ m/s}^2}} = \boxed{1.7 \text{ s}}$$

51. **REASONING** Once the man sees the block, the man must get out of the way in the time it takes for the block to fall through an additional 12.0 m. The velocity of the block at the instant that the man looks up can be determined from Equation 2.9. Once the velocity is known at that instant, Equation 2.8 can be used to find the time required for the block to fall through the additional distance.

SOLUTION When the man first notices the block, it is 14.0 m above the ground and its displacement from the starting point is $y = 14.0$ m $- 53.0$ m. Its velocity is given by Equation 2.9 $\left(v^2 = v_0^2 + 2ay\right)$. Since the block is moving down, its velocity has a negative value,

$$v = -\sqrt{v_0 + 2ay} = -\sqrt{(0 \text{ m/s})^2 + 2(-9.80 \text{ m/s}^2)(14.0 \text{ m} - 53.0 \text{ m})} = -27.7 \text{ m/s}$$

The block then falls the additional 12.0 m to the level of the man's head in a time t which satisfies Equation 2.8:

$$y = v_0 t + \frac{1}{2}at^2$$

where $y = -12.0$ m and $v_0 = -27.7$ m/s. Thus, t is the solution to the quadratic equation

$$4.90t^2 + 27.7t - 12.0 = 0$$

where the units have been suppressed for brevity. From the quadratic formula, we obtain

$$t = \frac{-27.7 \pm \sqrt{(27.7)^2 - 4(4.90)(-12.0)}}{2(4.90)} = 0.40 \text{ s} \quad \text{or} \quad -6.1 \text{ s}$$

The negative solution can be rejected as nonphysical, and the time it takes for the block to reach the level of the man is $\boxed{0.40 \text{ s}}$.

---

53. $\boxed{\text{WWW}}$ **REASONING AND SOLUTION**   The stone requires a time, t_1, to reach the bottom of the hole, a distance y below the ground. Assuming downward to be the positive direction, the variables are related by Equation 2.8 with $v_0 = 0$ m/s:

$$y = \frac{1}{2} a t_1^2 \tag{1}$$

The sound travels the distance y from the bottom to the top of the hole in a time t_2. Since the sound does not experience any acceleration, the variables y and t_2 are related by Equation 2.8 with $a = 0$ m/s^2 and v_{sound} denoting the speed of sound:

$$y = v_{sound} \, t_2 \tag{2}$$

Equating the right hand sides of Equations (1) and (2) and using the fact that the total elapsed time is $t = t_1 + t_2$, we have

$$\frac{1}{2} a t_1^2 = v_{sound} t_2 \quad \text{or} \quad \frac{1}{2} a t_1^2 = v_{sound} (t - t_1)$$

Rearranging gives

$$\frac{1}{2} a t_1^2 + v_{sound} t_1 - v_{sound} t = 0$$

Substituting values and suppressing units for brevity, we obtain the following quadratic equation for t_1:

$$4.90 t_1^2 + 343 t_1 - 514 = 0$$

From the quadratic formula, we obtain

$$t_1 = \frac{-343 \pm \sqrt{(343)^2 - 4(4.90)(-514)}}{2(4.90)} = 1.47 \text{ s} \quad \text{or} \quad -71.5 \text{ s}$$

The negative time corresponds to a nonphysical result and is rejected. The depth of the hole is then found using Equation 2.8 with the value of t_1 obtained above:

$$y = v_0 t_1 + \tfrac{1}{2} a t_1^2 = (0 \text{ m/s})(1.47 \text{ s}) + \tfrac{1}{2}(9.80 \text{ m/s}^2)(1.47 \text{ s})^2 = \boxed{10.6 \text{ m}}$$

---

57. **REASONING**   In order to construct the graph, the time for each segment of the trip must be determined.

SOLUTION   From the definition of average velocity, the time is $\Delta t = \Delta x / \overline{v}$. Therefore,

$$\Delta t_1 = \left(\frac{10.0 \text{ km}}{15.0 \text{ km/h}} \right) \left(\frac{60 \text{ min}}{1.0 \text{ h}} \right) = 40 \text{ min}$$

$$\Delta t_2 = \left(\frac{15.0 \text{ km}}{10.0 \text{ km/h}} \right) \left(\frac{60 \text{ min}}{1.0 \text{ h}} \right) = 90 \text{ min}$$

Thus, the second time interval ends $40 \text{ min} + 90 \text{ min} = 130 \text{ min}$ after the trip begins.

$$\Delta t_3 = \left(\frac{15.0 \text{ km}}{5.0 \text{ km/h}} \right) \left(\frac{60 \text{ min}}{1.0 \text{ h}} \right) = 180 \text{ min}$$

and the third time interval ends $130 \text{ min} + 180 \text{ min} = 310 \text{ min}$ after the trip begins.

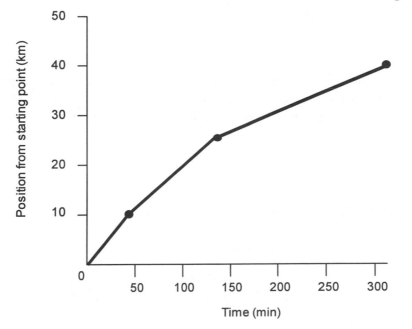

Note that the slope of each segment of the graph gives the average velocity during that interval.

---

61.   **REASONING**   The average acceleration is given by Equation 2.4: $\bar{a} = (v_C - v_A)/\Delta t$. The velocities v_A and v_C can be found from the slopes of the position-time graph for segments A and C.

SOLUTION   The average velocities in the segments A and C are

$$v_A = \frac{24 \text{ km} - 0 \text{ km}}{1.0 \text{ h} - 0 \text{ h}} = 24 \text{ km/h}$$

$$v_C = \frac{27 \text{ km} - 33 \text{ km}}{3.5 \text{ h} - 2.2 \text{ h}} = -5 \text{ km/h}$$

From the definition of average acceleration,

$$\bar{a} = \frac{\Delta v}{\Delta t} = \frac{v_C - v_A}{\Delta t} = \frac{(-5 \text{ km/h}) - (24 \text{ km/h})}{3.5 \text{ h} - 0 \text{ h}} = \boxed{-8.3 \text{ km/h}^2}$$

---

65.   **REASONING AND SOLUTION**   The speed of the penny as it hits the ground can be determined from Equation 2.9: $v^2 = v_0^2 + 2ay$. Since the penny is dropped from rest, $v_0 = 0$ m/s. Solving for v, with downward taken as the positive direction, we have

$$v = \sqrt{2(9.80 \text{ m/s}^2)(427 \text{ m})} = \boxed{91.5 \text{ m/s}}$$

---

67.   **REASONING**   Since the velocity and acceleration of the motorcycle point in the same direction, their numerical values will have the same algebraic sign. For convenience, we will choose them to be positive. The velocity, acceleration, and the time are related by Equation 2.4: $v = v_0 + at$.

SOLUTION

a.   Solving Equation 2.4 for t we have

$$t = \frac{v - v_0}{a} = \frac{(+31 \text{ m/s}) - (+21 \text{ m/s})}{+2.5 \text{ m/s}^2} = \boxed{4.0 \text{ s}}$$

b. Similarly,

$$t = \frac{v - v_0}{a} = \frac{(+61 \text{ m/s}) - (+51 \text{ m/s})}{+2.5 \text{ m/s}^2} = \boxed{4.0 \text{ s}}$$

73.   ***REASONING AND SOLUTION***
a. The total displacement traveled by the bicyclist for the entire trip is equal to the sum of the displacements traveled during each part of the trip. The displacement traveled during each part of the trip is given by Equation 2.2: $\Delta x = \overline{v}\Delta t$. Therefore,

$$\Delta x_1 = (7.2 \text{ m/s})(22 \text{ min})\left(\frac{60 \text{ s}}{1 \text{ min}}\right) = 9500 \text{ m}$$

$$\Delta x_2 = (5.1 \text{ m/s})(36 \text{ min})\left(\frac{60 \text{ s}}{1 \text{ min}}\right) = 11\,000 \text{ m}$$

$$\Delta x_3 = (13 \text{ m/s})(8.0 \text{ min})\left(\frac{60 \text{ s}}{1 \text{ min}}\right) = 6200 \text{ m}$$

The total displacement traveled by the bicyclist during the entire trip is then

$$\Delta x = 9500 \text{ m} + 11\,000 \text{ m} + 6200 \text{ m} = \boxed{2.67 \times 10^4 \text{ m}}$$

b. The average velocity can be found from Equation 2.2.

$$\overline{v} = \frac{\Delta x}{\Delta t} = \frac{2.67 \times 10^4 \text{ m}}{(22 \text{ min} + 36 \text{ min} + 8.0 \text{ min})}\left(\frac{1 \text{ min}}{60 \text{ s}}\right) = \boxed{6.74 \text{ m/s, due north}}$$

---

75.   ***REASONING AND SOLUTION***   The stone will reach the water (and hence the log) after falling for a time t, where t can be determined from Equation 2.8: $y = v_0 t + \frac{1}{2}at^2$. Since the stone is dropped from rest, $v_0 = 0$ m/s. Assuming that downward is positive and solving for t, we have

$$t = \sqrt{\frac{2y}{a}} = \sqrt{\frac{2(75 \text{ m})}{9.80 \text{ m/s}^2}} = 3.9 \text{ s}$$

During that time, the displacement of the log can be found from Equation 2.8. Since the log moves with constant velocity, $a = 0$ m/s^2, and v_0 is equal to the velocity of the log.

$$x = v_0 t = (5.0 \text{ m/s})(3.9 \text{ s}) = 2.0 \times 10^1 \text{ m}$$

Therefore, the horizontal distance between the log and the bridge when the stone is released is $\boxed{2.0 \times 10^1 \text{ m}}$.

79. **REASONING** Since the car is moving with a constant velocity, the displacement of the car in a time t can be found from Equation 2.8 with $a = 0$ m/s^2 and v_0 equal to the velocity of the car: $x_{car} = v_{car}t$. Since the train starts from rest with a constant acceleration, the displacement of the train in a time t is given by Equation 2.8 with $v_0 = 0$ m/s:

$$x_{train} = \frac{1}{2}a_{train}t^2$$

At a time t_1, when the car just reaches the front of the train, $x_{car} = L_{train} + x_{train}$, where L_{train} is the length of the train. Thus, at time t_1,

$$v_{car}t_1 = L_{train} + \frac{1}{2}a_{train}t_1^2 \tag{1}$$

At a time t_2, when the car is again at the rear of the train, $x_{car} = x_{train}$. Thus, at time t_2

$$v_{car}t_2 = \frac{1}{2}a_{train}t_2^2 \tag{2}$$

Equations (1) and (2) can be solved simultaneously for the speed of the car v_{car} and the acceleration of the train a_{train}.

SOLUTION
a. Solving Equation (2) for a_{train} we have

$$a_{train} = \frac{2v_{car}}{t_2} \tag{3}$$

Substituting this expression for a_{train} into Equation (1) and solving for v_{car}, we have

$$v_{car} = \frac{L_{train}}{t_1\left(1 - \dfrac{t_1}{t_2}\right)} = \frac{92 \text{ m}}{(14 \text{ s})\left(1 - \dfrac{14 \text{ s}}{28 \text{ s}}\right)} = \boxed{13 \text{ m/s}}$$

b. Direct substitution into Equation (3) gives the acceleration of the train:

$$a_{train} = \frac{2v_{car}}{t_2} = \frac{2\,(13 \text{ m/s})}{28 \text{ s}} = \boxed{0.93 \text{ m/s}^2}$$

CHAPTER 3 | KINEMATICS IN TWO DIMENSIONS

PROBLEMS

---

1. **REASONING AND SOLUTION** The horizontal and vertical components of the plane's velocity are related to the speed of the plane by the Pythagorean theorem: $v^2 = v_h^2 + v_v^2$. Solving for v_h we have

$$v_h = \sqrt{v^2 - v_v^2} = \sqrt{(245 \text{ m/s})^2 - (40.6 \text{ m/s})^2} = \boxed{242 \text{ m/s}}$$

---

5. **REASONING** The displacement of the elephant seal has two components; 460 m due east and 750 m downward. These components are mutually perpendicular; hence, the Pythagorean theorem can be used to determine their resultant.

SOLUTION From the Pythagorean theorem,

$$R^2 = (460 \text{ m})^2 + (750 \text{ m})^2$$

Therefore,

$$R = \sqrt{(460 \text{ m})^2 + (750 \text{ m})^2} = \boxed{8.8 \times 10^2 \text{ m}}$$

---

9. $\boxed{\text{WWW}}$ **REASONING AND SOLUTION** The escalator is the hypotenuse of a right triangle formed by the lower floor and the vertical distance between floors, as shown in the figure. The angle θ at which the escalator is inclined above the horizontal is related to the length L of the escalator and the vertical distance between the floors by the sine function:

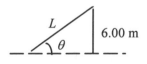

$$\sin\theta = \frac{6.00 \text{ m}}{L} \tag{1}$$

The length L of the escalator can be found from the right triangle formed by the components of the shopper's displacement up the escalator and to the right of the escalator.

From the Pythagorean theorem, we have

$$L^2 + (9.00 \text{ m})^2 = (16.0 \text{ m})^2$$

so that

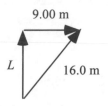

$$L = \sqrt{(16.0 \text{ m})^2 - (9.00 \text{ m})^2} = 13.2 \text{ m}$$

From Equation (1) we have

$$\theta = \sin^{-1}\left(\frac{6.00 \text{ m}}{13.2 \text{ m}}\right) = \boxed{27.0°}$$

---

15. ***REASONING AND SOLUTION***   As shown in Example 2, the time required for the package to hit the ground is given by $t = \sqrt{2y/a_y}$ and is independent of the plane's horizontal velocity. Thus, the time needed for the package to hit the ground is still $\boxed{14.6 \text{ s}}$.

---

17. $\boxed{\text{WWW}}$ ***REASONING***   Once the diver is airborne, he moves in the x direction with constant velocity while his motion in the y direction is accelerated (at the acceleration due to gravity). Therefore, the magnitude of the x component of his velocity remains constant at 1.20 m/s for all times t. The magnitude of the y component of the diver's velocity after he has fallen through a vertical displacement y can be determined from Equation 3.6b: $v_y^2 = v_{0y}^2 + 2a_y y$. Since the diver runs off the platform horizontally, $v_{0y} = 0$. Once the x and y components of the velocity are known for a particular vertical displacement y, the speed of the diver can be obtained from $v = \sqrt{v_x^2 + v_y^2}$.

SOLUTION   For convenience, we will take downward as the positive y direction. After the diver has fallen 10.0 m, the y component of his velocity is, from Equation 3.6b,

$$v_y = \sqrt{v_{0y}^2 + 2a_y y} = \sqrt{0^2 + 2(9.80 \text{ m/s}^2)(10.0 \text{ m})} = 14.0 \text{ m/s}$$

Therefore,

$$v = \sqrt{v_x^2 + v_y^2} = \sqrt{(1.20 \text{ m/s})^2 + (14.0 \text{ m/s})^2} = \boxed{14.1 \text{ m/s}}$$

---

21. ***REASONING AND SOLUTION***   The water exhibits projectile motion. The x component of the motion has zero acceleration while the y component is subject to the acceleration due to gravity. In order to reach the highest possible fire, the displacement of the hose from the building is x, where, according to Equation 3.5a (with $a_x = 0$),

$$x = v_{0x}t = (v_0 \cos\theta)t$$

with t equal to the time required for the water the reach its maximum vertical displacement. The time t can be found by considering the vertical motion. From Equation 3.3b,

$$v_y = v_{0y} + a_y t$$

When the water has reached its maximum vertical displacement, $v_y = 0$. Taking up and to the right as the positive directions, we find that

$$t = \frac{-v_{0y}}{a_y} = \frac{-v_0 \sin\theta}{a_y}$$

and

$$x = (v_0 \cos\theta)\left(\frac{-v_0 \sin\theta}{a_y}\right)$$

Therefore, we have

$$x = -\frac{v_0^2 \cos\theta \sin\theta}{a_y} = -\frac{(25.0 \text{ m/s})^2 \cos 35.0° \sin 35.0°}{-9.80 \text{ m/s}^2} = \boxed{30.0 \text{ m}}$$

---

25. **REASONING**   Since the magnitude of the velocity of the fuel tank is given by $v = \sqrt{v_x^2 + v_y^2}$, it is necessary to know the velocity components v_x and v_y just before impact. At the instant of release, the empty fuel tank has the same velocity as that of the plane. Therefore, the magnitudes of the initial velocity components of the fuel tank are given by $v_{0x} = v_0 \cos\theta$ and $v_{0y} = v_0 \sin\theta$, where v_0 is the speed of the plane at the instant of release. Since the x motion has zero acceleration, the x component of the velocity of the plane remains equal to v_{0x} for all later times while the tank is airborne. The y component of the velocity of the tank after it has undergone a vertical displacement y is given by Equation 3.6b.

SOLUTION
a. Taking up as the positive direction, the velocity components of the fuel tank just before it hits the ground are

$$v_x = v_{0x} = v\cos\theta = (135 \text{ m/s}) \cos 15° = 1.30 \times 10^2 \text{ m/s}$$

From Equation 3.6b, we have

$$v_y = -\sqrt{v_{0y}^2 + 2a_y y} = -\sqrt{(v_0 \sin\theta)^2 + 2a_y y}$$

$$= -\sqrt{\left[(135 \text{ m/s}) \sin 15.0°\right]^2 + \left[2(-9.80 \text{ m/s}^2)(-2.00 \times 10^3 \text{ m})\right]} = -201 \text{ m/s}$$

Therefore, the magnitude of the velocity of the fuel tank just before impact is

$$v = \sqrt{v_x^2 + v_y^2} = \sqrt{(1.30 \times 10^2 \text{ m/s})^2 + (201 \text{ m/s})^2} = \boxed{239 \text{ m/s}}$$

The velocity vector just before impact is inclined at an angle ϕ above the horizontal. This angle is

$$\phi = \tan^{-1}\left(\frac{201 \text{ m/s}}{1.30 \times 10^2 \text{ m/s}}\right) = \boxed{57.1°}$$

b. As shown in Conceptual Example 9, once the fuel tank in part a rises and falls to the same altitude at which it was released, its motion is identical to the fuel tank in part b. Therefore, the velocity of the fuel tank in part b just before impact is $\boxed{\text{239 m/s at an angle of 57.1° above the horizontal}}$.

---

29. **REASONING** The upward direction is chosen as positive. Since the ballast bag is released from rest relative to the balloon, its initial velocity relative to the ground is equal to the velocity of the balloon relative to the ground, so that $v_{0y} = 3.0$ m/s. Time required for the ballast to reach the ground can be found by solving Equation 3.5b for t.

SOLUTION Using Equation 3.5b, we have

$$\tfrac{1}{2}a_y t^2 + v_{0y}t - y = 0 \quad \text{or} \quad \tfrac{1}{2}(-9.80 \text{ m/s}^2)t^2 + (3.0 \text{ m/s})t - (-9.5 \text{ m}) = 0$$

This equation is quadratic in t, and t may be found from the quadratic formula. Using the quadratic formula, suppressing the units, and discarding the negative root, we find

$$t = \frac{-3.0 \pm \sqrt{(3.0)^2 - 4(-4.90)(9.5)}}{2(-4.90)} = \boxed{1.7 \text{ s}}$$

---

33. $\boxed{\text{WWW}}$ **REASONING**   The horizontal distance covered by stone **1** is equal to the distance covered by stone **2** after it passes point **P** in the following diagram. Thus, the distance Δx between the points where the stones strike the ground is equal to x_2, the horizontal distance covered by stone **2** when it reaches **P**. In the diagram, we assume up and to the right are positive.

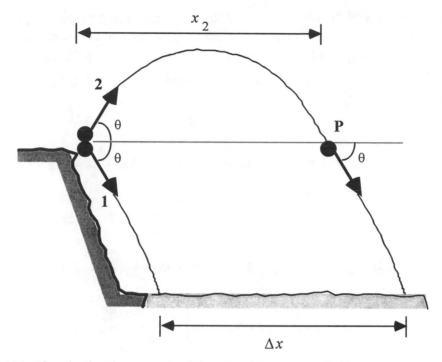

SOLUTION   If t_P is the time required for stone **2** to reach **P**, then

$$x_2 = v_{0x}t_P = (v_0 \cos\theta)t_P$$

For the vertical motion of stone **2**, $v_y = v_0 \sin\theta + a_y t$.   Solving for t gives

$$t = \frac{v_y - v_0 \sin\theta}{a_y}$$

When stone **2** reaches **P**, $v_y = -v_0 \sin\theta$, so the time required to reach **P** is

$$t_P = \frac{-2v_0 \sin\theta}{a_y}$$

Then,

$$x_2 = v_{0x}t_P = (v_0 \cos\theta)\left(\frac{-2v_0 \sin\theta}{a_y}\right)$$

$$x_2 = \frac{-2v_0^2 \sin\theta \cos\theta}{a_y} = \frac{-2(13.0 \text{ m/s})^2 \sin 30.0^\circ \cos 30.0^\circ}{-9.80 \text{ m/s}^2} = \boxed{14.9 \text{ m}}$$

37. **_REASONING AND SOLUTION_** The horizontal displacement of the jumper is given by Equation 3.5a with $a_x = 0$: $x = v_{0x}t = (v_0 \cos\theta)t$. The vertical component of the jumper's velocity at any time t is given by Equation 3.3b: $v_y = v_{0y} + a_y t$. At the instant that the jumper lands, $v_y = -v_{0y}$. Therefore, the vertical component of the jumper's velocity is

$$-v_{0y} = v_{0y} + a_y t$$

Solving for t and assuming that upward is the positive direction gives

$$t = \frac{-2v_{0y}}{a_y} = \frac{-2v_0 \sin\theta}{a_y}$$

Substituting this expression for t into the expression for x gives

$$x = (v_0 \cos\theta)t = v_0 \cos\theta\left(\frac{-2v_0 \sin\theta}{a_y}\right)$$

or

$$x = \frac{-2v_0^2 \sin\theta \cos\theta}{a_y}$$

Solving for v_0 gives

$$v_0 = \sqrt{\frac{-x\,a_y}{2\cos\theta \sin\theta}} = \sqrt{\frac{-(8.7 \text{ m})(-9.80 \text{ m}/\text{s}^2)}{2\cos 23° \sin 23°}} = \boxed{11 \text{ m}/\text{s}}$$

---

41. **_REASONING_** The angle θ can be found from

$$\theta = \tan^{-1}\left(\frac{2400 \text{ m}}{x}\right) \tag{1}$$

where x is the horizontal displacement of the flare. Since $a_x = 0$, it follows that $x = (v_0 \cos 30.0°)t$. The flight time t is determined by the vertical motion. In particular, the time t can be found from Equation 3.5b. Once the time is known, x can be calculated.

SOLUTION From Equation 3.5b, assuming upward is the positive direction, we have

$$y = -(v_0 \sin 30.0°)t + \tfrac{1}{2}a_y t^2$$

which can be rearranged to give the following equation that is quadratic in t:

$$\tfrac{1}{2}a_y t^2 - (v_0 \sin 30.0°)t - y = 0$$

Using $y = -2400$ m and $a_y = -9.80$ m/s^2 and suppressing the units, we obtain the quadratic equation

$$4.9t^2 + 120t - 2400 = 0$$

Using the quadratic formula, we obtain $t = 13$ s. Therefore, we find that

$$x = (v_0 \cos 30.0°)t = (240 \text{ m/s})(\cos 30.0°)(13 \text{ s}) = 2700 \text{ m}$$

Equation (1) then gives

$$\theta = \tan^{-1}\left(\frac{2400 \text{ m}}{2700 \text{ m}}\right) = \boxed{42°}$$

43. **REASONING**   Since the horizontal motion is not accelerated, we know that the x component of the velocity remains constant at 340 m/s. Thus, we can use Equation 3.5a (with $a_x = 0$) to determine the time that the bullet spends in the building before it is embedded in the wall. Since we know the vertical displacement of the bullet after it enters the building, we can use the flight time in the building and Equation 3.5b to find the y component of the velocity of the bullet as it enters the window. Then, Equation 3.6b can be used (with $v_{0y} = 0$) to determine the vertical displacement y of the bullet as it passes between the buildings. We can determine the distance H by adding the magnitude of y to the vertical distance of 0.50 m within the building.

Once we know the vertical displacement of the bullet as it passes between the buildings, we can determine the time t_1 required for the bullet to reach the window using Equation 3.4b. Since the motion in the x direction is not accelerated, the distance D can then be found from $D = v_{0x}t_1$.

SOLUTION   Assuming that the direction to the right is positive, we find that the time that the bullet spends in the building is (according to Equation 3.5a)

$$t = \frac{x}{v_{0x}} = \frac{6.9 \text{ m}}{340 \text{ m/s}} = 0.0203 \text{ s}$$

The vertical displacement of the bullet after it enters the building is, taking down as the negative direction, equal to -0.50 m. Therefore, the vertical component of the velocity of the bullet as it passes through the window is, from Equation 3.5b,

$$v_{0y(\text{window})} = \frac{y - \frac{1}{2}a_y t^2}{t} = \frac{y}{t} - \frac{1}{2}a_y t = \frac{-0.50 \text{ m}}{0.0203 \text{ s}} - \frac{1}{2}(-9.80 \text{ m/s}^2)(0.0203 \text{ s}) = -24.5 \text{ m/s}$$

The vertical displacement of the bullet as it travels between the buildings is (according to Equation 3.6b with $v_{0y} = 0$)

$$y = \frac{v_y^2}{2a_y} = \frac{(-24.5 \text{ m/s})^2}{2(-9.80 \text{ m/s}^2)} = -30.6 \text{ m}$$

Therefore, the distance H is

$$H = (30.6 \text{ m}) + (0.50 \text{ m}) = \boxed{31 \text{ m}}$$

The time for the bullet to reach the window, according to Equation 3.4b, is

$$t_1 = \frac{2y}{v_{0y} + v_y} = \frac{2y}{v_y} = \frac{2(-30.6 \text{ m})}{(-24.5 \text{ m/s})} = 2.50 \text{ s}$$

Hence, the distance D is given by

$$D = v_{0x}t_1 = (340 \text{ m/s})(2.50 \text{ s}) = \boxed{850 \text{ m}}$$

---

47. **REASONING**   The velocity $\mathbf{v}_{AB}$ of train A relative to train B is the vector sum of the velocity $\mathbf{v}_{AG}$ of train A relative to the ground and the velocity $\mathbf{v}_{GB}$ of the ground relative to train B, as indicated by Equation 3.7: $\mathbf{v}_{AB} = \mathbf{v}_{AG} + \mathbf{v}_{GB}$. The values of $\mathbf{v}_{AG}$ and $\mathbf{v}_{BG}$ are given in the statement of the problem. We must also make use of the fact that $\mathbf{v}_{GB} = -\mathbf{v}_{BG}$.

SOLUTION
a. Taking east as the positive direction, the velocity of A relative to B is, according to Equation 3.7,

$$\mathbf{v}_{AB} = \mathbf{v}_{AG} + \mathbf{v}_{GB} = \mathbf{v}_{AG} - \mathbf{v}_{BG} = (+13 \text{ m/s}) - (-28 \text{ m/s}) = \boxed{+41 \text{ m/s}}$$

The positive sign indicates that the direction of $\mathbf{v}_{AB}$ is $\boxed{\text{due east}}$.

b. Similarly, the velocity of B relative to A is

$$\mathbf{v}_{BA} = \mathbf{v}_{BG} + \mathbf{v}_{GA} = \mathbf{v}_{BG} - \mathbf{v}_{AG} = (-28 \text{ m/s}) - (+13 \text{ m/s}) = \boxed{-41 \text{ m/s}}$$

The negative sign indicates that the direction of $\mathbf{v}_{BA}$ is $\boxed{\text{due west}}$.

51. **REASONING** The velocity $\mathbf{v}_{SG}$ of the swimmer relative to the ground is the vector sum of the velocity $\mathbf{v}_{SW}$ of the swimmer relative to the water and the velocity $\mathbf{v}_{WG}$ of the water relative to the ground as shown at the right: $\mathbf{v}_{SG} = \mathbf{v}_{SW} + \mathbf{v}_{WG}$.

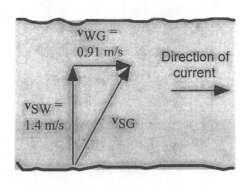

The component of $\mathbf{v}_{SG}$ that is parallel to the width of the river determines how fast the swimmer is moving across the river; this parallel component is v_{SW}. The time for the swimmer to cross the river is equal to the width of the river divided by the magnitude of this velocity component.

The component of $\mathbf{v}_{SG}$ that is parallel to the direction of the current determines how far the swimmer is carried down stream; this component is v_{WG}. Since the motion occurs with constant velocity, the distance that the swimmer is carried downstream while crossing the river is equal to the magnitude of v_{WG} multiplied by the time it takes for the swimmer to cross the river.

SOLUTION
a. The time t for the swimmer to cross the river is

$$t = \frac{\text{width}}{v_{SW}} = \frac{2.8 \times 10^3 \text{ m}}{1.4 \text{ m/s}} = \boxed{2.0 \times 10^3 \text{ s}}$$

b. The distance x that the swimmer is carried downstream while crossing the river is

$$x = v_{WG}t = (0.91 \text{ m/s})(2.0 \times 10^3 \text{ s}) = \boxed{1.8 \times 10^3 \text{ m}}$$

---

57. **REASONING** The velocity $\mathbf{v}_{PM}$ of the puck relative to Mario is the vector sum of the velocity $\mathbf{v}_{PI}$ of the puck relative to the ice and the velocity $\mathbf{v}_{IM}$ of the ice relative to Mario as indicated by Equation 3.7: $\mathbf{v}_{PM} = \mathbf{v}_{PI} + \mathbf{v}_{IM}$. The values of $\mathbf{v}_{MI}$ and $\mathbf{v}_{PI}$ are given in the statement of the problem. In order to use the data, we must make use of the fact that $\mathbf{v}_{IM} = -\mathbf{v}_{MI}$, with the result that $\mathbf{v}_{PM} = \mathbf{v}_{PI} - \mathbf{v}_{MI}$.

SOLUTION The first two rows of the following table give the east/west and north/south components of the vectors $\mathbf{v}_{PI}$ and $-\mathbf{v}_{MI}$. The third row gives the components of their resultant $\mathbf{v}_{PM} = \mathbf{v}_{PI} - \mathbf{v}_{MI}$. Due east and due north have been taken as positive.

Vector	East/West Component	North/South Component
$\mathbf{v}_{PI}$	$-(11.0\ \text{m/s})\sin 22° = -4.1\ \text{m/s}$	$-(11.0\ \text{m/s})\cos 22° = -10.2\ \text{m/s}$
$-\mathbf{v}_{MI}$	0	$+7.0\ \text{m/s}$
$\mathbf{v}_{PM} = \mathbf{v}_{PI} - \mathbf{v}_{MI}$	$-4.1\ \text{m/s}$	$-3.2\ \text{m/s}$

Now that the components of $\mathbf{v}_{PM}$ are known, the Pythagorean theorem can be used to find the magnitude.

$$v_{PM} = \sqrt{(-4.1\ \text{m/s})^2 + (-3.2\ \text{m/s})^2} = \boxed{5.2\ \text{m/s}}$$

The direction of $\mathbf{v}_{PM}$ is found from

$$\phi = \tan^{-1}\left(\frac{4.1\ \text{m/s}}{3.2\ \text{m/s}}\right) = \boxed{52° \text{ west of south}}$$

59. $\boxed{\text{WWW}}$ *REASONING* The velocity $\mathbf{v}_{OW}$ of the object relative to the water is the vector sum of the velocity $\mathbf{v}_{OS}$ of the object relative to the ship and the velocity $\mathbf{v}_{SW}$ of the ship relative to the water, as indicated by Equation 3.7: $\mathbf{v}_{OW} = \mathbf{v}_{OS} + \mathbf{v}_{SW}$. The value of $\mathbf{v}_{SW}$ is given in the statement of the problem. We can find the value of $\mathbf{v}_{OS}$ from the fact that we know the position of the object relative to the ship at two different times. The initial position is $\mathbf{r}_{OS1}$, and the final position is $\mathbf{r}_{OS2}$. Since the object moves with constant velocity,

$$\mathbf{v}_{OS} = \frac{\Delta\mathbf{r}_{OS}}{\Delta t} = \frac{\mathbf{r}_{OS2} - \mathbf{r}_{OS1}}{\Delta t} \qquad (1)$$

SOLUTION The first two rows of the following table give the east/west and north/south components of the vectors $\mathbf{r}_{OS2}$ and $-\mathbf{r}_{OS1}$. The third row of the table gives the components of $\Delta\mathbf{r}_{OS} = \mathbf{r}_{OS2} - \mathbf{r}_{OS1}$. Due east and due north have been taken as positive.

Vector	East/West Component	North/South Component
$\mathbf{r}_{OS2}$	$-(1120\ \text{m})\cos 57.0°$ $= -6.10 \times 10^2\ \text{m}$	$-(1120\ \text{m})\sin 57.0°$ $= -9.39 \times 10^2\ \text{m}$
$-\mathbf{r}_{OS1}$	$-(2310\ \text{m})\cos 32.0°$ $= -1.96 \times 10^3\ \text{m}$	$+(2310\ \text{m})\sin 32.0°$ $= 1.22 \times 10^3\ \text{m}$
$\Delta \mathbf{r}_{OS} = \mathbf{r}_{OS2} - \mathbf{r}_{OS1}$	$-2.57 \times 10^3\ \text{m}$	$2.81 \times 10^2\ \text{m}$

Now that the components of $\Delta \mathbf{r}_{OS}$ are known, the Pythagorean theorem can be used to find the magnitude.

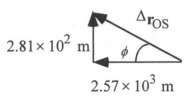

$$\Delta r_{OS} = \sqrt{(-2.57 \times 10^3\ \text{m})^2 + (2.81 \times 10^2\ \text{m})^2} = 2.59 \times 10^3\ \text{m}$$

The direction of $\Delta \mathbf{r}_{OS}$ is found from

$$\phi = \tan^{-1}\left(\frac{2.81 \times 10^2\ \text{m}}{2.57 \times 10^3\ \text{m}}\right) = 6.24°$$

Therefore, from Equation (1),

$$\mathbf{v}_{OS} = \frac{\Delta \mathbf{r}_{OS}}{\Delta t} = \frac{\mathbf{r}_{OS2} - \mathbf{r}_{OS1}}{\Delta t} = \frac{2.59 \times 10^3\ \text{m}}{360\ \text{s}} = 7.19\ \text{m/s, } 6.24°\text{ north of west}$$

Now that $\mathbf{v}_{OS}$ is known, we can find $\mathbf{v}_{OW}$, as indicated by Equation 3.7: $\mathbf{v}_{OW} = \mathbf{v}_{OS} + \mathbf{v}_{SW}$. The following table summarizes the vector addition:

Vector	East/West Component	North/South Component
$\mathbf{v}_{OS}$	$-(7.19\ \text{m/s})\cos 6.24° = -7.15\ \text{m/s}$	$(7.19\ \text{m/s})\sin 6.24° = 0.782\ \text{m/s}$
$\mathbf{v}_{SW}$	$+4.20\ \text{m/s}$	$0\ \text{m/s}$
$\mathbf{v}_{OW} = \mathbf{v}_{OS} + \mathbf{v}_{SW}$	$-2.95\ \text{m/s}$	$0.782\ \text{m/s}$

Now that the components of $\mathbf{v}_{OW}$ are known, the Pythagorean theorem can be used to find the magnitude.

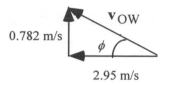

$$v_{OW} = \sqrt{(-2.95 \text{ m/s})^2 + (0.782 \text{ m/s})^2} = \boxed{3.05 \text{ m/s}}$$

The direction of $\mathbf{v}_{OW}$ is found from

$$\phi = \tan^{-1}\left(\frac{0.782 \text{ m/s}}{2.95 \text{ m/s}}\right) = \boxed{14.8° \text{ north of west}}$$

---

63. ***REASONING***   The time that the ball spends in the air is determined by its vertical motion. The time required for the ball to reach the lake can be found by solving Equation 3.5b for t. The motion of the golf ball is characterized by constant velocity in the x direction and accelerated motion (due to gravity) in the y direction. Thus, the x component of the velocity of the golf ball is constant, and the y component of the velocity at any time t can be found from Equation 3.3b. Once the x and y components of the velocity are known for a particular time t, the speed can be obtained from $v = \sqrt{v_x^2 + v_y^2}$.

SOLUTION
a. Since the ball rolls off the cliff horizontally, $v_{0y} = 0$. If the origin is chosen at top of the cliff and upward is assumed to be the positive direction, then the vertical component of the ball's displacement is $y = -15.5$ m. Thus, Equation 3.5b gives

$$t = \sqrt{\frac{2y}{a_y}} = \sqrt{\frac{2(-15.5 \text{ m})}{(-9.80 \text{ m/s}^2)}} = \boxed{1.78 \text{ s}}$$

b. Since there is no acceleration in the x direction, $v_x = v_{0x} = 11.4$ m/s. The y component of the velocity of the ball just before it strikes the water is, according to Equation 3.3b,

$$v_y = v_{0y} + a_y t = \left[0 + (-9.80 \text{ m/s}^2)(1.78 \text{ s})\right] = -17.4 \text{ m/s}$$

The speed of the ball just before it strikes the water is, therefore,

$$v = \sqrt{v_x^2 + v_y^2} = \sqrt{(11.4 \text{ m/s})^2 + (-17.4 \text{ m/s})^2} = \boxed{20.8 \text{ m/s}}$$

---

65. ***REASONING***   The speed of the fish at any time t is given by $v = \sqrt{v_x^2 + v_y^2}$, where v_x and v_y are the x and y components of the velocity at that instant. Since the horizontal motion of the fish has zero acceleration, $v_x = v_{0x}$ for all times t. Since the fish is dropped by the

eagle, v_{0x} is equal to the horizontal speed of the eagle and $v_{0y} = 0$. The y component of the velocity of the fish for any time t is given by Equation 3.3b with $v_{0y} = 0$. Thus, the speed at any time t is given by $v = \sqrt{v_{0x}^2 + (a_y t)^2}$.

SOLUTION

a. The initial speed of the fish is $v_0 = \sqrt{v_{0x}^2 + v_{0y}^2} = \sqrt{v_{0x}^2 + 0^2} = v_{0x}$. When the fish's speed doubles, $v = 2v_{0x}$. Therefore,

$$2v_{0x} = \sqrt{v_{0x}^2 + (a_y t)^2} \qquad \text{or} \qquad 4v_{0x}^2 = v_{0x}^2 + (a_y t)^2$$

Assuming that downward is positive and solving for t, we have

$$t = \sqrt{3}\,\frac{v_{0x}}{a_y} = \sqrt{3}\left(\frac{6.0 \text{ m/s}}{9.80 \text{ m/s}^2}\right) = \boxed{1.1 \text{ s}}$$

b. When the fish's speed doubles again, $v = 4v_{0x}$. Therefore,

$$4v_{0x} = \sqrt{v_{0x}^2 + (a_y t)^2} \qquad \text{or} \qquad 16v_{0x}^2 = v_{0x}^2 + (a_y t)^2$$

Solving for t, we have

$$t = \sqrt{15}\,\frac{v_{0x}}{a_y} = \sqrt{15}\left(\frac{6.0 \text{ m/s}}{9.80 \text{ m/s}^2}\right) = 2.37 \text{ s}$$

Therefore, the additional time for the speed to double again is $(2.4 \text{ s}) - (1.1 \text{ s}) = \boxed{1.3 \text{ s}}$.

---

69. **REASONING** The velocity $\mathbf{v}_{PG}$ of the plane relative to the ground is the vector sum of the velocity $\mathbf{v}_{PA}$ of the plane relative to the air and the velocity $\mathbf{v}_{AG}$ of the air relative to the ground, as indicated by Equation 3.7: $\mathbf{v}_{PG} = \mathbf{v}_{PA} + \mathbf{v}_{AG}$, or $\mathbf{v}_{AG} = \mathbf{v}_{PG} - \mathbf{v}_{PA}$. We are given $\mathbf{v}_{PA}$ and can find $\mathbf{v}_{PG}$ from the given displacement and travel time. Thus,

$$\mathbf{v}_{PG} = \frac{81.0 \times 10^3 \text{ m}}{9.00 \times 10^2 \text{ s}} = 90.0 \text{ m/s, } 45° \text{ west of south}$$

SOLUTION The first two rows of the following table give the east/west and north/south components of the vectors $\mathbf{v}_{PG}$ and $-\mathbf{v}_{PA}$. The third row gives the components of $\mathbf{v}_{AG} = \mathbf{v}_{PG} - \mathbf{v}_{PA}$. Due east and due north have been chosen as the positive directions.

Vector	East/West Component	North/South Component
$\mathbf{v}_{PG}$	$-(90.0 \text{ m/s}) \sin 45° = -63.6 \text{ m/s}$	$-(90.0 \text{ m/s}) \cos 45° = -63.6 \text{ m/s}$
$-\mathbf{v}_{PA}$	0 m/s	+57.8 m/s
$\mathbf{v}_{AG} = \mathbf{v}_{PG} - \mathbf{v}_{PA}$	-63.6 m/s	-5.8 m/s

The magnitude of the velocity of the wind with respect to the ground can be obtained using the Pythagorean theorem:

$$v_{AG} = \sqrt{(-63.6 \text{ m/s})^2 + (-5.8 \text{ m/s})^2} = \boxed{63.9 \text{ m/s}}$$

The direction of $\mathbf{v}_{AG}$ is found from

$$\phi = \tan^{-1}\left(\frac{63.6 \text{ m/s}}{5.8 \text{ m/s}}\right) = \boxed{85° \text{ west of south}}$$

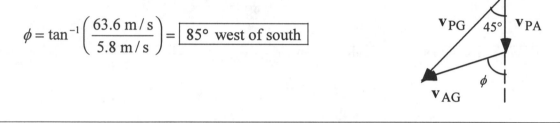

---

71. **REASONING AND SOLUTION**   In the absence of air resistance, the bullet exhibits projectile motion. The x component of the motion has zero acceleration while the y component of the motion is subject to the acceleration due to gravity. The horizontal distance traveled by the bullet is given by Equation 3.5a (with $a_x = 0$):

$$x = v_{0x}t = (v_0 \cos\theta)t$$

with t equal to the time required for the bullet to reach the target. The time t can be found by considering the vertical motion. From Equation 3.3b,

$$v_y = v_{0y} + a_y t$$

When the bullet reaches the target, $v_y = -v_{0y}$. Assuming that up and to the right are the positive directions, we have

$$t = \frac{-2v_{0y}}{a_y} = \frac{-2v_0 \sin\theta}{a_y} \qquad \text{and} \qquad x = (v_0 \cos\theta)\left(\frac{-2v_0 \sin\theta}{a_y}\right)$$

Using the fact that $2\sin\theta \cos\theta = \sin 2\theta$, we have

$$x = -\frac{2v_0^2 \cos\theta \sin\theta}{a_y} = -\frac{v_0^2 \sin 2\theta}{a_y}$$

Thus, we find that

$$\sin 2\theta = -\frac{x\, a_y}{v_0^2} = -\frac{(91.4 \text{ m}) (-9.80 \text{ m/s}^2)}{(427 \text{ m/s})^2} = 4.91 \times 10^{-3}$$

and

$$2\theta = 0.281° \quad \text{or} \quad 2\theta = 180.000° - 0.281° = 179.719°$$

Therefore,

$$\theta = \boxed{0.141° \text{ and } 89.860°}$$

CHAPTER 4 | FORCES AND NEWTON'S LAWS OF MOTION

3. **REASONING** According to Newton's second law of motion, the net force applied to the fist is equal to the mass of the fist multiplied by its acceleration. The data in the problem gives the final velocity of the fist and the time it takes to acquire that velocity. The average acceleration can be obtained directly from these data using the definition of average acceleration given in Equation 2.4.

SOLUTION The magnitude of the average net force applied to the fist is, therefore,

$$\Sigma \overline{F} = ma = m\left(\frac{\Delta v}{\Delta t}\right) = (0.70 \text{ kg})\left(\frac{8.0 \text{ m/s} - 0 \text{ m/s}}{0.15 \text{ s}}\right) = \boxed{37 \text{ N}}$$

5. **REASONING** The net force acting on the ball can be calculated using Newton's second law. Before we can use Newton's second law, however, we must use Equation 2.9 from the equations of kinematics to determine the acceleration of the ball.

SOLUTION According to Equation 2.9, the acceleration of the ball is given by

$$a = \frac{v^2 - v_0^2}{2x}$$

Thus, the magnitude of the net force on the ball is given by

$$\Sigma F = ma = m\left(\frac{v^2 - v_0^2}{2x}\right) = (0.058 \text{ kg})\left[\frac{(45 \text{ m/s})^2 - (0 \text{ m/s})^2}{2(0.44 \text{ m})}\right] = \boxed{130 \text{ N}}$$

9. $\boxed{\text{WWW}}$ **REASONING** Let due east be chosen as the positive direction. Then, when both forces point due east, Newton's second law gives

$$\underbrace{F_A + F_B}_{\Sigma F} = ma_1 \tag{1}$$

where $a_1 = 0.50 \text{ m/s}^2$. When F_A points due east and F_B points due west, Newton's second law gives

$$\underbrace{F_A - F_B}_{\Sigma F} = ma_2 \tag{2}$$

where $a_2 = 0.40 \text{ m/s}^2$. These two equations can be used to find the magnitude of each force.

SOLUTION
a. Adding Equations 1 and 2 gives

$$F_A = \frac{m(a_1 + a_2)}{2} = \frac{(8.0 \text{ kg})(0.50 \text{ m/s}^2 + 0.40 \text{ m/s}^2)}{2} = \boxed{3.6 \text{ N}}$$

b. Subtracting Equation 2 from Equation 1 gives

$$F_B = \frac{m(a_1 - a_2)}{2} = \frac{(8.0 \text{ kg})(0.50 \text{ m/s}^2 - 0.40 \text{ m/s}^2)}{2} = \boxed{0.40 \text{ N}}$$

---

13. **REASONING**   According to Newton's second law ($\Sigma \mathbf{F} = m\mathbf{a}$), the acceleration of the object is given by $\mathbf{a} = \Sigma\mathbf{F}/m$, where $\Sigma\mathbf{F}$ is the net force that acts on the object. We must first find the net force that acts on the object, and then determine the acceleration using Newton's second law.

SOLUTION The following table gives the x and y components of the two forces that act on the object. The third row of that table gives the components of the net force.

Force	x-Component	y-Component
$\mathbf{F}_1$	40.0 N	0 N
$\mathbf{F}_2$	$(60.0 \text{ N}) \cos 45.0° = 42.4 \text{ N}$	$(60.0 \text{ N}) \sin 45.0° = 42.4 \text{ N}$
$\Sigma\mathbf{F} = \mathbf{F}_1 + \mathbf{F}_2$	82.4 N	42.4 N

The magnitude of $\Sigma\mathbf{F}$ is given by the Pythagorean theorem as

$$\Sigma F = \sqrt{(82.4 \text{ N})^2 + (42.4)^2} = 92.7 \text{ N}$$

The angle θ that $\Sigma\mathbf{F}$ makes with the $+x$ axis is

$$\theta = \tan^{-1}\left(\frac{42.4 \text{ N}}{82.4 \text{ N}}\right) = 27.2°$$

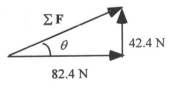

According to Newton's second law, the magnitude of the acceleration of the object is

$$a = \frac{\Sigma F}{m} = \frac{92.7 \text{ N}}{3.00 \text{ kg}} = \boxed{30.9 \text{ m/s}^2}$$

Since Newton's second law is a vector equation, we know that the direction of the right hand side must be equal to the direction of the left hand side. In other words, the direction of the acceleration a is the same as the direction of the net force $\Sigma \mathbf{F}$. Therefore, the direction of the acceleration of the object is $\boxed{27.2° \text{ above the } +x \text{ axis}}$.

---

17. $\boxed{\text{WWW}}$ *REASONING* We first determine the acceleration of the boat. Then, using Newton's second law, we can find the net force $\Sigma \mathbf{F}$ that acts on the boat. Since two of the three forces are known, we can solve for the unknown force $\mathbf{F_W}$ once the net force $\Sigma \mathbf{F}$ is known.

SOLUTION Let the direction due east be the positive x direction and the direction due north be the positive y direction. The x and y components of the initial velocity of the boat are then

$$v_{0\,x} = (2.00 \text{ m/s}) \cos 15.0° = 1.93 \text{ m/s}$$

$$v_{0\,y} = (2.00 \text{ m/s}) \sin 15.0° = 0.518 \text{ m/s}$$

Thirty seconds later, the x and y velocity components of the boat are

$$v_x = (4.00 \text{ m/s}) \cos 35.0° = 3.28 \text{ m/s}$$

$$v_y = (4.00 \text{ m/s}) \sin 35.0° = 2.29 \text{ m/s}$$

Therefore, according to Equations 3.3a and 3.3b, the x and y components of the acceleration of the boat are

$$a_x = \frac{v_x - v_{0x}}{t} = \frac{3.28 \text{ m/s} - 1.93 \text{ m/s}}{30.0 \text{ s}} = 4.50 \times 10^{-2} \text{ m/s}^2$$

$$a_y = \frac{v_y - v_{0y}}{t} = \frac{2.29 \text{ m/s} - 0.518 \text{ m/s}}{30.0 \text{ s}} = 5.91 \times 10^{-2} \text{ m/s}^2$$

Thus, the x and y components of the net force that act on the boat are

$$\Sigma F_x = ma_x = (325 \text{ kg}) (4.50 \times 10^{-2} \text{ m/s}^2) = 14.6 \text{ N}$$

$$\Sigma F_y = ma_y = (325 \text{ kg}) (5.91 \times 10^{-2} \text{ m/s}^2) = 19.2 \text{ N}$$

The following table gives the x and y components of the net force $\Sigma \mathbf{F}$ and the two known forces that act on the boat.   The fourth row of that table gives the components of the unknown force $\mathbf{F_W}$.

Force	x-Component	y-Component
$\Sigma \mathbf{F}$	14.6 N	19.2 N
$\mathbf{F_1}$	$(31.0 \text{ N}) \cos 15.0° = 29.9$ N	$(31.0 \text{ N}) \sin 15.0° = 8.02$ N
$\mathbf{F_2}$	$-(23.0 \text{ N}) \cos 15.0° = -22.2$ N	$-(23.0 \text{ N}) \sin 15.0° = -5.95$ N
$\mathbf{F_W} = \Sigma \mathbf{F} - \mathbf{F_1} - \mathbf{F_2}$	14.6 N $-$ 29.9 N $+$ 22.2 N $=$ 6.9 N	19.2 N $-$ 8.02 N $+$ 5.95 N $=$ 17.1 N

The magnitude of $\mathbf{F_W}$ is given by the Pythagorean theorem as

$$F_W = \sqrt{(6.9 \text{ N})^2 + (17.1 \text{ N})^2} = \boxed{18.4 \text{ N}}$$

The angle θ that $\mathbf{F_W}$ makes with the x axis is

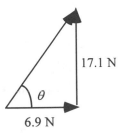

$$\theta = \tan^{-1}\left(\frac{17.1 \text{ N}}{6.9 \text{ N}}\right) = 68°$$

17.1 N

6.9 N

Therefore, the direction of $\mathbf{F_W}$ is $\boxed{68°, \text{ north of east}}$.

---

19. ***REASONING AND SOLUTION***

a.   According to Equation 4.5, the weight of the space traveler of mass $\boxed{m = 115 \text{ kg}}$ on earth is

$$W = mg = (115 \text{ kg})(9.80 \text{ m/s}^2) = \boxed{1.13 \times 10^3 \text{ N}}$$

b.   In interplanetary space where there are no nearby planetary objects, the gravitational force exerted on the space traveler is zero and $g = 0 \text{ m/s}^2$.   Therefore, the weight is $\boxed{W = 0 \text{ N}}$.   Since the mass of an object is an intrinsic property of the object and is independent of its location in the universe, the mass of the space traveler is still $\boxed{m = 115 \text{ kg}}$.

23. *REASONING AND SOLUTION*
a.   Combining Equations 4.4 and 4.5, we see that the acceleration due to gravity on the surface of Saturn can be calculated as follows:

$$g_{Saturn} = G\frac{M_{Saturn}}{r_{Saturn}^2} = \left(6.67\times10^{-11}\ \text{N}\cdot\text{m}^2/\text{kg}^2\right)\frac{\left(5.67\times10^{26}\ \text{kg}\right)}{\left(6.00\times10^7\ \text{m}\right)^2} = \boxed{10.5\ \text{m/s}^2}$$

b.   The ratio of the person's weight on Saturn to that on earth is

$$\frac{W_{Saturn}}{W_{earth}} = \frac{mg_{Saturn}}{mg_{earth}} = \frac{g_{Saturn}}{g_{earth}} = \frac{10.5\ \text{m/s}^2}{9.80\ \text{m/s}^2} = \boxed{1.07}$$

---

27. *REASONING AND SOLUTION*   According to Equations 4.4 and 4.5, the weight of an object of mass m at a distance r from the *center* of the earth is

$$mg = \frac{GM_E m}{r^2}$$

In a circular orbit that is 3.59×10^7 m above the surface of the earth (radius $= 6.38\times10^6$ m, mass $= 5.98\times10^{24}$ kg), the total distance from the center of the earth is $r = 3.59\times10^7\text{m} + 6.38\times10^6$ m. Thus the acceleration g due to gravity is

$$g = \frac{GM_E}{r^2} = \frac{(6.67\times10^{-11}\text{N}\cdot\text{m}^2/\text{kg}^2)(5.98\times10^{24}\text{kg})}{(3.59\times10^7\text{m} + 6.38\times10^6\text{m})^2} = \boxed{0.223\ \text{m/s}^2}$$

---

31. ⬛WWW⬛   *REASONING AND SOLUTION*   There are two forces that act on the balloon; they are, the combined weight of the balloon and its load, Mg, and the upward buoyant force F_B.   If we take upward as the positive direction, then, initially when the balloon is motionless, Newton's second law gives $F_B - Mg = 0$. If an amount of mass m is dropped overboard so that the balloon has an upward acceleration, Newton's second law for this situation is

$$F_B - (M - m)g = (M - m)a$$

But $F_B = mg$, so that

$$Mg - (M - m)g = mg = (M - m)a$$

Solving for the mass m that should be dropped overboard, we obtain

$$m = \frac{Ma}{g+a} = \frac{(310\ \text{kg})(0.15\ \text{m/s}^2)}{9.80\ \text{m/s}^2 + 0.15\ \text{m/s}^2} = \boxed{4.7\ \text{kg}}$$

35. ***REASONING AND SOLUTION***   According to Equation 3.3b, the acceleration of the astronaut is $a_y = (v_y - v_{0y})/t = v_y/t$. The apparent weight and the true weight of the astronaut are related according to Equation 4.6. Direct substitution gives

$$\underbrace{F_N}_{\substack{\text{Apparent}\\\text{weight}}} = \underbrace{mg}_{\substack{\text{True}\\\text{weight}}} + ma_y = m\,(g + a_y) = m\left(g + \frac{v_y}{t}\right)$$

$$= (57 \text{ kg})\left(9.80 \text{ m/s}^2 + \frac{45 \text{ m/s}}{15 \text{ s}}\right) = \boxed{7.3 \times 10^2 \text{ N}}$$

39. ***REASONING*** In order to start the crate moving, an external agent must supply a force that is at least as large as the maximum value $f_s^{MAX} = \mu_s F_N$, where μ_s is the coefficient of static friction (see Equation 4.7). Once the crate is moving, the magnitude of the frictional force is very nearly constant at the value $f_k = \mu_k F_N$, where μ_k is the coefficient of kinetic friction (see Equation 4.8). In both cases described in the problem statement, there are only two vertical forces that act on the crate; they are the upward normal force $\mathbf{F}_N$, and the downward pull of gravity (the weight) mg. Furthermore, the crate has no vertical acceleration in either case. Therefore, if we take upward as the positive direction, Newton's second law in the vertical direction gives $F_N - mg = 0$, and we see that, in both cases, the magnitude of the normal force is $F_N = mg$.

SOLUTION
a. Therefore, the applied force needed to start the crate moving is

$$f_s^{MAX} = \mu_s mg = (0.760)(60.0 \text{ kg})(9.80 \text{ m/s}^2) = \boxed{447 \text{ N}}$$

b. When the crate moves in a straight line at constant speed, its velocity does not change, and it has zero acceleration. Thus, Newton's second law in the horizontal direction becomes $\mathbf{P} - \mathbf{f}_k = 0$, where P is the required pushing force. Thus, the applied force required to keep the crate sliding across the dock at a constant speed is

$$P = f_k = \mu_k mg = (0.410)(60.0 \text{ kg})(9.80 \text{ m/s}^2) = \boxed{241 \text{ N}}$$

43. ***REASONING***   If we assume that kinetic friction is the only horizontal force that acts on the skater, then, since kinetic friction is a resistive force, it acts opposite to the direction of motion and the skater slows down. According to Newton's second law ($\Sigma \mathbf{F} = m\mathbf{a}$), the magnitude of the deceleration is $a = f_k/m$.

The magnitude of the frictional force that acts on the skater is, according to Equation 4.8, $f_k = \mu_k F_N$ where μ_k is the coefficient of kinetic friction between the ice and the skate blades. There are only two vertical forces that act on the skater; they are the upward normal force $\mathbf{F_N}$ and the downward pull of gravity (the weight) mg. Since the skater has no vertical acceleration, Newton's second law in the vertical direction gives (if we take upward as the positive direction) $F_N - mg = 0$. Therefore, the magnitude of the normal force is $F_N = mg$ and the magnitude of the deceleration is given by

$$a = \frac{f_k}{m} = \frac{\mu_k F_N}{m} = \frac{\mu_k mg}{m} = \mu_k g$$

SOLUTION
a.   Direct substitution into the previous expression gives

$$a = \mu_k g = (0.100)(9.80 \text{ m/s}^2) = \boxed{0.980 \text{ m/s}^2}$$

Since the skater is slowing down,

$$\boxed{\text{the direction of the acceleration must be opposite to the direction of motion}}.$$

b.   The displacement through which the skater will slide before he comes to rest can be obtained from Equation 2.9 ($v^2 - v_0^2 = 2ax$). Since the skater comes to rest, $v = 0$ m/s. If we take the direction of motion of the skater as the positive direction, then, solving for x, we obtain

$$x = \frac{-v_0^2}{2a} = \frac{-(7.60 \text{ m/s})^2}{2(-0.980 \text{ m/s}^2)} = \boxed{29.5 \text{ m}}$$

___

47.   ***REASONING AND SOLUTION***   There are three vertical forces that act on the lantern. The two upward forces of tension exerted by the wires, and the downward pull of gravity (the weight). If we let upward be the positive direction, and let T represent the tension in *one* of the wires, then Equation 4.9b gives

$$\Sigma F_y = 2T - mg = 0$$

Solving for T gives

$$T = \frac{mg}{2} = \frac{(12.0 \text{ kg})(9.80 \text{ m/s}^2)}{2} = \boxed{58.8 \text{ N}}$$

49. **REASONING AND SOLUTION**   The figure at the right shows the forces that act on the wine bottle. Newton's second law applied in the horizontal and vertical directions gives

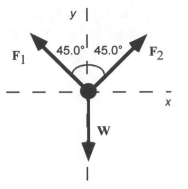

$$\Sigma F_y = F_1 \cos 45.0° + F_2 \cos 45.0° - W = 0 \qquad (1)$$

$$\Sigma F_x = F_2 \sin 45.0° - F_1 \sin 45.0° = 0 \qquad (2)$$

From Equation (2), we see that $F_1 = F_2$. According to Equation (1), we have

$$F_1 = \frac{W}{2 \cos 45.0°} = \frac{mg}{2 \cos 45.0°}$$

Therefore,

$$F_1 = F_2 = \frac{(1.40 \text{ kg}) (9.80 \text{ m/s}^2)}{2 \cos 45.0°} = \boxed{9.70 \text{ N}}$$

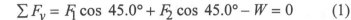

53. **REASONING**   In order for the object to move with constant velocity, the net force on the object must be zero. Therefore, the north/south component of the third force must be equal in magnitude and opposite in direction to the 80.0 N force, while the east/west component of the third force must be equal in magnitude and opposite in direction to the 60.0 N force. Therefore, the third force has components: 80.0 N due south and 60.0 N due east. We can use the Pythagorean theorem and trigonometry to find the magnitude and direction of this third force.

SOLUTION   The magnitude of the third force is

$$F_3 = \sqrt{(80.0 \text{ N})^2 + (60.0 \text{ N})^2} = \boxed{1.00 \times 10^2 \text{ N}}$$

The direction of F_3 is specified by the angle θ where

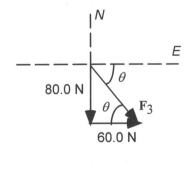

$$\theta = \tan^{-1} \left(\frac{80.0 \text{ N}}{60.0 \text{ N}} \right) = \boxed{53.1°, \text{ south of east}}$$

57. **REASONING**   There are four forces that act on the chandelier; they are the forces of tension T in each of the three wires, and the downward force of gravity mg. Under the influence of these forces, the chandelier is at rest and, therefore, in equilibrium. Consequently, the sum of the x components as well as the sum of the y components of the forces must each be zero.   The figure below shows the free-body diagram for the chandelier and the force components for a suitable system of x, y axes. Note that the free-body diagram only shows one of the forces of tension; the second and third tension forces are not shown.

The triangle at the right shows the geometry of one of the cords, where ℓ is the length of the cord, and d is the distance from the ceiling.

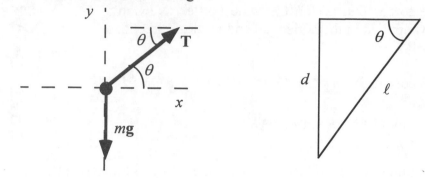

We can use the forces in the y direction to find the magnitude T of the tension in any one wire.

SOLUTION   Remembering that there are three tension forces, we see from the free-body diagram that

$$3T \sin\theta = mg \qquad \text{or} \qquad T = \frac{mg}{3\sin\theta} = \frac{mg}{3(d/\ell)} = \frac{mg\ell}{3d}$$

Therefore, the magnitude of the tension in any one of the cords is

$$T = \frac{(44 \text{ kg})(9.80 \text{ m/s}^2)(2.0 \text{ m})}{3(1.5 \text{ m})} = \boxed{1.9 \times 10^2 \text{N}} \quad (90 \text{ N})$$

---

61. **REASONING**   The following figure shows the crate on the incline and the free body diagram for the crate. The diagram at the far right shows all the forces resolved into components that are parallel and perpendicular to the surface of the incline. We can analyze the motion of the crate using Newton's second law. The coefficient of friction can be determined from the resulting equations.

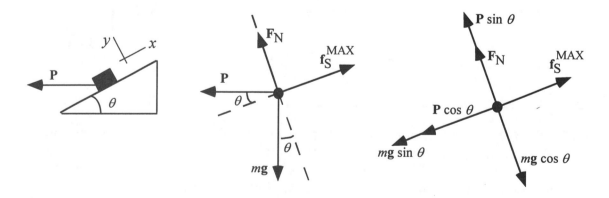

SOLUTION Since the crate is at rest, it is in equilibrium and its acceleration is zero in all directions. If we take the direction down the incline as positive, Newton's second law indicates that

$$\Sigma F_x = P\cos\theta + mg\sin\theta - f_s^{MAX} = 0$$

According to Equation 4.7, $f_s^{MAX} = \mu_s F_N$. Therefore, we have

$$P\cos\theta + mg\sin\theta - \mu_s F_N = 0 \qquad\qquad (1)$$

The expression for the normal force can be found from analyzing the forces that are perpendicular to the incline. Taking up to be positive, we have

$$\Sigma F_y = P\sin\theta + F_N - mg\cos\theta = 0 \qquad \text{or} \qquad F_N = mg\cos\theta - P\sin\theta$$

Equation (1) then becomes

$$P\cos\theta + mg\sin\theta - \mu_s(mg\cos\theta - P\sin\theta) = 0$$

Solving for the coefficient of static friction, we find that

$$\mu_s = \frac{P\cos\theta + mg\sin\theta}{mg\cos\theta - P\sin\theta} = \frac{(535\text{ N})\cos 20.0° + (225\text{ kg})(9.80\text{ m/s}^2)\sin 20.0°}{(225\text{ kg})(9.80\text{ m/s}^2)\cos 20.0° - (535\text{ N})\sin 20.0°} = \boxed{0.665}$$

---

65. **REASONING** If we assume that the acceleration is constant, we can use Equation 2.4 ($v = v_0 + at$) to find the acceleration of the car. Once the acceleration is known, Newton's second law ($\Sigma \mathbf{F} = m\mathbf{a}$) can be used to find the magnitude and direction of the net force that produces the deceleration of the car.

SOLUTION The average acceleration of the car is, according to Equation 2.4,

$$a = \frac{v - v_0}{t} = \frac{17.0\text{ m/s} - 27.0\text{ m/s}}{8.00\text{ s}} = -1.25\text{ m/s}^2$$

where the minus sign indicates that the direction of the acceleration is opposite to the direction of motion; therefore, the acceleration points due west.

According to Newton's Second law, the net force on the car is

$$\Sigma F = ma = (1380\text{ kg})(-1.25\text{ m/s}^2) = -1730\text{ N}$$

The magnitude of the net force is $\boxed{1730 \text{ N}}$. From Newton's second law, we know that the direction of the force is the same as the direction of the acceleration, so the force also points $\boxed{\text{due west}}$.

---

69. $\boxed{\textbf{WWW}}$ *REASONING*   The speed of the skateboarder at the bottom of the ramp can be found by solving Equation 2.9 ($v^2 = v_0^2 + 2ax$ where x is the distance that the skater moves down the ramp) for v. The figure at the right shows the free-body diagram for the skateboarder. The net force ΣF, which accelerates the skateboarder down the ramp, is the component of the weight that is parallel to the incline: $\Sigma F = mg \sin\theta$. Therefore, we know from Newton's second law that the acceleration of the skateboarder down the ramp is

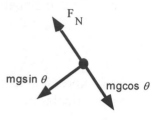

$$a = \frac{\Sigma F}{m} = \frac{mg \sin\theta}{m} = g \sin\theta$$

SOLUTION   Thus, the speed of the skateboarder at the bottom of the ramp is

$$v = \sqrt{v_0^2 + 2ax} = \sqrt{v_0^2 + 2gx \sin\theta} = \sqrt{(2.6 \text{ m/s})^2 + 2(9.80 \text{ m/s}^2)(6.0 \text{ m}) \sin 18°} = \boxed{6.6 \text{ m/s}}$$

---

71.   *REASONING*   We can use the appropriate equation of kinematics to find the acceleration of the bullet. Then Newton's second law can be used to find the average net force on the bullet.

SOLUTION   According to Equation 2.4, the acceleration of the bullet is

$$a = \frac{v - v_0}{t} = \frac{715 \text{ m/s} - 0 \text{ m/s}}{2.50 \times 10^{-3} \text{ s}} = 2.86 \times 10^5 \text{ m/s}^2$$

Therefore, the net average force on the bullet is

$$\Sigma F = ma = (15 \times 10^{-3} \text{ kg})(2.86 \times 10^5 \text{ m/s}^2) = \boxed{4290 \text{ N}}$$

---

75.   *REASONING*   The shortest time to pull the person from the cave corresponds to the maximum acceleration, a_y, that the rope can withstand. We first determine this acceleration and then use kinematic Equation 3.5b ($y = v_{0y}t + \frac{1}{2}a_y t^2$) to find the time t.

SOLUTION   As the person is being pulled from the cave, there are two forces that act on him; they are the tension T in the rope that points vertically upward, and the weight of the person mg that points vertically downward.   Thus, if we take upward as the positive direction, Newton's second law gives $\Sigma F_y = T - mg = ma_y$. Solving for a_y, we have

$$a_y = \frac{T}{m} - g = \frac{T}{W/g} - g = \frac{569 \text{ N}}{(5.20 \times 10^2 \text{ N})/(9.80 \text{ m/s}^2)} - 9.80 \text{ m/s}^2 = 0.92 \text{ m/s}^2$$

Therefore, from Equation 3.5b with $v_{0y} = 0$ m/s, we have $y = \frac{1}{2} a_y t^2$. Solving for t, we find

$$t = \sqrt{\frac{2y}{a_y}} = \sqrt{\frac{2(35.1 \text{ m})}{0.92 \text{ m/s}^2}} = \boxed{8.7 \text{ s}}$$

---

79. **REASONING**   The box comes to a halt because the kinetic frictional force and the component of its weight parallel to the incline oppose the motion and cause the box to slow down. The distance that the box travels up the incline can be can be found by solving Equation 2.9 ($v^2 = v_0^2 + 2ax$) for x. Before we use this approach, however, we must first determine the acceleration of the box as it travels along the incline.

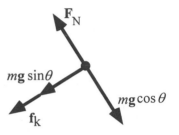

SOLUTION   The figure above shows the free-body diagram for the box. It shows the resolved components of the forces that act on the box. If we take the direction up the incline as the positive x direction, then, Newton's second law gives

$$\Sigma F_x = -mg\sin\theta - f_k = ma_x \qquad \text{or} \qquad -mg\sin\theta - \mu_k F_N = ma_x$$

where we have used Equation 4.8, $f_k = \mu_k F_N$. In the y direction we have

$$\Sigma F_y = F_N - mg\cos\theta = 0 \qquad \text{or} \qquad F_N = mg\cos\theta$$

since there is no acceleration in the y direction. Therefore, the equation for the motion in the x direction becomes

$$-mg\sin\theta - \mu_k mg\cos\theta = ma_x \qquad \text{or} \qquad a_x = -g(\sin\theta + \mu_k \cos\theta)$$

According to Equation 2.9, with this value for the acceleration and the fact that $v = 0$ m/s, the distance that the box slides up the incline is

$$x = -\frac{v_0^2}{2a} = \frac{v_0^2}{2g(\sin\theta + \mu_k \cos\theta)} = \frac{(1.50 \text{ m/s})^2}{2(9.80 \text{ m/s}^2)[\sin 15.0° + (0.180)\cos 15.0°]} = \boxed{0.265 \text{ m}}$$

83. ***REASONING AND SOLUTION***
The penguin comes to a halt on the horizontal surface because the kinetic frictional force opposes the motion and causes it to slow down. The time required for the penguin to slide to a halt ($v = 0$ m/s) after entering the horizontal patch of ice is, according to Equation 2.4,

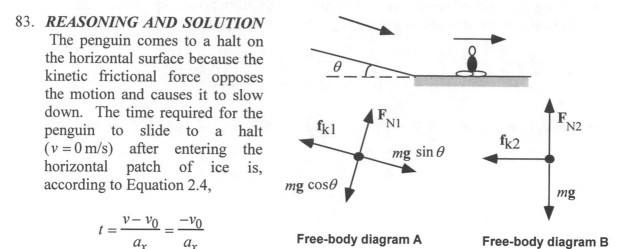

Free-body diagram A   **Free-body diagram B**

$$t = \frac{v - v_0}{a_x} = \frac{-v_0}{a_x}$$

We must, therefore, determine the acceleration of the penguin as it slides along the horizontal patch.

For the penguin sliding on the horizontal patch of ice, we find from free-body diagram B and Newton's second law in the x direction (motion to the right is taken as positive) that

$$\sum F_x = -f_{k2} = ma_x \qquad \text{or} \qquad a_x = \frac{-f_{k2}}{m} = \frac{-\mu_k F_{N2}}{m}$$

In the y direction in free-body diagram B, we have $\sum F_y = F_{N2} - mg = 0$, or $F_{N2} = mg$. Therefore, the acceleration of the penguin is

$$a_x = \frac{-\mu_k mg}{m} = -\mu_k g \qquad\qquad (1)$$

Equation (1) indicates that, in order to find the acceleration a_x, we must find the coefficient of kinetic friction.

We are told in the problem statement that the coefficient of kinetic friction between the penguin and the ice is the same for the incline as for the horizontal patch. Therefore, we can use the motion of the penguin on the incline to determine the coefficient of friction and use it in Equation (1).

For the penguin sliding down the incline, we find from free-body diagram A and Newton's second law (taking the direction of motion as positive) that

$$\sum F_x = mg\sin\theta - f_{k1} = ma_x = 0 \qquad \text{or} \qquad f_{k1} = mg\sin\theta \qquad (2)$$

Here, we have used the fact that the penguin slides down the incline with a constant velocity, so that it has zero acceleration. From Equation 4.8, we know that $f_{k1} = \mu_k F_{N1}$. Applying Newton's second law in the direction perpendicular to the incline, we have

$$\Sigma F_y = F_{N1} - mg\cos\theta = 0 \qquad \text{or} \qquad F_{N1} = mg\cos\theta$$

Therefore, $f_{k1} = \mu_k mg\cos\theta$, so that according to Equation (2), we find

$$f_{k1} = \mu_k mg\cos\theta = mg\sin\theta$$

Solving for the coefficient of kinetic friction, we have

$$\mu_k = \frac{\sin\theta}{\cos\theta} = \tan\theta$$

Finally, the time required for the penguin to slide to a halt after entering the horizontal patch of ice is

$$t = \frac{-v_0}{a_x} = \frac{-v_0}{-\mu_k g} = \frac{v_0}{g\tan\theta} = \frac{1.4 \text{ m/s}}{(9.80 \text{ m/s}^2)\tan 6.9°} = \boxed{1.2 \text{ s}}$$

---

87. ***REASONING AND SOLUTION***
a.   According to Equation 4.4, the weight of an object of mass m on the surface of Mars would be given by

$$W = \frac{GM_M m}{R_M^2}$$

where M_M is the mass of Mars and R_M is the radius of Mars. On the surface of Mars, the weight of the object can be given as $W = mg$ (see Equation 4.5), so

$$mg = \frac{GM_M m}{R_M^2} \qquad \text{or} \qquad g = \frac{GM_M}{R_M^2}$$

Substituting values, we have

$$g = \frac{(6.67\times10^{-11}\text{N}\cdot\text{m}^2/\text{kg}^2)(6.46\times10^{23} \text{ kg})}{(3.39\times10^6 \text{ m})^2} = \boxed{3.75 \text{ m/s}^2}$$

b.   According to Equation 4.5,

$$W = mg = (65 \text{ kg})(3.75 \text{ m/s}^2) = \boxed{2.4\times10^2 \text{ N}}$$

---

91. ***REASONING AND SOLUTION***   Four forces act on the sled. They are the pulling force P, the force of kinetic friction $\mathbf{f}_k$, the weight mg of the sled, and the normal force $\mathbf{F}_N$ exerted on the sled by the surface on which it slides. The following figures show free-body

diagrams for the sled. In the diagram on the right, the forces have been resolved into their x and y components.

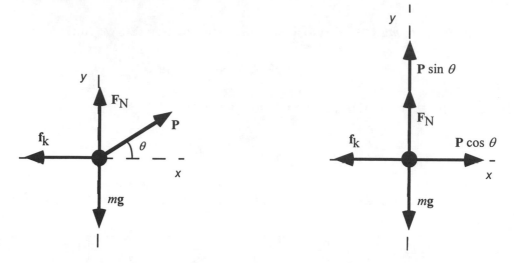

Since the sled is pulled at constant velocity, its acceleration is zero, and Newton's second law in the direction of motion is (with right chosen as the positive direction)

$$\Sigma F_x = P\cos\theta - f_k = ma_x = 0$$

From Equation 4.8, we know that $f_k = \mu_k F_N$, so that the above expression becomes

$$P\cos\theta - \mu_k F_N = 0 \qquad (1)$$

In the vertical direction,

$$\Sigma F_y = P\sin\theta + F_N - mg = ma_y = 0 \qquad (2)$$

Solving Equation (2) for the normal force, and substituting into Equation (1), we obtain

$$P\cos\theta - \mu_k\left(mg - P\sin\theta\right) = 0$$

Solving for μ_k, the coefficient of kinetic friction, we find

$$\mu_k = \frac{P\cos\theta}{mg - P\sin\theta} = \frac{(80.0\text{ N})\cos 30.0°}{(20.0\text{ kg})(9.80\text{ m/s}^2) - (80.0\text{N})\sin 30.0°} = \boxed{0.444}$$

---

95. ***REASONING AND SOLUTION***

a. Each cart has the same mass and acceleration; therefore, the net force acting on any one of the carts is, according to Newton's second law

$$\Sigma F = ma = (26\text{ kg})(0.050\text{ m/s}^2) = \boxed{1.3\text{ N}}$$

b.   The fifth cart must essentially push the sixth, seventh, eight, ninth and tenth cart. In other words, it must exert on the sixth cart a total force of

$$\sum F = ma = 5(26 \text{ kg})(0.050 \text{ m/s}^2) = \boxed{6.5 \text{ N}}$$

---

99. ***REASONING AND SOLUTION***   The system is shown in the drawing. We will let $m_1 = 21.0$ kg, and $m_2 = 45.0$ kg . Then, m_1 will move upward, and m_2 will move downward.   There are two forces that act on each object; they are the tension T in the cord and the weight mg of the object.   The forces are shown in the free-body diagrams at the far right.

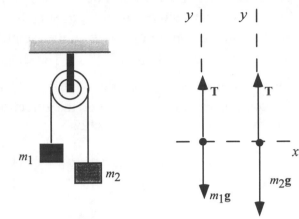

We will take up as the positive direction. If the acceleration of m_1 is a , then the acceleration of m_2 must be $-a$.

From Newton's second law, we have for m_1

$$\sum F_y = T - m_1 g = m_1 a \qquad (1)$$

and for m_2

$$\sum F_y = T - m_2 g = -m_2 a \qquad (2)$$

a.   Eliminating T between these two equations, we obtain

$$a = \frac{m_2 - m_1}{m_2 + m_1} g = \left(\frac{45.0 \text{ kg} - 21.0 \text{ kg}}{45.0 \text{ kg} + 21.0 \text{ kg}} \right) (9.80 \text{ m/s}^2) = \boxed{3.56 \text{ m/s}^2}$$

b.   Eliminating a between Equations (1) and (2), we find

$$T = \frac{2 m_1 m_2}{m_1 + m_2} g = \left[\frac{2(21.0 \text{ kg})(45.0 \text{ kg})}{21.0 \text{ kg} + 45.0 \text{ kg}} \right] (9.80 \text{ m/s}^2) = \boxed{281 \text{ N}}$$

---

103. ***REASONING AND SOLUTION***   Three forces act on the man. They are two upward forces of tension of equal magnitude T, and the force of gravity $m\mathbf{g}$. Therefore, if we take up as the positive direction, Newton's second law gives

$$\sum F = 2T - mg = ma$$

Solving for the acceleration a, we find

$$a = \frac{2T - mg}{m} = \frac{2T}{m} - g = \frac{2(358 \text{ N})}{72.0 \text{ kg}} - 9.80 \text{ m/s}^2 = \boxed{0.14 \text{ m/s}^2}$$

---

107. **REASONING AND SOLUTION** The free-body diagram is shown at the right. The forces that act on the picture are the pressing force P, the normal force $\mathbf{F_N}$ exerted on the picture by the wall, the weight mg of the picture, and the force of static friction $\mathbf{f_s}^{MAX}$. The maximum magnitude for the frictional force is given by Equation 4.7: $f_s^{MAX} = \mu_s F_N$. The picture is in equilibrium, and, if we take the directions to the right and up as positive, we have in the x direction

$$\Sigma F_x = P - F_N = 0 \qquad \text{or} \qquad P = F_N$$

and in the y direction

$$\Sigma F_y = f_s^{MAX} - mg = 0 \qquad \text{or} \qquad f_s^{MAX} = mg$$

Therefore,

$$f_s^{MAX} = \mu_s F_N = mg$$

But since $F_N = P$, we have

$$\mu_s P = mg$$

Solving for P, we have

$$P = \frac{mg}{\mu_s} = \frac{(1.10 \text{ kg})(9.80 \text{ m/s}^2)}{0.660} = \boxed{16.3 \text{ N}}$$

---

111. **REASONING** When the bicycle is coasting straight down the hill, the forces that act on it are the normal force $\mathbf{F_N}$ exerted by the surface of the hill, the force of gravity mg, and the force of air resistance R. When the bicycle climbs the hill, there is one additional force; it is the applied force that is required for the bicyclist to climb the hill at constant speed. We can use our knowledge of the motion of the bicycle down the hill to find R. Once R is known, we can analyze the motion of the bicycle as it climbs the hill.

SOLUTION The figure to the left below shows the free-body diagram for the forces during the downhill motion. The hill is inclined at an angle θ above the horizontal. The figure to the right shows these forces resolved into components parallel to and perpendicular to the line of motion.

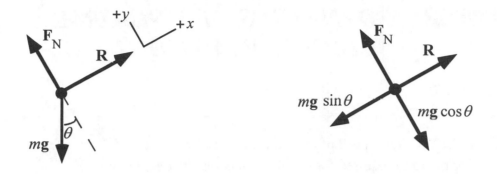

Since the bicyclist is traveling at a constant velocity, his acceleration is zero. Therefore, according to Newton's second law, we have $\Sigma F_x = 0$ and $\Sigma F_y = 0$. Taking the direction up the hill as positive, we have $\Sigma F_x = R - mg \sin \theta = 0$, or

$$R = mg \sin \theta = (80.0 \text{ kg})(9.80 \text{ m/s}^2) \sin 15.0° = 203 \text{ N}$$

When the bicyclist climbs the same hill at constant speed, an applied force P must push the system up the hill, and the force of air resistance will oppose the motion by pointing down the hill.

The figure at the right shows the resolved forces that act on the system during the uphill motion.

Using the same sign convention as above, we have $\Sigma F_x = P - mg \sin \theta - R = 0$, or

$$P = R + mg \sin \theta = 203 \text{ N} + 203 \text{ N} = \boxed{406 \text{ N}}$$

CHAPTER 5 | DYNAMICS OF UNIFORM CIRCULAR MOTION

PROBLEMS

1. **REASONING** The speed of the plane is given by Equation 5.1: $v = 2\pi r / T$, where T is the period or the time required for the plane to complete one revolution.

SOLUTION Solving Equation 5.1 for T we have

$$T = \frac{2\pi r}{v} = \frac{2\pi(2850 \text{ m})}{110 \text{ m/s}} = \boxed{160 \text{ s}}$$

---

5. **REASONING AND SOLUTION** In each case, the magnitude of the centripetal acceleration is given by Equation 5.2, $a_c = v^2 / r$. Therefore,

$$\frac{a_{cA}}{a_{cB}} = \frac{v_A^2 / r_A}{v_B^2 / r_B}$$

Since each boat experiences the same centripetal acceleration, $a_{cA} / a_{cB} = 1$. Solving for the ratio of the speeds gives

$$\frac{v_A}{v_B} = \sqrt{\frac{r_A}{r_B}} = \sqrt{\frac{120 \text{ m}}{240 \text{ m}}} = \boxed{0.71}$$

---

9. **REASONING** The magnitude of the centripetal acceleration of any point on the helicopter blade is given by Equation 5.2, $a_C = v^2 / r$, where r is the radius of the circle on which that point moves. From Equation 5.1: $v = 2\pi r / T$. Combining these two expressions, we obtain

$$a_C = \frac{4\pi^2 r}{T^2}$$

All points on the blade move with the same period T.

SOLUTION The ratio of the centripetal acceleration at the end of the blade (point 1) to that which exists at a point located 3.0 m from the center of the circle (point 2) is

$$\frac{a_{C1}}{a_{C2}} = \frac{4\pi^2 r_1 / T^2}{4\pi^2 r_2 / T^2} = \frac{r_1}{r_2} = \frac{6.7 \text{ m}}{3.0 \text{ m}} = \boxed{2.2}$$

---

11. ***REASONING*** In Example 3, it was shown that the magnitudes of the centripetal acceleration for the two cases are

$$[\text{Radius} = 33 \text{ m}] \qquad a_{\text{C}} = 35 \text{ m}/\text{s}^2$$
$$[\text{Radius} = 24 \text{ m}] \qquad a_{\text{C}} = 48 \text{ m}/\text{s}^2$$

According to Newton's second law, the centripetal force is $F_{\text{C}} = ma_{\text{C}}$ (see Equation 5.3).

SOLUTION a. Therefore, when the sled undergoes the turn of radius 33 m,

$$F_{\text{C}} = ma_{\text{C}} = (350 \text{ kg})(35 \text{ m}/\text{s}^2) = \boxed{1.2 \times 10^4 \text{ N}}$$

b. Similarly, when the radius of the turn is 24 m,

$$F_{\text{C}} = ma_{\text{C}} = (350 \text{ kg})(48 \text{ m}/\text{s}^2) = \boxed{1.7 \times 10^4 \text{ N}}$$

---

17. $\boxed{\text{WWW}}$ ***REASONING*** Let v_0 be the initial speed of the ball as it begins its projectile motion. Then, the centripetal force is given by Equation 5.3: $F_{\text{C}} = mv_0^2/r$. We are given the values for m and r; however, we must determine the value of v_0 from the details of the projectile motion after the ball is released.

In the absence of air resistance, the x component of the projectile motion has zero acceleration, while the y component of the motion is subject to the acceleration due to gravity. The horizontal distance traveled by the ball is given by Equation 3.5a (with $a_x = 0$):

$$x = v_{0x}t = (v_0 \cos\theta)t$$

with t equal to the flight time of the ball while it exhibits projectile motion. The time t can be found by considering the vertical motion. From Equation 3.3b,

$$v_y = v_{0y} + a_y t$$

After a time t, $v_y = -v_{0y}$. Assuming that up and to the right are the positive directions, we have

$$t = \frac{-2v_{0y}}{a_y} = \frac{-2v_0 \sin\theta}{a_y}$$

and

$$x = (v_0 \cos\theta)\left(\frac{-2v_0 \sin\theta}{a_y}\right)$$

Using the fact that $2\sin\theta\cos\theta = \sin 2\theta$, we have

$$x = -\frac{2v_0^2 \cos\theta \sin\theta}{a_y} = -\frac{v_0^2 \sin 2\theta}{a_y} \qquad (1)$$

Equation (1) (with upward and to the right chosen as the positive directions) can be used to determine the speed v_0 with which the ball begins its projectile motion. Then Equation 5.3 can be used to find the centripetal force.

SOLUTION   Solving equation (1) for v_0, we have

$$v_0 = \sqrt{\frac{-x\, a_y}{\sin 2\theta}} = \sqrt{\frac{-(86.75 \text{ m})(-9.80 \text{ m/s}^2)}{\sin 2(41°)}} = 29.3 \text{ m/s}$$

Then, from Equation 5.3,

$$F_C = \frac{mv_0^2}{r} = \frac{(7.3 \text{ kg})(29.3 \text{ m/s})^2}{1.8 \text{ m}} = \boxed{3500 \text{ N}}$$

---

21.   **REASONING AND SOLUTION**   Equation 5.4 gives the relationship between the speed v the angle of banking, and the radius of curvature. Solving for v, we obtain

$$v = \sqrt{r\,g\,\tan\theta} = \sqrt{(120 \text{ m})(9.80 \text{ m/s}^2)\tan 18°} = \boxed{2.0\times10^1 \text{ m/s}}$$

---

25. $\boxed{\text{WWW}}$ **REASONING**   Refer to Figure 5.10 in the text. The horizontal component of the lift **L** is the centripetal force that holds the plane in the circle. Thus,

$$L\sin\theta = \frac{mv^2}{r} \qquad (1)$$

The vertical component of the lift supports the weight of the plane; therefore,

$$L\cos\theta = mg \qquad (2)$$

Dividing the first equation by the second gives

$$\tan\theta = \frac{v^2}{rg} \qquad (3)$$

Equation (3) can be used to determine the angle θ of banking. Once θ is known, then the magnitude of L can be found from either equation (1) or equation (2).

SOLUTION Solving equation (3) for θ gives

$$\theta = \tan^{-1}\left[\frac{(123 \text{ m/s})^2}{(3810 \text{ m})(9.80 \text{ m/s}^2)}\right] = 22.1°$$

The lifting force is, from equation (2),

$$L = \frac{mg}{\cos\theta} = \frac{(2.00\times10^5 \text{ kg})(9.80 \text{ m/s}^2)}{\cos 22.1°} = \boxed{2.12\times10^6 \text{ N}}$$

---

27. $\boxed{\text{WWW}}$ **REASONING** Equation 5.5 gives the orbital speed for a satellite in a circular orbit around the earth. It can be modified to determine the orbital speed around any planet **P** by replacing the mass of the earth M_E by the mass of the planet M_P: $v = \sqrt{GM_P/r}$.

SOLUTION The ratio of the orbital speeds is, therefore,

$$\frac{v_2}{v_1} = \frac{\sqrt{GM_P/r_2}}{\sqrt{GM_P/r_1}} = \sqrt{\frac{r_1}{r_2}}$$

Solving for v_2 gives

$$v_2 = v_1\sqrt{\frac{r_1}{r_2}} = (1.70\times10^4 \text{ m/s})\sqrt{\frac{5.25\times10^6 \text{ m}}{8.60\times10^6 \text{ m}}} = \boxed{1.33\times10^4 \text{ m/s}}$$

---

33. **REASONING** The true weight of the satellite when it is at rest on the planet's surface can be found from Equation 4.4: $W = (GM_P m)/r^2$ where M_P and m are the masses of the planet and the satellite, respectively, and r is the radius of the planet. However, before we can use Equation 4.4, we must determine the mass M_P of the planet.

The mass of the planet can be found by replacing M_E by M_P in Equation 5.6 and solving for M_P. When using Equation 5.6, we note that r corresponds to the radius of the circular orbit *relative to the center of the planet*.

SOLUTION The period of the satellite is $T = 2.00 \text{ h} = 7.20\times10^3 \text{ s}$. From Equation 5.6,

$$M_{\mathrm{P}} = \frac{4\pi^2 r^3}{GT^2} = \frac{4\pi^2 \left[(4.15 \times 10^6 \text{ m}) + (4.1 \times 10^5 \text{ m}) \right]^3}{(6.67 \times 10^{-11} \text{ N} \cdot \text{m}^2 / \text{kg}^2)(7.20 \times 10^3 \text{ s})^2} = 1.08 \times 10^{24} \text{ kg}$$

Using Equation 4.4, we have

$$W = \frac{GM_{\mathrm{P}}m}{r^2} = \frac{(6.67 \times 10^{-11} \text{ N} \cdot \text{m}^2 / \text{kg}^2)(1.08 \times 10^{24} \text{ kg})(5850 \text{ kg})}{(4.15 \times 10^6 \text{ m})^2} = \boxed{2.45 \times 10^4 \text{ N}}$$

---

37. **REASONING**   This situation is similar to the loop-the-loop trick discussed in Section 5.7 of the text. When the plane passes over the top of a vertical circle of radius r such that the passengers experience apparent weightlessness, the centripetal force is provided entirely by the true weight mg. Thus, $mg = mv^2/r$.

SOLUTION   Solving the above expression for r gives

$$r = \frac{v^2}{g} = \frac{(215 \text{ m/s})^2}{9.80 \text{ m/s}^2} = \boxed{4.72 \times 10^3 \text{ m}}$$

---

39. **REASONING**   As the motorcycle passes over the top of the hill, it will experience a centripetal force, the magnitude of which is given by Equation 5.3: $F_{\mathrm{c}} = mv^2/r$. The centripetal force is provided by the net force on the cycle + driver system. At that instant, the net force on the system is composed of the normal force, which points upward, and the weight, which points downward. Taking the direction toward the center of the circle (downward) as the positive direction, we have $F_{\mathrm{c}} = mg - F_{\mathrm{N}}$. This expression can be solved for F_{N}, the normal force.

SOLUTION
a. The magnitude of the centripetal force is

$$F_{\mathrm{c}} = \frac{mv^2}{r} = \frac{(342 \text{ kg})(25.0 \text{ m/s})^2}{126 \text{ m}} = \boxed{1.70 \times 10^3 \text{ N}}$$

b. The magnitude of the normal force is

$$F_{\mathrm{N}} = mg - F_{\mathrm{c}} = (343 \text{ kg})(9.80 \text{ m/s}^2) - 1.70 \times 10^3 \text{ N} = \boxed{1.66 \times 10^3 \text{ N}}$$

---

43. **REASONING AND SOLUTION**   The magnitude of the centripetal force on the ball is given by Equation 5.3: $F_{\mathrm{c}} = mv^2/r$. Solving for v, we have

$$v = \sqrt{\frac{F_c r}{m}} = \sqrt{\frac{(0.028 \text{ N})(0.25 \text{ m})}{0.015 \text{ kg}}} = \boxed{0.68 \text{ m/s}}$$

---

47. ***REASONING AND SOLUTION*** Since the magnitude of the centripetal acceleration is given by Equation 5.2, $a_C = v^2/r$, we can solve for r and find that

$$r = \frac{v^2}{a_C} = \frac{(98.8 \text{ m/s})^2}{3.00(9.80 \text{ m/s}^2)} = \boxed{332 \text{ m}}$$

---

49. ***REASONING AND SOLUTION***   Since the tension serves the same purpose as the normal force at point 1 in Figure 5.21, we have, using the equation for the situation at point 1 with F_{N1} replaced by T,

$$\frac{mv^2}{r} = T - mg$$

Solving for T gives

$$T = \frac{mv^2}{r} + mg = m\left(\frac{v^2}{r} + g\right) = (2100 \text{ kg})\left[\frac{(7.6 \text{ m/s})^2}{15 \text{ m}} + (9.80 \text{ m/s}^2)\right] = \boxed{2.9 \times 10^4 \text{ N}}$$

---

53. $\boxed{\text{WWW}}$ ***REASONING***   If the effects of gravity are not ignored in Example 5, the plane will make an angle θ with the vertical as shown in figure **A** below. The figure **B** shows the forces that act on the plane, and figure **C** shows the horizontal and vertical components of these forces.

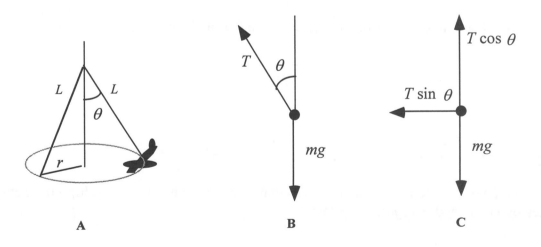

From figure **C** we see that the resultant force in the horizontal direction is the horizontal component of the tension in the guideline and provides the centripetal force. Therefore,

$$T \sin \theta = \frac{mv^2}{r}$$

From figure **A**, the radius r is related to the length L of the guideline by $\quad r = L \sin\theta$; therefore,

$$T \sin \theta = \frac{mv^2}{L \sin \theta} \qquad (1)$$

The resultant force in the vertical direction is zero: $T\cos\theta - mg = 0$, so that

$$T \cos\theta = mg \qquad (2)$$

From equation (2) we have

$$T = \frac{mg}{\cos \theta} \qquad (3)$$

Equation (3) contains two unknown, T and θ. First we will solve equations (1) and (3) simultaneously to determine the value(s) of the angle θ. Once θ is known, we can calculate the tension using equation (3).

SOLUTION Substituting equation (3) into equation (1):

$$\left(\frac{mg}{\cos \theta} \right) \sin \theta = \frac{mv^2}{L \sin \theta}$$

Thus,

$$\frac{\sin^2 \theta}{\cos \theta} = \frac{v^2}{gL} \qquad (4)$$

Using the fact that $\cos^2 \theta + \sin^2 \theta = 1$, equation (4) can be written

$$\frac{1 - \cos^2 \theta}{\cos \theta} = \frac{v^2}{gL}$$

or

$$\frac{1}{\cos \theta} - \cos \theta = \frac{v^2}{gL}$$

This can be put in the form of an equation that is quadratic in $\cos \theta$. Multiplying both sides by $\cos \theta$ and rearranging yields:

$$\cos^2 \theta + \frac{v^2}{gL} \cos \theta - 1 = 0 \qquad (5)$$

Equation (5) is of the form

$$ax^2 + bx + c = 0 \tag{6}$$

with $x = \cos\theta$, $a = 1$, $b = v^2/(gL)$, and $c = -1$. The solution to equation (6) is found from the quadratic formula:

$$x = \frac{-b \pm \sqrt{b^2 - 4ac}}{2a}$$

When $v = 19.0$ m/s, $b = 2.17$. The positive root from the quadratic formula gives $x = \cos\theta = 0.391$. Substitution into equation (3) yields

$$T = \frac{mg}{\cos\theta} = \frac{(0.900 \text{ kg})(9.80 \text{ m/s}^2)}{0.391} = \boxed{23 \text{ N}}$$

When $v = 38.0$ m/s, $b = 8.67$. The positive root from the quadratic formula gives $x = \cos\theta = 0.114$. Substitution into equation (3) yields

$$T = \frac{mg}{\cos\theta} = \frac{(0.900 \text{ kg})(9.80 \text{ m/s}^2)}{0.114} = \boxed{77 \text{ N}}$$

CHAPTER 6 | *WORK AND ENERGY*

1. ***REASONING AND SOLUTION*** We will assume that the tug-of-war rope remains parallel to the ground, so that the force that moves team B is in the same direction as the displacement. According to Equation 6.1, the work done by team A is

$$W = (F \cos\theta)s = (1100 \text{ N})(\cos 0°)(2.0 \text{ m}) = \boxed{2.2 \times 10^3 \text{ J}}$$

5. ***REASONING AND SOLUTION*** Solving Equation 6.1 for the angle θ, we obtain

$$\theta = \cos^{-1}\left(\frac{W}{Fs}\right) = \cos^{-1}\left[\frac{1.10 \times 10^3 \text{ J}}{(30.0 \text{ N})(50.0 \text{ m})}\right] = \boxed{42.8°}$$

9. ***REASONING AND SOLUTION*** According to Equation 6.1, the work done by the husband and wife are, respectively,

$$[\text{Husband}] \qquad W_\text{H} = (F_\text{H} \cos \theta_\text{H})s$$

$$[\text{Wife}] \qquad W_\text{W} = (F_\text{W} \cos \theta_\text{W})s$$

Since both the husband and the wife do the same amount of work,

$$(F_\text{H} \cos \theta_\text{H})s = (F_\text{W} \cos \theta_\text{W})s$$

Since the displacement has the same magnitude s in both cases, the magnitude of the force exerted by the wife is

$$F_\text{W} = F_\text{H} \frac{\cos \theta_\text{H}}{\cos \theta_\text{W}} = (67 \text{ N})\frac{\cos 58°}{\cos 38°} = \boxed{45 \text{ N}}$$

13. ***REASONING*** The work done to launch either object can be found from Equation 6.3, the work-energy theorem, $W = \text{KE}_\text{f} - \text{KE}_0 = \frac{1}{2}mv_\text{f}^2 - \frac{1}{2}mv_0^2$.

SOLUTION
a. The work required to launch the hammer is

$$W = \tfrac{1}{2}mv_\text{f}^2 - \tfrac{1}{2}mv_0^2 = \tfrac{1}{2}m\left(v_\text{f}^2 - v_0^2\right) = \tfrac{1}{2}(7.3 \text{ kg})\left[(29 \text{ m/s})^2 - (0 \text{ m/s})^2\right] = \boxed{3.1 \times 10^3 \text{ J}}$$

b. Similarly, the work required to launch the bullet is

$$W = \tfrac{1}{2}m\left(v_f^2 - v_0^2\right) = \tfrac{1}{2}(0.0026 \text{ kg})\left[(410 \text{ m/s})^2 - (0 \text{ m/s})^2\right] = \boxed{2.2 \times 10^2 \text{ J}}$$

---

15. **WWW** **REASONING AND SOLUTION** The work required to bring each car up to speed is, from the work-energy theorem, $W = KE_f - KE_0 = \tfrac{1}{2}mv_f^2 - \tfrac{1}{2}mv_0^2$. Therefore,

$$W_B = \tfrac{1}{2}m\left(v_f^2 - v_0^2\right) = \tfrac{1}{2}(1.20 \times 10^3 \text{ kg})\left[(40.0 \text{ m/s})^2 - (0 \text{ m/s})^2\right] = 9.60 \times 10^5 \text{ J}$$

$$W_B = \tfrac{1}{2}m\left(v_f^2 - v_0^2\right) = \tfrac{1}{2}(2.00 \times 10^3 \text{ kg})\left[(40.0 \text{ m/s})^2 - (0 \text{ m/s})^2\right] = 1.60 \times 10^6 \text{ J}$$

The *additional* work required to bring car B up to speed is, therefore,

$$W_B - W_A = (1.6 \times 10^6 \text{ J}) - (9.60 \times 10^5 \text{ J}) = \boxed{6.4 \times 10^5 \text{ J}}$$

---

19. **REASONING** According to the work-energy theorem, the kinetic energy of the sled increases in each case because work is done on the sled. The work-energy theorem is given by Equation 6.3: $W = KE_f - KE_0 = \tfrac{1}{2}mv_f^2 - \tfrac{1}{2}mv_0^2$. The work done on the sled is given by Equation 6.1: $W = (F\cos\theta)s$. The work done in each case can, therefore, be expressed as

$$W_1 = (F\cos 0°)s = \tfrac{1}{2}mv_f^2 - \tfrac{1}{2}mv_0^2 = \Delta KE_1$$

and

$$W_2 = (F\cos 62°)s = \tfrac{1}{2}mv_f^2 - \tfrac{1}{2}mv_0^2 = \Delta KE_2$$

The fractional increase in the kinetic energy of the sled when $\theta = 0°$ is

$$\frac{\Delta KE_1}{KE_0} = \frac{(F\cos 0°)s}{KE_0} = 0.38$$

Therefore,

$$Fs = (0.38)\, KE_0 \tag{1}$$

The fractional increase in the kinetic energy of the sled when $\theta = 62°$ is

$$\frac{\Delta KE_2}{KE_0} = \frac{(F\cos 62°)s}{KE_0} = \frac{Fs}{KE_0}(\cos 62°) \tag{2}$$

Equation (1) can be used to substitute in Equation (2) for Fs.

SOLUTION   Combining Equations (1) and (2), we have

$$\frac{\Delta KE_2}{KE_0} = \frac{Fs}{KE_0}(\cos 62°) = \frac{(0.38)\,KE_0}{KE_0}(\cos 62°) = (0.38)(\cos 62°) = 0.18$$

Thus, the sled's kinetic energy would increase by $\boxed{18\,\%}$.

---

23. $\boxed{\text{WWW}}$ **REASONING**   When the satellite goes from the first to the second orbit, its kinetic energy changes. The net work that the external force must do to change the orbit can be found from the work-energy theorem: $W = KE_f - KE_0 = \frac{1}{2}mv_f^2 - \frac{1}{2}mv_0^2$. The speeds v_f and v_0 can be obtained from Equation 5.5 for the speed of a satellite in a circular orbit of radius r. Given the speeds, the work energy theorem can be used to obtain the work.

SOLUTION   According to Equation 5.5, $v = \sqrt{GM_E/r}$. Substituting into the work-energy theorem, we have

$$W = \tfrac{1}{2}mv_f^2 - \tfrac{1}{2}mv_0^2 = \tfrac{1}{2}m\left(v_f^2 - v_0^2\right) = \tfrac{1}{2}m\left[\left(\sqrt{\frac{GM_E}{r_f}}\right)^2 - \left(\sqrt{\frac{GM_E}{r_0}}\right)^2\right] = \frac{GM_E m}{2}\left(\frac{1}{r_f} - \frac{1}{r_0}\right)$$

Therefore,

$$W = \frac{(6.67 \times 10^{-11}\ \text{N} \cdot \text{m}^2/\text{kg}^2)(5.98 \times 10^{24}\ \text{kg})(6200\ \text{kg})}{2}$$

$$\times\left(\frac{1}{7.0 \times 10^6\ \text{m}} - \frac{1}{3.3 \times 10^7\ \text{m}}\right) = \boxed{1.4 \times 10^{11}\ \text{J}}$$

---

27.   **REASONING**   During each portion of the trip, the work done by the resistive force is given by Equation 6.1, $W = (F\cos\theta)s$. Since the resistive force always points opposite to the displacement of the bicyclist, $\theta = 180°$; hence, on each part of the trip, $W = (F\cos 180°)s = -Fs$. The work done by the resistive force during the round trip is the algebraic sum of the work done during each portion of the trip.

SOLUTION
a. The work done during the round trip is, therefore,

$$W_{\text{total}} = W_1 + W_2 = -F_1 s_1 - F_2 s_2$$

$$= -(3.0 \text{ N})(5.0 \times 10^3 \text{ m}) - (3.0 \text{ N})(5.0 \times 10^3 \text{ m}) = \boxed{-3.0 \times 10^4 \text{ J}}$$

b. Since the work done by the resistive force over the closed path is *not* zero, we can conclude that $\boxed{\text{the resistive force is } not \text{ a conservative force}}$.

---

31. ***REASONING*** The only nonconservative force that acts on Rocket man is the force generated by the propulsion unit. Thus, Equation 6.7b can be used to determine the net work done by this force.

SOLUTION From Equation 6.7b, we have

$$W_{nc} = \Delta KE + \Delta PE = \left(\tfrac{1}{2}mv_f^2 - \tfrac{1}{2}mv_0^2 \right) + \left(mgh_f - mgh_0 \right)$$

Since rocket man starts from rest, $v_0 = 0$ m/s. If we take $h_0 = 0$ m on the ground, we have

$$W_{nc} = \frac{1}{2}mv_f^2 + mgh_f = m\left(\tfrac{1}{2}v_f^2 + gh_f \right)$$

$$= (136 \text{ kg})\left[\tfrac{1}{2}(5.0 \text{ m/s})^2 + (9.80 \text{ m/s}^2)(16 \text{ m}) \right] = \boxed{2.3 \times 10^4 \text{ J}}$$

---

35. ***REASONING*** No forces other than gravity act on the rock since air resistance is being ignored. Thus, the net work done by nonconservative forces is zero, $W_{nc} = 0$ J. Consequently, the principle of conservation of mechanical energy holds, so the total mechanical energy remains constant as the rock falls.

If we take $h = 0$ m at ground level, the gravitational potential energy at any height h is, according to Equation 6.5, $PE = mgh$. The kinetic energy of the rock is given by Equation 6.2: $KE = \frac{1}{2}mv^2$. In order to use Equation 6.2, we must have a value for v^2 at each desired height h. The quantity v^2 can be found from Equation 2.9 with $v_0 = 0$ m/s, since the rock is released from rest. At a height h, the rock has fallen through a distance $(20.0 \text{ m}) - h$, and according to Equation 2.9, $v^2 = 2ay = 2a[(20.0 \text{ m}) - h]$. Therefore, the kinetic energy at any height h is given by $KE = ma[(20.0 \text{ m}) - h]$. The total energy of the rock at any height h is the sum of the kinetic energy and potential energy at the particular height.

SOLUTION The calculations are performed below for $h = 10.0$ m. The table that follows also shows the results for $h = 20.0$ m and $h = 0$ m.

$$PE = mgh = (2.00 \text{ kg})(9.80 \text{ m/s}^2)(10.0 \text{ m}) = \boxed{196 \text{ J}}$$

$$KE = ma\left[(20.0 \text{ m}) - h\right] = (2.00 \text{ kg})(9.80 \text{ m/s}^2)(20.0 \text{ m} - 10.0 \text{ m}) = \boxed{196 \text{ J}}$$

$$E = KE + PE = 196 \text{ J} + 196 \text{ J} = \boxed{392 \text{ J}}$$

h (m)	KE (J)	PE (J)	E (J)
20.0	0	392	392
10.0	196	196	392
0	392	0	392

Note that in each case, the value of E is the same, because mechanical energy is conserved.

---

39. $\boxed{\textbf{WWW}}$ *REASONING* Gravity is the only force acting on the vaulters, since friction and air resistance are being ignored. Therefore, the net work done by the nonconservative forces is zero, and the principle of conservation of mechanical energy holds.

SOLUTION Let E_{2f} represent the total mechanical energy of the second vaulter at the ground, and E_{20} represent the total mechanical energy of the second vaulter at the bar. Then, the principle of mechanical energy is written for the second vaulter as

$$\underbrace{\frac{1}{2}mv_{2f}^2 + mgh_{2f}}_{E_{2f}} = \underbrace{\frac{1}{2}mv_{20}^2 + mgh_{20}}_{E_{20}}$$

Since the mass m of the vaulter appears in every term of the equation, m can be eliminated algebraically. The quantity $h_{20} = h$, where h is the height of the bar. Furthermore, when the vaulter is at ground level, $h_{2f} = 0$ m. Solving for v_{20} we have

$$v_{20} = \sqrt{v_{2f}^2 - 2gh} \tag{1}$$

In order to use Equation (1), we must first determine the height h of the bar. The height h can be determined by applying the principle of conservation of mechanical energy to the first vaulter on the ground and at the bar. Using notation similar to that above, we have

$$\underbrace{\frac{1}{2}mv_{1f}^2 + mgh_{1f}}_{E_{1f}} = \underbrace{\frac{1}{2}mv_{10}^2 + mgh_{10}}_{E_{10}}$$

where E_{10} corresponds to the total mechanical energy of the first vaulter at the bar. The height of the bar is, therefore,

$$h = h_{10} = \frac{v_{1f}^2 - v_{10}^2}{2g} = \frac{(8.90 \text{ m/s})^2 - (1.00 \text{ m/s})^2}{2(9.80 \text{ m/s}^2)} = 3.99 \text{ m}$$

The speed at which the second vaulter clears the bar is, from Equation (1),

$$v_{20} = \sqrt{v_{2f}^2 - 2gh} = \sqrt{(9.00 \text{ m/s})^2 - 2(9.80 \text{ m/s}^2)(3.99 \text{ m})} = \boxed{1.7 \text{ m/s}}$$

---

41. **REASONING**   Friction and air resistance are being ignored. The normal force from the slide is perpendicular to the motion, so it does no work. Thus, no net work is done by nonconservative forces, and the principle of conservation of mechanical energy applies.

SOLUTION   Applying the principle of conservation of mechanical energy to the swimmer at the top and the bottom of the slide, we have

$$\underbrace{\tfrac{1}{2}mv_f^2 + mgh_f}_{E_f} = \underbrace{\tfrac{1}{2}mv_0^2 + mgh_0}_{E_0}$$

If we let h be the height of the bottom of the slide above the water, $h_f = h$, and $h_0 = H$. Since the swimmer starts from rest, $v_0 = 0$ m/s, and the above expression becomes

$$\tfrac{1}{2}v_f^2 + gh = gH$$

Solving for H, we obtain

$$H = h + \frac{v_f^2}{2g}$$

Before we can calculate H, we must find v_f and h. Since the velocity in the horizontal direction is constant,

$$v_f = \frac{\Delta x}{\Delta t} = \frac{5.00 \text{ m}}{0.500 \text{ s}} = 10.0 \text{ m/s}$$

The vertical displacement of the swimmer after leaving the slide is, from Equation 3.5b (with down being negative),

$$y = \tfrac{1}{2}a_y t^2 = \tfrac{1}{2}(-9.80 \text{ m/s}^2)(0.500 \text{ s})^2 = -1.23 \text{ m}$$

Therefore, $h = 1.23$ m. Using these values of v_f and h in the above expression for H, we find

$$H = h + \frac{v_f^2}{2g} = 1.23 \text{ m} + \frac{(10.0 \text{ m/s})^2}{2(9.80 \text{ m/s}^2)} = \boxed{6.33 \text{ m}}$$

45. ***REASONING AND SOLUTION*** The work-energy theorem can be used to determine the change in kinetic energy of the car according to Equation 6.8:

$$W_{nc} = \underbrace{\left(\tfrac{1}{2}mv_f^2 + mgh_f\right)}_{E_f} - \underbrace{\left(\tfrac{1}{2}mv_0^2 + mgh_0\right)}_{E_0}$$

The nonconservative forces are the forces of friction and the force due to the chain mechanism. Thus, we can rewrite Equation 6.8 as

$$W_{friction} + W_{chain} = \left(KE_f + mgh_f\right) - \left(KE_0 + mgh_0\right)$$

We will measure the heights from ground level. Solving for the change in kinetic energy of the car, we have

$$KE_f - KE_0 = W_{friction} + W_{chain} - mgh_f + mgh_0$$

$$= -2.00 \times 10^4 \text{ J} + 3.00 \times 10^4 \text{ J} - (375 \text{ kg})(9.80 \text{ m/s}^2)(20.0 \text{ m})$$
$$+ (375 \text{ kg})(9.80 \text{ m/s}^2)(5.00 \text{ m}) = \boxed{-4.51 \times 10^4 \text{ J}}$$

---

53. ***REASONING AND SOLUTION***   According to the work-energy theorem as given in Equation 6.8, we have

$$W_{nc} = \left(\tfrac{1}{2}mv_f^2 + mgh_f\right) - \left(\tfrac{1}{2}mv_0^2 + mgh_0\right)$$

The metal piece starts at rest and is at rest just as it barely strikes the bell, so that $v_f = v_0 = 0$ m/s. In addition, $h_f = h$ and $h_0 = 0$ m, while $W_{nc} = 0.25\left(\tfrac{1}{2}Mv^2\right)$, where M and v are the mass and speed of the hammer. Thus, the work-energy theorem becomes

$$0.25\left(\tfrac{1}{2}Mv^2\right) = mgh$$

Solving for the speed of the hammer, we find

$$v = \sqrt{\frac{2mgh}{0.25M}} = \sqrt{\frac{2(0.400 \text{ kg})(9.80 \text{ m/s}^2)(5.00 \text{ m})}{0.25\,(9.00 \text{ kg})}} = \boxed{4.17 \text{ m/s}}$$

57. ***REASONING AND SOLUTION*** One kilowatt·hour is the amount of work or energy generated when one kilowatt of power is supplied for a time of one hour. From Equation 6.10a, we know that $W = \overline{P}t$. Using the fact that $1 \text{ kW} = 1.0 \times 10^3$ J/s and that $1 \text{h} = 3600$ s, we have

$$1.0 \text{ kWh} = (1.0 \times 10^3 \text{ J/s})(1 \text{ h}) = (1.0 \times 10^3 \text{ J/s})(3600 \text{ s}) = \boxed{3.6 \times 10^6 \text{ J}}$$

---

61. ***REASONING AND SOLUTION*** In the drawings below, the positive direction is to the right. When the boat is not pulling a skier, the engine must provide a force F_1 to overcome the resistive force of the water F_R. Since the boat travels with constant speed, these forces must be equal in magnitude and opposite in direction.

$$-F_R + F_1 = ma = 0$$

or

$$F_R = F_1 \qquad (1)$$

When the boat is pulling a skier, the engine must provide a force F_2 to balance the resistive force of the water F_R and the tension in the tow rope T.

$$-F_R + F_2 - T = ma = 0$$

or

$$F_2 = F_R + T \qquad (2)$$

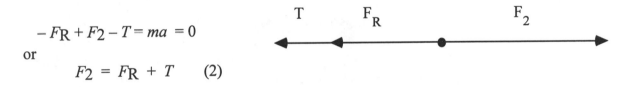

The average power is given by $\overline{P} = F\overline{v}$, according to Equation 6.11 in the text. Since the boat moves with the same speed in both cases, $v_1 = v_2$, and we have

$$\frac{\overline{P}_1}{F_1} = \frac{\overline{P}_2}{F_2} \qquad \text{or} \qquad F_2 = F_1 \frac{\overline{P}_2}{\overline{P}_1}$$

Using Equations (1) and (2), this becomes

$$F_1 + T = F_1 \frac{\overline{P}_2}{\overline{P}_1}$$

Solving for T gives

$$T = F_1 \left(\frac{\overline{P}_2}{\overline{P}_1} - 1 \right)$$

The force F_1 can be determined from $\overline{P}_1 = F_1 v_1$, thereby giving

$$T = \frac{\overline{P}_1}{v_1}\left(\frac{\overline{P}_2}{\overline{P}_1} - 1\right) = \frac{7.50 \times 10^4 \text{ W}}{12 \text{ m/s}}\left(\frac{8.30 \times 10^4 \text{ W}}{7.50 \times 10^4 \text{ W}} - 1\right) = \boxed{6.7 \times 10^2 \text{ N}}$$

---

63. ***REASONING***   The area under the force-versus-displacement graph over any displacement interval is equal to the work done over that displacement interval.

SOLUTION
a. Since the area under the force-versus-displacement graph from $s = 0$ to 0.50 m is greater for bow 1, it follows that $\boxed{\text{bow 1 requires more work to draw the bow fully}}$.

b. The work required to draw bow 1 is equal to the area of the triangular region under the force-versus-displacement curve for bow 1. Since the area of a triangle is equal one-half times the base of the triangle times the height of the triangle, we have

$$W_1 = \frac{1}{2}(0.50\text{m})(350 \text{ N}) = 88 \text{ J}$$

For bow 2, we note that each small square under the force-versus-displacement graph has an area of

$$(0.050 \text{ m})(40.0 \text{ N}) = 2.0 \text{ J}$$

We estimate that there are approximately 31.3 squares under the force-versus-displacement graph for bow 2; therefore, the total work done is

$$(31.3 \text{ squares})\left(\frac{2.0 \text{ J}}{\text{square}}\right) = 63 \text{ J}$$

Therefore, the additional work required to stretch bow 1 as compared to bow 2 is

$$88 \text{ J} - 63 \text{ J} = \boxed{25 \text{ J}}$$

---

69. ***REASONING AND SOLUTION***   Since we are ignoring friction and air resistance, the net work done by the nonconservative forces is zero, and the principle of conservation of mechanical energy holds. Let E_0 be the total mechanical energy of the vaulter at take-off and E_f be the total mechanical energy as he clears the bar. The principle of conservation of mechanical energy gives

$$\underbrace{\frac{1}{2}mv_f^2 + mgh_f}_{E_f} = \underbrace{\frac{1}{2}mv_0^2 + mgh_0}_{E_0}$$

For maximum height h_f, the vaulter just clears the bar with zero speed, $v_f = 0$ m/s. If we let $h_0 = 0$ m at ground level, we have

$$mgh_f = \frac{1}{2}mv_0^2$$

Solving for h_f, we find

$$h_f = \frac{v_0^2}{2g} = \frac{(9.00 \text{ m/s})^2}{2(9.80 \text{ m/s}^2)} = \boxed{4.13 \text{ m}}$$

---

71. ***REASONING AND SOLUTION***   The work done by the retarding force is given by Equation 6.1: $W = (F\cos\theta)s$. Since the force is a retarding force, it must point opposite to the direction of the displacement, so that $\theta = 180°$. Thus, we have

$$W = (F\cos\theta)s = (3.0\times10^3 \text{ N})(\cos 180°)(850 \text{ m}) = \boxed{-2.6\times10^6 \text{ J}}$$

The work done by this force is $\boxed{\text{negative}}$, because the retarding force is directed opposite to the direction of the displacement of the truck.

---

75. ***REASONING*** The work-energy theorem can be used to determine the net work done on the car by the nonconservative forces of friction and air resistance. The work-energy theorem is, according to Equation 6.8,

$$W_{nc} = \left(\tfrac{1}{2}mv_f^2 + mgh_f\right) - \left(\tfrac{1}{2}mv_0^2 + mgh_0\right)$$

The nonconservative forces are the force of friction, the force of air resistance, and the force provided by the engine. Thus, we can rewrite Equation 6.8 as

$$W_{friction} + W_{air} + W_{engine} = \left(\tfrac{1}{2}mv_f^2 + mgh_f\right) - \left(\tfrac{1}{2}mv_0^2 + mgh_0\right)$$

This expression can be solved for $W_{friction} + W_{air}$.

SOLUTION   We will measure the heights from sea level, where $h_0 = 0$ m. Since the car starts from rest, $v_0 = 0$ m/s. Thus, we have

$$W_{friction} + W_{air} = m\left(\tfrac{1}{2}v_f^2 + gh_f\right) - W_{engine}$$

$$= (1.50\times10^3 \text{ kg})\left[\tfrac{1}{2}(27.0 \text{ m/s})^2 + (9.80 \text{ m/s}^2)(2.00\times10^2 \text{ m})\right] - (4.70\times10^6 \text{ J})$$

$$= \boxed{-1.21\times10^6 \text{ J}}$$

79.   **WWW**   *REASONING*   After the wheels lock, the only nonconservative force acting on the truck is friction. The work done by this conservative force can be determined from the work-energy theorem. According to Equation 6.8,

$$W_{nc} = E_f - E_0 = \left(\tfrac{1}{2}mv_f^2 + mgh_f\right) - \left(\tfrac{1}{2}mv_0^2 + mgh_0\right) \tag{1}$$

where W_{nc} is the work done by friction. According to Equation 6.1, $W = (F\cos\theta)s$; since the force of kinetic friction points opposite to the displacement, $\theta = 180°$. According to Equation 4.8, the kinetic frictional force has a magnitude of $f_k = \mu_k F_N$, where μ_k is the coefficient of kinetic friction and F_N is the magnitude of the normal force. Thus,

$$W_{nc} = (f_k \cos\theta)s = \mu_k F_N (\cos 180°)s = -\mu_k F_N s \tag{2}$$

Since the truck is sliding down an incline, we refer to the free-body diagram in order to determine the magnitude of the normal force F_N. The free-body diagram at the right shows the three forces that act on the truck. They are the normal force $\mathbf{F_N}$, the force of kinetic friction $\mathbf{f_k}$, and the weight $m\mathbf{g}$. The weight has been resolved into its components, and these vectors are shown as dashed arrows. From the free body diagram and the fact that the truck does not accelerate perpendicular to the incline, we can see that the magnitude of the normal force is given by

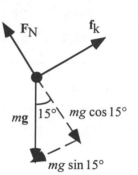

$$F_N = mg\cos 15°$$

Therefore, Equation (2) becomes

$$W_{nc} = -\mu_k\left(mg\cos 15°\right)s \tag{3}$$

Combining Equations (1), (2), and (3), we have

$$-\mu_k\left(mg\cos 15°\right)s = \left(\tfrac{1}{2}mv_f^2 + mgh_f\right) - \left(\tfrac{1}{2}mv_0^2 + mgh_0\right) \tag{4}$$

This expression may be solved for s.

SOLUTION   When the car comes to a stop, $v_f = 0$ m/s. If we take $h_f = 0$ m at ground level, then, from the drawing at the right, we see that $h_0 = s \sin 15°$. Equation (4) then gives

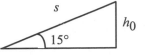

$$s = \frac{v_0^2}{2g\left(\mu_k \cos 15° - \sin 15°\right)} = \frac{(11.1 \text{ m/s})^2}{2(9.80 \text{ m/s}^2)\left[(0.750)(\cos 15°) - \sin 15°\right]} = \boxed{13.5 \text{ m}}$$

CHAPTER 7 | *IMPULSE AND MOMENTUM*

PROBLEMS

---

1. ***REASONING AND SOLUTION*** The impulse is given directly by Equation 7.1:

$$\textbf{Impulse} = \overline{\textbf{F}}\,\Delta t = (+1400 \text{ N})(7.9 \times 10^{-3} \text{ s}) = \boxed{+11 \text{ N} \cdot \text{s}}$$

The plus sign indicates that the direction of the impulse is the same as that of $\overline{\textbf{F}}$.

---

5. ***REASONING AND SOLUTION*** The impulse can be found from Equation 7.4, the impulse-momentum theorem:

$$\underbrace{\overline{\textbf{F}}\,\Delta t}_{\text{Impulse}} = \underbrace{m\textbf{v}_f}_{\substack{\text{Final} \\ \text{momentum}}} - \underbrace{m\textbf{v}_0}_{\substack{\text{Initial} \\ \text{momentum}}}$$

$$= m(\textbf{v}_f - \textbf{v}_0) = (0.35 \text{ kg})\left[(-21 \text{ m/s}) - (+4.0 \text{ m/s})\right] = \boxed{-8.7 \text{ kg} \cdot \text{m/s}}$$

The minus sign indicates that the direction of the impulse is the same as that of the final velocity.

---

9. **WWW** ***REASONING*** The impulse applied to the golf ball by the floor can be found from Equation 7.4, the impulse-momentum theorem: $\overline{\textbf{F}}\,\Delta t = m\textbf{v}_f - m\textbf{v}_0$. Only the vertical component of the ball's momentum changes during impact with the floor. In order to use Equation 7.4 directly, we must first find the vertical components of the initial and final velocities. We begin, then, by finding these velocity components.

SOLUTION The figures below show the initial and final velocities of the golf ball.

Before impact   *After impact*

If we take up as the positive direction, then the vertical components of the initial and final velocities are, respectively, $\textbf{v}_{0y} = -v_0 \cos 30.0°$ and $\textbf{v}_{fy} = v_f \cos 30.0°$. Then, from Equation 7.4 the magnitude of the impulse is

$$\overline{F}\Delta t = m(v_{f_y} - v_{0y}) = m\left[(v_f \cos 30.0°) - (-v_0 \cos 30.0°)\right]$$

Since $v_0 = v_f = 45$ m/s,

$$\overline{F}\Delta t = 2mv_0 \cos 30.0° = 2(0.047 \text{ kg})(45 \text{ m/s})(\cos 30.0°) = \boxed{3.7 \text{ N·s}}$$

---

17. **REASONING** Batman and the boat with the criminal constitute the system. Gravity acts on this system as an external force; however, gravity acts vertically, and we are concerned only with the horizontal motion of the system. If we neglect air resistance and friction, there are no external forces that act horizontally; therefore, the total linear momentum in the horizontal direction is conserved. When Batman collides with the boat, the horizontal component of his velocity is zero, so the statement of conservation of linear momentum in the horizontal direction can be written as

$$\underbrace{(m_1 + m_2)v_f}_{\substack{\text{Total horizontal momentum}\\\text{after collision}}} = \underbrace{m_1 v_{01} + 0}_{\substack{\text{Total horizontal momentum}\\\text{before collision}}}$$

Here, m_1 is the mass of the boat, and m_2 is the mass of Batman. This expression can be solved for v_f, the velocity of the boat after Batman lands in it.

SOLUTION Solving for v_f gives

$$v_f = \frac{m_1 v_{01}}{m_1 + m_2} = \frac{(510 \text{ kg})(+11 \text{ m/s})}{510 \text{ kg} + 91 \text{ kg}} = \boxed{+9.3 \text{ m/s}}$$

The plus sign indicates that the boat continues to move in its initial direction of motion; it does not recoil.

---

21. **REASONING AND SOLUTION** The comet piece and Jupiter constitute an isolated system, since no external forces act on them. Therefore, the head-on collision obeys the conservation of linear momentum:

$$\underbrace{(m_{\text{comet}} + m_{\text{Jupiter}})v_f}_{\substack{\text{Total momentum}\\\text{after collision}}} = \underbrace{m_{\text{comet}} v_{\text{comet}} + m_{\text{Jupiter}} v_{\text{Jupiter}}}_{\substack{\text{Total momentum}\\\text{before collision}}}$$

where v_f is the common velocity of the comet piece and Jupiter after the collision. We assume initially that Jupiter is moving in the $+x$ direction so $v_{\text{Jupiter}} = +1.3 \times 10^4$ m/s. The comet piece must be moving in the opposite direction so $v_{\text{comet}} = -6.0 \times 10^4$ m/s. The final velocity v_f of Jupiter and the comet piece can be written as $v_f = v_{\text{Jupiter}} + \Delta v$ where Δv is the change in velocity of Jupiter due to the collision. Substituting this expression into the conservation of momentum equation gives

$$(m_{\text{comet}} + m_{\text{Jupiter}})(v_{\text{Jupiter}} + \Delta v) = m_{\text{comet}} v_{\text{comet}} + m_{\text{Jupiter}} v_{\text{Jupiter}}$$

Multiplying out the left side of this equation, algebraically canceling the term $m_{\text{Jupiter}} v_{\text{Jupiter}}$ from both sides of the equation, and solving for Δv yields

$$\Delta v = \frac{m_{\text{comet}}(v_{\text{comet}} - v_{\text{Jupiter})}}{m_{\text{comet}} + m_{\text{Jupiter}}}$$

$$= \frac{(4.0 \times 10^{12} \text{ kg})(-6.0 \times 10^4 \text{ m/s} - 1.3 \times 10^4 \text{ m/s})}{4.0 \times 10^{12} \text{ kg} + 1.9 \times 10^{27} \text{ kg}} = -1.5 \times 10^{-10} \text{ m/s}$$

The change in Jupiter's speed is $\boxed{1.5 \times 10^{-10} \text{ m/s}}$.

___

23. ***REASONING***   We will define the system to be the platform, the two people and the ball. Since the ball travels nearly horizontally, the effects of gravity are negligible. Momentum is conserved. Since the initial momentum of the system is zero, it must remain zero as the ball is thrown and caught. While the ball is in motion, the platform will recoil in such a way that the total momentum of the system remains zero. As the ball is caught, the system must come to rest so that the total momentum remains zero. The distance that the platform moves before coming to rest again can be determined by using the expressions for momentum conservation and the kinematic description for this situation.

SOLUTION   While the ball is in motion, we have

$$MV + mv = 0 \tag{1}$$

where M is the combined mass of the platform and the two people, V is the recoil velocity of the platform, m is the mass of the ball, and v is the velocity of the ball.

The distance that the platform moves is given by

$$x = Vt \tag{2}$$

where t is the time that the ball is in the air. The time that the ball is in the air is given by

$$t = \frac{L}{v - V} \tag{3}$$

where L is the length of the platform, and the quantity $(v - V)$ is the velocity of the ball relative to the platform. Remember, both the ball and the platform are moving while the ball is in the air. Combining equations (2) and (3) gives

$$x = \left(\frac{V}{v - V}\right) L \tag{4}$$

From equation (1) the ratio of the velocities is $V/v = -(m/M)$. Equation (4) then gives

$$x = -\frac{(V/v)L}{1 - (V/v)} = \frac{(-m/M)L}{1 + (m/M)} = -\frac{mL}{M + m} = -\frac{(6.0 \text{ kg})(2.0 \text{ m})}{118 \text{ kg} + 6.0 \text{ kg}} = -0.097 \text{ m}$$

The minus sign indicates that displacement of the platform is in the opposite direction to the displacement of the ball. The distance moved by the platform is the magnitude of this displacement, or $\boxed{0.097 \text{ m}}$.

---

27. **REASONING** Since all of the collisions are elastic, the total mechanical energy of the ball is conserved. However, since gravity affects its vertical motion, its linear momentum is *not* conserved. If h_f is the maximum height of the ball on its final bounce, conservation of energy gives

$$\underbrace{\frac{1}{2}mv_f^2 + mgh_f}_{E_f} = \underbrace{\frac{1}{2}mv_0^2 + mgh_0}_{E_0}$$

$$h_f = \frac{v_0^2 - v_f^2}{2g} + h_0$$

SOLUTION In order to use this expression, we must obtain the values for the velocities v_0 and v_f. The initial velocity has only a horizontal component, $v_0 = v_{0x}$. The final velocity also has only a horizontal component since the ball is at the top of its trajectory, $v_f = v_{fx}$. No forces act in the horizontal direction so the momentum of the ball in this direction is conserved, hence $v_0 = v_f$. Therefore

$$h_f = h_0 = \boxed{3.00 \text{ m}}$$

---

31. $\boxed{\text{WWW}}$ **REASONING** The system consists of the two balls. The total linear momentum of the two-ball system is conserved because the net external force acting on it is zero. The principle of conservation of linear momentum applies whether or not the collision is elastic.

$$\underbrace{m_1 v_{f1} + m_2 v_{f2}}_{\substack{\text{Total momentum} \\ \text{after collision}}} = \underbrace{m_1 v_{01} + 0}_{\substack{\text{Total momentum} \\ \text{before collision}}}$$

When the collision is elastic, the kinetic energy is also conserved during the collision

$$\underbrace{\tfrac{1}{2}m_1 v_{f1}^2 + \tfrac{1}{2}m_2 v_{f2}^2}_{\substack{\text{Total kinetic energy} \\ \text{after collision}}} = \underbrace{\tfrac{1}{2}m_1 v_{01}^2 + 0}_{\substack{\text{Total kinetic energy} \\ \text{before collision}}}$$

SOLUTION

a. The final velocities for an elastic collision are determined by simultaneously solving the above equations for the final velocities. The procedure is discussed in Example 7 in the text, and leads to Equations 7.8a and 7.8b. According to Equation 7.8:

$$v_{f1} = \left(\frac{m_1 - m_2}{m_1 + m_2}\right)v_{01} \qquad \text{and} \qquad v_{f2} = \left(\frac{2m_1}{m_1 + m_2}\right)v_{01}$$

Let the initial direction of motion of the 5.00-kg ball define the positive direction. Substituting the values given in the text, these equations give

$$[5.00\text{-kg ball}] \qquad v_{f1} = \left(\frac{5.00\text{ kg} - 7.50\text{ kg}}{5.00\text{ kg} + 7.50\text{ kg}}\right)(2.00\text{ m/s}) = \boxed{-0.400\text{ m/s}}$$

$$[7.50\text{-kg ball}] \qquad v_{f2} = \left(\frac{2(5.00\text{ kg})}{5.00\text{ kg} + 7.50\text{ kg}}\right)(2.00\text{ m/s}) = \boxed{+1.60\text{ m/s}}$$

The signs indicate that, after the collision, the 5.00-kg ball reverses its direction of motion, while the 7.50-kg ball moves in the direction in which the 5.00-kg ball was initially moving.

b. When the collision is completely inelastic, the balls stick together, giving a composite body of mass $m_1 + m_2$ which moves with a velocity v_f. The statement of conservation of linear momentum then becomes

$$\underbrace{(m_1 + m_2)v_f}_{\substack{\text{Total momentum} \\ \text{after collision}}} = \underbrace{m_1 v_{01} + 0}_{\substack{\text{Total momentum} \\ \text{before collision}}}$$

The final velocity of the two balls after the collision is, therefore,

$$v_f = \frac{m_1 v_{01}}{m_1 + m_2} = \frac{(5.00\text{ kg})(2.00\text{ m/s})}{5.00\text{ kg} + 7.50\text{ kg}} = \boxed{+0.800\text{ m/s}}$$

35. **REASONING** The two skaters constitute the system. Since the net external force acting on the system is zero, the total linear momentum of the system is conserved. In the x direction (the east/west direction), conservation of linear momentum gives $P_{fx} = P_{0x}$, or

$$(m_1 + m_2)v_f \cos\theta = m_1 v_{01}$$

Note that since the skaters hold onto each other, they move away with a common velocity v_f. In the y direction, $P_{fy} = P_{0y}$, or

$$(m_1 + m_2)v_f \sin\theta = m_2 v_{02}$$

These equations can be solved simultaneously to obtain both the angle θ and the velocity v_f.

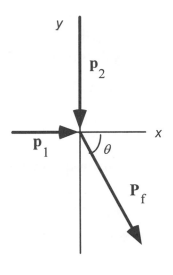

SOLUTION

a. Division of the equations above gives

$$\theta = \tan^{-1}\left(\frac{m_2 v_{02}}{m_1 v_{01}}\right) = \tan^{-1}\left[\frac{(70.0\ \text{kg})(7.00\ \text{m/s})}{(50.0\ \text{kg})(3.00\ \text{m/s})}\right] = \boxed{73.0°}$$

b. Solution of the first of the momentum equations gives

$$v_f = \frac{m_1 v_{01}}{(m_1 + m_2)\cos\theta} = \frac{(50.0\ \text{kg})(3.00\ \text{m/s})}{(50.0\ \text{kg} + 70.0\ \text{kg})(\cos 73.0°)} = \boxed{4.28\ \text{m/s}}$$

---

39. **REASONING** The two balls constitute the system. The tension in the wire is the only nonconservative force that acts on the ball. The tension does no work since it is perpendicular to the displacement of the ball. Since $W_{nc} = 0$, the principle of conservation of mechanical energy holds and can be used to find the speed of the 1.50-kg ball just before the collision. Momentum is conserved during the collision, so the principle of conservation of momentum can be used to find the velocities of both balls just after the collision. Once the collision has occurred, energy conservation can be used to determine how high each ball rises.

SOLUTION

a. Applying the principle of energy conservation to the 1.50-kg ball, we have

$$\underbrace{\tfrac{1}{2}mv_f^2 + mgh_f}_{E_f} = \underbrace{\tfrac{1}{2}mv_0^2 + mgh_0}_{E_0}$$

If we measure the heights from the lowest point in the swing, $h_f = 0$, and the expression above simplifies to

$$\tfrac{1}{2}mv_f^2 = \tfrac{1}{2}mv_0^2 + mgh_0$$

Solving for v_f, we have

$$v_f = \sqrt{v_0^2 + 2gh_0} = \sqrt{(5.00 \text{ m/s})^2 + 2(9.80 \text{ m/s}^2)(0.300 \text{ m})} = \boxed{5.56 \text{ m/s}}$$

b. If we assume that the collision is elastic, then the velocities of both balls just after the collision can be obtained from Equations 7.8a and 7.8b:

$$v_{f1} = \left(\frac{m_1 - m_2}{m_1 + m_2}\right)v_{01} \qquad \text{and} \qquad v_{f2} = \left(\frac{2m_1}{m_1 + m_2}\right)v_{01}$$

Since v_{01} corresponds to the speed of the 1.50-kg ball just before the collision, it is equal to the quantity v_f calculated in part (a). With the given values of $m_1 = 1.50$ kg and $m_2 = 4.60$ kg, and the value of $v_{01} = 5.56$ m/s obtained in part (a), Equations 7.8a and 7.8b yield the following values:

$$\boxed{v_{f1} = -2.83 \text{ m/s}} \qquad \text{and} \qquad \boxed{v_{f2} = +2.73 \text{ m/s}}$$

The minus sign in v_{f1} indicates that the first ball reverses its direction as a result of the collision.

c. If we apply the conservation of mechanical energy to either ball after the collision we have

$$\underbrace{\tfrac{1}{2}mv_f^2 + mgh_f}_{E_f} = \underbrace{\tfrac{1}{2}mv_0^2 + mgh_0}_{E_0}$$

where v_0 is the speed of the ball just after the collision, and h_f is the final height to which the ball rises. For either ball, $h_0 = 0$, and when either ball has reached its maximum height, $v_f = 0$. Therefore, the expression of energy conservation reduces to

$$gh_f = \tfrac{1}{2}v_0^2 \qquad \text{or} \qquad h_f = \frac{v_0^2}{2g}$$

Thus, the heights to which each ball rises after the collision are

$$[1.50\text{-kg ball}] \qquad h_f = \frac{v_0^2}{2g} = \frac{(2.83 \text{ m/s})^2}{2(9.80 \text{ m/s}^2)} = \boxed{0.409 \text{ m}}$$

$$[4.60\text{-kg ball}] \qquad h_f = \frac{v_0^2}{2g} = \frac{(2.73 \text{ m/s})^2}{2(9.80 \text{ m/s}^2)} = \boxed{0.380 \text{ m}}$$

---

41.   ***REASONING AND SOLUTION***   The location of the center of mass for a two-body system is given by Equation 7.10:

$$x_{cm} = \frac{m_1 x_1 + m_2 x_2}{m_1 + m_2}$$

where the subscripts "1" and "2" refer to the earth and the moon, respectively.   For convenience, we will let the center of the earth be coincident with the origin so that $x_1 = 0$ and $x_2 = d$, the center-to-center distance between the earth and the moon.   Direct calculation then gives

$$x_{cm} = \frac{m_2 d}{m_1 + m_2} = \frac{(7.35 \times 10^{22} \text{ kg})(3.85 \times 10^8 \text{ m})}{5.98 \times 10^{24} \text{ kg} + 7.35 \times 10^{22} \text{ kg}} = \boxed{4.67 \times 10^6 \text{ m}}$$

---

45.   ☐WWW   ***REASONING***   The system consists of the lumberjack and the log.   For this system, the sum of the external forces is zero.   This is because the weight of the system is balanced by the corresponding normal force (provided by the buoyant force of the water) and the water is assumed to be frictionless.   The lumberjack and the log, then, constitute an isolated system, and the principle of conservation of linear momentum holds.

SOLUTION
a.   The total linear momentum of the system before the lumberjack begins to move is zero, since all parts of the system are at rest.   Momentum conservation requires that the total momentum remains zero during the motion of the lumberjack.

$$\underbrace{m_1 v_{f1} + m_2 v_{f2}}_{\substack{\text{Total momentum} \\ \text{just before the jump}}} = \underbrace{0}_{\text{Initial momentum}}$$

Here the subscripts "1" and "2" refer to the first log and lumberjack, respectively.   Let the direction of motion of the lumberjack be the positive direction.   Then, solving for v_{1f} gives

$$v_{f1} = -\frac{m_2 v_{f2}}{m_1} = -\frac{(98 \text{ kg})(+3.6 \text{ m/s})}{230 \text{ kg}} = \boxed{-1.5 \text{ m/s}}$$

The minus sign indicates that the first log recoils as the lumberjack jumps off.

b.   Now the system is composed of the lumberjack, just before he lands on the second log, and the second log.   Gravity acts on the system, but for the short time under consideration while the lumberjack lands, the effects of gravity in changing the linear momentum of the system are negligible. Therefore, to a very good approximation, we can say that the linear momentum of the system is very nearly conserved. In this case, the initial momentum is not zero as it was in part (a); rather the initial momentum of the system is the momentum of the lumberjack just before he lands on the second log. Therefore,

$$\underbrace{m_1 v_{f1} + m_2 v_{f2}}_{\substack{\text{Total momentum} \\ \text{just after lumberjack lands}}} = \underbrace{m_1 v_{01} + m_2 v_{02}}_{\text{Initial momentum}}$$

In this expression, the subscripts "1" and "2" now represent the second log and lumberjack, respectively. Since the second log is initially at rest, $v_{01} = 0$. Furthermore, since the lumberjack and the second log move with a common velocity, $v_{f1} = v_{f2} = v_f$. The statement of momentum conservation then becomes

$$m_1 v_f + m_2 v_f = m_2 v_{02}$$

Solving for v_f, we have

$$v_f = \frac{m_2 v_{02}}{m_1 + m_2} = \frac{(98 \text{ kg})(+3.6 \text{ m/s})}{230 \text{ kg} + 98 \text{ kg}} = \boxed{+1.1 \text{ m/s}}$$

The positive sign indicates that the system moves in the same direction as the original direction of the lumberjack's motion.

---

47. ***REASONING AND SOLUTION***   Bonzo and Ender constitute an isolated system. Therefore, the principle of conservation of linear momentum holds. Since both members of the system are initially stationary, the initial linear momentum is zero, and must remain zero during and after any interaction between Bonzo and Ender.

a. Since the total momentum of the system is conserved, and the total momentum of the system is zero, Bonzo and Ender must have equal but opposite linear momenta. $\boxed{\text{Bonzo}}$ must have the larger mass, since $\boxed{\text{he flies off with the smaller recoil velocity}}$.

b. Conservation of linear momentum gives $0 = m_{\text{Bonzo}} v_{\text{Bonzo}} + m_{\text{Ender}} v_{\text{Ender}}$. Solving for the ratio of the masses, we have

$$\frac{m_{\text{Bonzo}}}{m_{\text{Ender}}} = -\frac{v_{\text{Ender}}}{v_{\text{Bonzo}}} = -\frac{-2.5 \text{ m/s}}{1.5 \text{ m/s}} = \boxed{1.7}$$

---

51. ***REASONING***   The two-stage rocket constitutes the system. The forces that act to cause the separation during the explosion are, therefore, forces that are internal to the system. Since

no external forces act on this system, it is isolated and the principle of conservation of linear momentum applies:

$$\underbrace{m_1 v_{f1} + m_2 v_{f2}}_{\substack{\text{Total momentum} \\ \text{after separation}}} = \underbrace{(m_1 + m_2) v_0}_{\substack{\text{Total momentum} \\ \text{before separation}}}$$

where the subscripts "1" and "2" refer to the lower and upper stages, respectively. This expression can be solved for v_{f1}.

SOLUTION   Solving for v_{f1} gives

$$v_{f1} = \frac{(m_1 + m_2) v_0 - m_2 v_{f2}}{m_1}$$

$$= \frac{[2400 \text{ kg} + 1200 \text{ kg}](4900 \text{ m/s}) - (1200 \text{ kg})(5700 \text{ m/s})}{2400 \text{ kg}} = \boxed{+4500 \text{ m/s}}$$

Since v_{f1} is positive $\boxed{\text{its direction is the same as the rocket before the explosion}}$.

---

53. **REASONING** No net external force acts on the plate parallel to the floor; therefore, the component of the momentum of the plate that is parallel to the floor is conserved as the plate breaks and flies apart. Initially, the total momentum parallel to the floor is zero. After the collision with the floor, the component of the total momentum parallel to the floor must remain zero. The drawing in the text shows the pieces in the plane parallel to the floor just after the collision. Clearly, the linear momentum in the plane parallel to the floor has two components; therefore the linear momentum of the plate must be conserved in each of these two mutually perpendicular directions. Using the drawing in the text, with the positive directions taken to be up and to the right, we have

$[x \text{ direction}]$ $\quad\quad\quad\quad -m_1 v_1 (\sin 25.0°) + m_2 v_2 (\cos 45.0°) = 0$

$[y \text{ direction}]$ $\quad\quad\quad\quad m_1 v_1 (\cos 25.0°) + m_2 v_2 (\sin 45.0°) - m_3 v_3 = 0$

These equations can be solved simultaneously for the masses m_1 and m_2.

SOLUTION Using the values given in the drawing for the velocities after the plate breaks, we have, suppressing units,

$$-1.27 m_1 + 1.27 m_2 = 0 \tag{1}$$

$$2.72 m_1 + 1.27 m_2 - 3.99 = 0 \tag{2}$$

Subtracting (1) from (2) gives $\boxed{m_1 = 1.00 \text{ kg}}$. Substituting this value into either (1) or (2) then yields $\boxed{m_2 = 1.00 \text{ kg}}$.

---

57. $\boxed{\text{WWW}}$ **REASONING**   The cannon and the shell constitute the system. Since no external force hinders the motion of the system after the cannon is unbolted, conservation of linear momentum applies in that case. If we assume that the burning gun powder imparts the same kinetic energy to the system in each case, we have sufficient information to develop a mathematical description of this situation, and solve it for the velocity of the shell fired by the loose cannon.

SOLUTION   For the case where the cannon is unbolted, momentum conservation gives

$$\underbrace{m_1 v_{f1} + m_2 v_{f2}}_{\substack{\text{Total momentum} \\ \text{after shell is fired}}} = \underbrace{0}_{\substack{\text{Initial momentum} \\ \text{of system}}} \tag{1}$$

where the subscripts "1" and "2" refer to the cannon and shell, respectively. In both cases, the burning gun power imparts the same kinetic energy to the system. When the cannon is bolted to the ground, only the shell moves and the kinetic energy imparted to the system is

$$KE = \tfrac{1}{2} m_{shell} v_{shell}^2 = \tfrac{1}{2}(85.0 \text{ kg})(551 \text{ m/s})^2 = 1.29 \times 10^7 \text{ J}$$

The kinetic energy imparted to the system when the cannon is unbolted has the same value and can be written using the same notation as equation (1):

$$KE = \tfrac{1}{2} m_1 v_{f1}^2 + \tfrac{1}{2} m_2 v_{f2}^2 \tag{2}$$

Solving equation (1) for v_{f1}, the velocity of the cannon after the shell is fired, and substituting the resulting expression into Equation (2) gives

$$KE = \frac{m_2^2 v_{f2}^2}{2m_1} + \tfrac{1}{2} m_2 v_{f2}^2 \tag{3}$$

Solving equation (3) for v_{f2} gives

$$v_{f2} = \sqrt{\frac{2KE}{m_2\left(\dfrac{m_2}{m_1}+1\right)}} = \sqrt{\frac{2(1.29 \times 10^7 \text{ J})}{(85 \text{ kg})\left(\dfrac{85 \text{ kg}}{5.80 \times 10^3 \text{ kg}}+1\right)}} = \boxed{+547 \text{ m/s}}$$

CHAPTER 8 | *ROTATIONAL KINEMATICS*

1. **REASONING AND SOLUTION** Since there are 2π radians per revolution, and it is stated in the problem that there are 100 grads in one-quarter of a circle, we find that the number of grads in one radian is

$$(1.00 \text{ rad})\left(\frac{1 \text{ rev}}{2\pi \text{ rad}}\right)\left(\frac{100 \text{ grad}}{0.250 \text{ rev}}\right) = \boxed{63.7 \text{ grad}}$$

3. **REASONING AND SOLUTION** Equation 8.2 gives the desired result. Since the answer is to be expressed in rad/s, we must convert revolutions to radians.

$$\bar{\omega} = \frac{\Delta\theta}{\Delta t} = \left(\frac{3.5 \text{ rev}}{1.7 \text{ s}}\right)\left(\frac{2\pi \text{ rad}}{1 \text{ rev}}\right) = \boxed{13 \text{ rad/s}}$$

7. **REASONING AND SOLUTION** Equation 8.4 gives the desired result. Assuming $t_0 = 0$ s, the final angular velocity is

$$\omega = \omega_0 + \alpha t = 0 \text{ rad/s} + (328 \text{ rad/s}^2)(1.50 \text{ s}) = \boxed{492 \text{ rad/s}}$$

11. **REASONING** The time required for the bullet to travel the distance d is equal to the time required for the discs to undergo an angular displacement of 0.240 rad. The time can be found from Equation 8.2; once the time is known, the speed of the bullet can be found using Equation 2.2.

    SOLUTION From the definition of average angular velocity:

$$\bar{\omega} = \frac{\Delta\theta}{\Delta t}$$

The required time is

$$\Delta t = \frac{\Delta\theta}{\bar{\omega}} = \frac{0.240 \text{ rad}}{95.0 \text{ rad/s}} = 2.53 \times 10^{-3} \text{ s}$$

Note that $\bar{\omega} = \omega$ because the angular speed is constant. The (constant) speed of the bullet can then be determined from the definition of average speed:

$$\bar{v} = \frac{\Delta x}{\Delta t} = \frac{d}{\Delta t} = \frac{0.850 \text{ m}}{2.53 \times 10^{-3} \text{ s}} = \boxed{336 \text{ m/s}}$$

17. ***REASONING AND SOLUTION***
    a. From Equation 8.7 we obtain

$$\theta = \omega_0 t + \tfrac{1}{2}\alpha t^2 = (5.00 \text{ rad/s})(4.00 \text{ s}) + \tfrac{1}{2}(2.50 \text{ rad/s}^2)(4.00 \text{ s})^2 = \boxed{4.00 \times 10^1 \text{ rad}}$$

    b. From Equation 8.4, we obtain

$$\omega = \omega_0 + \alpha t = 5.00 \text{ rad/s} + (2.50 \text{ rad/s}^2)(4.00 \text{ s}) = \boxed{15.0 \text{ rad/s}}$$

19. $\boxed{\text{WWW}}$ ***REASONING AND SOLUTION***
     a. Since the flywheel comes to rest, its final angular velocity is zero. Furthermore, if the initial angular velocity ω_0 is assumed to be a positive number and the flywheel decelerates, the angular acceleration must be a negative number. Solving Equation 8.8 for θ, we obtain

$$\theta = \frac{\omega^2 - \omega_0^2}{2\alpha} = \frac{(0 \text{ rad/s})^2 - (220 \text{ rad/s})^2}{2(-2.0 \text{ rad/s}^2)} = \boxed{1.2 \times 10^4 \text{ rad}}$$

     b. The time required for the flywheel to come to rest can be found from Equation 8.4. Solving for t, we obtain

$$t = \frac{\omega - \omega_0}{\alpha} = \frac{0 \text{ rad/s} - 220 \text{ rad/s}}{-2.0 \text{ rad/s}^2} = \boxed{1.1 \times 10^2 \text{ s}}$$

23. ***REASONING***   The time required for the change in the angular velocity to occur can be found by solving Equation 8.4 for t. In order to use Equation 8.4, however, we must know the initial angular velocity ω_0. Equation 8.6 can be used to find the initial angular velocity.

     SOLUTION   From Equation 8.6 we have

$$\theta = \frac{1}{2}(\omega_0 + \omega)t$$

Solving for ω_0 gives

$$\omega_0 = \frac{2\theta}{t} - \omega$$

Since the angular displacement θ is zero, $\omega_0 = -\omega$. Solving Equation 8.4 for t, and using the fact that $\omega_0 = -\omega$, gives

$$t = \frac{2\omega}{\alpha} = \frac{2(-25.0\ \text{rad/s})}{-4.00\ \text{rad/s}^2} = \boxed{12.5\ \text{s}}$$

---

27. $\boxed{\text{WWW}}$ ***REASONING*** The angular displacement of the child when he catches the horse is, from Equation 8.2, $\theta_c = \omega_c t$. In the same time, the angular displacement of the horse is, from Equation 8.7 with $\omega_0 = 0\ \text{rad/s}$, $\theta_h = \frac{1}{2}\alpha t^2$. If the child is to catch the horse $\theta_c = \theta_h + (\pi/2)$.

SOLUTION Using the above conditions yields

$$\tfrac{1}{2}\alpha t^2 - \omega_c t + \tfrac{\pi}{2} = 0$$

or (suppressing the units)

$$\tfrac{1}{2}(0.0100)t^2 - 0.250t + \tfrac{\pi}{2} = 0$$

The quadratic formula yields $t = 7.37$ s and 42.6 s; therefore, the shortest time needed to catch the horse is $\boxed{t = 7.37\ \text{s}}$.

---

31. ***REASONING AND SOLUTION*** Equation 8.9 gives the desired result.

$$v_T = r\omega = (2.00 \times 10^{-3}\ \text{m})(7.85 \times 10^4\ \text{rad/s}) = \boxed{157\ \text{m/s}}$$

---

35. ***REASONING AND SOLUTION***
    a. From Equation 8.9, and the fact that 1 revolution $= 2\pi$ radians, we obtain

$$v_T = r\omega = (0.0568\ \text{m})\left(3.50\ \frac{\text{rev}}{\text{s}}\right)\left(\frac{2\pi\ \text{rad}}{1\ \text{rev}}\right) = \boxed{1.25\ \text{m/s}}$$

   b. Since the disk rotates at *constant* tangential speed,

$$v_{T1} = v_{T2} \qquad \text{or} \qquad \omega_1 r_1 = \omega_2 r_2$$

Solving for ω_2, we obtain

$$\omega_2 = \frac{\omega_1 r_1}{r_2} = \frac{(3.50\ \text{rev/s})(0.0568\ \text{m})}{0.0249\ \text{m}} = \boxed{7.98\ \text{rev/s}}$$

---

39. ***REASONING***   The top of the racket has both tangential and centripetal acceleration components given by Equations 8.10 and 8.11, respectively: $a_T = r\alpha$ and $a_c = r\omega^2$. The total acceleration of the top of the racket is the resultant of these two components. Since these acceleration components are mutually perpendicular, their resultant can be found by using the Pythagorean theorem.

SOLUTION   Employing the Pythagorean theorem, we obtain

$$a = \sqrt{a_T^2 + a_c^2} = \sqrt{(r\alpha)^2 + (r\omega^2)^2} = r\sqrt{\alpha^2 + \omega^4}$$

Therefore,

$$a = (1.5 \text{ m})\sqrt{(160 \text{ rad/s}^2)^2 + (14 \text{ rad/s})^4} = \boxed{380 \text{ m/s}^2}$$

---

41. ***REASONING***   The tangential acceleration $\mathbf{a}_T$ of the speedboat can be found by using Newton's second law, $\mathbf{F}_T = m\mathbf{a}_T$, where $\mathbf{F}_T$ is the net tangential force. Once the tangential acceleration of the boat is known, Equation 2.4 can be used to find the tangential speed of the boat 2.0 s into the turn. With the tangential speed and the radius of the turn known, Equation 5.2 can then be used to find the centripetal acceleration of the boat.

SOLUTION
a. From Newton's second law, we obtain

$$a_T = \frac{F_T}{m} = \frac{550 \text{ N}}{220 \text{ kg}} = \boxed{2.5 \text{ m/s}^2}$$

b. The tangential speed of the boat 2.0 s into the turn is, according to Equation 2.4,

$$v_T = v_{0T} + a_T t = 5.0 \text{ m/s} + (2.5 \text{ m/s}^2)(2.0 \text{ s}) = 1.0 \times 10^1 \text{ m/s}$$

The centripetal acceleration of the boat is then

$$a_c = \frac{v_T^2}{r} = \frac{(1.0 \times 10^1 \text{ m/s})^2}{32 \text{ m}} = \boxed{3.1 \text{ m/s}^2}$$

---

45. ***REASONING***   The tangential acceleration and the centripetal acceleration of a point at a distance r from the rotation axis are given by Equations 8.10 and 8.11, respectively: $a_T = r\alpha$ and $a_c = r\omega^2$. After the drill has rotated through the angle in question, $a_c = 2a_T$, or

$$r\omega^2 = 2r\alpha$$

This expression can be used to find the angular acceleration α. Once the angular acceleration is known, Equation 8.8 can be used to find the desired angle.

SOLUTION Solving the expression obtained above for α gives

$$\alpha = \frac{\omega^2}{2}$$

Solving Equation 8.8 for θ (with $\omega_0 = 0$ rad/s since the drill starts from rest), and using the expression above for the angular acceleration α gives

$$\theta = \frac{\omega^2}{2\alpha} = \frac{\omega^2}{2(\omega^2/2)} = \left(\frac{\omega^2}{2}\right)\left(\frac{2}{\omega^2}\right) = \boxed{1.00 \text{ rad}}$$

Note that since both Equations 8.10 and 8.11 require that the angles be expressed in radians, the final result for θ is in radians.

---

47. $\boxed{\text{WWW}}$ **REASONING AND SOLUTION** From Equation 2.4, the linear acceleration of the motorcycle is

$$a = \frac{v - v_0}{t} = \frac{22.0 \text{ m/s} - 0 \text{ m/s}}{9.00 \text{ s}} = 2.44 \text{ m/s}^2$$

Since the tire rolls without slipping, the linear acceleration equals the tangential acceleration of a point on the outer edge of the tire: $a = a_T$. Solving Equation 8.13 for α gives

$$\alpha = \frac{a_T}{r} = \frac{2.44 \text{ m/s}^2}{0.280 \text{ m}} = \boxed{8.71 \text{ rad/s}^2}$$

---

51. **REASONING AND SOLUTION** According to Equation 8.9, the crank handle has an angular speed of

$$\omega = \frac{v_T}{r} = \frac{1.20 \text{ m/s}}{\frac{1}{2}(0.400 \text{ m})} = 6.00 \text{ rad/s}$$

The crank barrel must have the same angular speed as the handle; therefore, the tangential speed of a point on the barrel is, again using Equation 8.9,

$$v_T = r\omega = (5.00 \times 10^{-2} \text{ m})(6.00 \text{ rad/s}) = 0.300 \text{ m/s}$$

If we assume that the rope does not slip, then all points on the rope must move with the same speed as the tangential speed of any point on the barrel. Therefore the linear speed with which the bucket moves down the well is $\boxed{v = 0.300 \text{ m/s}}$.

---

57. ***REASONING*** Since the car is traveling with a constant speed, its tangential acceleration must be zero. The radial or centripetal acceleration of the car can be found from Equation 5.2. Since the tangential acceleration is zero, the total acceleration of the car is equal to its radial acceleration.

SOLUTION
a. Using Equation 5.2, we find that the car's radial acceleration, and therefore its total acceleration, is

$$a = a_R = \frac{v_T^2}{r} = \frac{(75.0 \text{ m/s})^2}{625 \text{ m}} = \boxed{9.00 \text{ m/s}^2}$$

b. The direction of the car's total acceleration is the same as the direction of its radial acceleration. That is, the direction is $\boxed{\text{radially inward}}$.

---

59. ***REASONING AND SOLUTION*** Since the angular speed of the fan decreases, the sign of the angular acceleration must be opposite to the sign for the angular velocity. Taking the angular velocity to be positive, the angular acceleration, therefore, must be a negative quantity. Using Equation 8.4 we obtain

$$\omega_0 = \omega - \alpha t = 83.8 \text{ rad/s} - (-42.0 \text{ rad/s}^2)(1.75 \text{ s}) = \boxed{157.3 \text{ rad/s}}$$

---

63. $\boxed{\text{WWW}}$ ***REASONING*** Assuming that the belt does not slip on the platter or the shaft pulley, the tangential speed of points on the platter and shaft pulley must be equal; therefore, $r_s \omega_s = r_p \omega_p$.

SOLUTION Solving the above expression for ω_s gives

$$\omega_s = \frac{r_p \omega_p}{r_s} = \frac{(3.49 \text{ rad/s})(0.102 \text{ m})}{1.27 \times 10^{-2} \text{ m}} = \boxed{28.0 \text{ rad/s}}$$

67. ***REASONING AND SOLUTION***   By inspection, the distance traveled by the "axle" or the center of the moving quarter is $d = 2\pi(2r) = 4\pi r$, where r is the radius of the quarter. The distance d traveled by the "axle" of the moving quarter must be equal to the circular arc length s along the outer edge of the quarter. This arc length is $s = r\theta$, where θ is the angle through which the quarter rotates. Thus,

$$4\pi r = r\theta$$

so that $\theta = 4\pi$ rad. This is equivalent to

$$(4\pi\ \text{rad})\left(\frac{1\ \text{rev}}{2\pi\ \text{rad}}\right) = \boxed{2\ \text{revolutions}}$$

CHAPTER 9 | *ROTATIONAL DYNAMICS*

PROBLEMS

---

1. ***REASONING AND SOLUTION*** Solving Equation 9.1 for the lever arm ℓ, we obtain

$$\ell = \frac{\tau}{F} = \frac{3.0 \, \text{N} \cdot \text{m}}{12 \, \text{N}} = \boxed{0.25 \, \text{m}}$$

---

5. ***REASONING*** The torque on either wheel is given by Equation 9.1, $\tau = F\ell$, where F is the magnitude of the force and ℓ is the lever arm. Regardless of how the force is applied, the lever arm will be proportional to the radius of the wheel.

    SOLUTION The ratio of the torque produced by the force in the truck to the torque produced in the car is

$$\frac{\tau_{\text{truck}}}{\tau_{\text{car}}} = \frac{F\ell_{\text{truck}}}{F\ell_{\text{car}}} = \frac{Fr_{\text{truck}}}{Fr_{\text{car}}} = \frac{r_{\text{truck}}}{r_{\text{car}}} = \frac{0.25 \, \text{m}}{0.19 \, \text{m}} = \boxed{1.3}$$

---

9. ***REASONING*** Since the meter stick does not move, it is in equilibrium. The forces and the torques, therefore, each add to zero. We can determine the location of the 6.00 N force, by using the condition that the sum of the torques must vectorially add to zero.

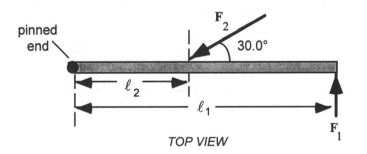

TOP VIEW

SOLUTION If we take counterclockwise torques as positive, then the torque of the first force about the pin is

$$\tau_1 = F_1 \ell_1 = (2.00 \, \text{N})(1.00 \, \text{m}) = 2.00 \, \text{N} \cdot \text{m}$$

The torque due to the second force is

$$\tau_2 = -F_2 (\sin 30.0°) \, \ell_2 = -(6.00 \, \text{N})(\sin 30.0°) \, \ell_2 = -(3.00 \, \text{N}) \, \ell_2$$

The torque τ_2 is negative because the force F_2 tends to produce a clockwise rotation about the pinned end. Since the net torque is zero, we have

$$2.00 \text{ N} \cdot \text{m} + [-(3.00 \text{ N})\ell_2] = 0$$

Thus,

$$\ell_2 = \frac{2.00 \text{ N} \cdot \text{m}}{3.00 \text{ N}} = \boxed{0.667 \text{ m}}$$

---

13. **REASONING** The minimum value for the coefficient of static friction between the ladder and the ground, so that the ladder does not slip, is given by Equation 4.7:

$$f_s^{\text{MAX}} = \mu_s F_N$$

SOLUTION From Example 4, the magnitude of the force of static friction is $G_x = 727$ N. The magnitude of the normal force applied to the ladder by the ground is $G_y = 1230$ N. The minimum value for the coefficient of static friction between the ladder and the ground is

$$\mu_s = \frac{f_s^{\text{MAX}}}{F_N} = \frac{G_x}{G_y} = \frac{727 \text{ N}}{1230 \text{ N}} = \boxed{0.591}$$

---

15. **REASONING** The figure at the right shows the door and the forces that act upon it. Since the door is uniform, the center of gravity, and, thus, the location of the weight **W**, is at the geometric center of the door.

Let **U** represent the force applied to the door by the upper hinge, and **L** the force applied to the door by the lower hinge. Taking forces that point to the right and forces that point up as positive, we have

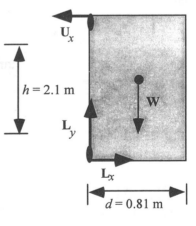

$$\sum F_x = L_x - U_x = 0 \quad \text{or} \quad L_x = U_x \qquad (1)$$

$$\sum F_y = L_y - W = 0 \quad \text{or} \quad L_y = W = 140 \text{ N} \qquad (2)$$

Taking torques about an axis perpendicular to the plane of the door and through the lower hinge, with counterclockwise torques being positive, gives

$$\sum \tau = -W\left(\frac{d}{2}\right) + U_x h = 0 \qquad (3)$$

Equations (1), (2), and (3) can be used to find the force applied to the door by the hinges; Newton's third law can then be used to find the force applied by the door to the hinges.

SOLUTION

a. Solving Equation (3) for U_x gives

$$U_x = \frac{Wd}{2h} = \frac{(140 \text{ N})(0.81 \text{ m})}{2(2.1 \text{ m})} = 27 \text{ N}$$

From Equation (1), $L_x = 27$ N.

Therefore, the upper hinge exerts on the door a horizontal force of $\boxed{27 \text{ N to the left}}$.

b. From Equation (1), $L_x = 27$ N, we conclude that the bottom hinge exerts on the door a horizontal force of $\boxed{27 \text{ N to the right}}$.

c. From Newton's third law, the door applies a force of $\boxed{27 \text{ N}}$ on the upper hinge. This force $\boxed{\text{points horizontally to the right}}$, according to Newton's third law.

d. Let **D** represent the force applied by the door to the lower hinge. Then, from Newton's third law, the force exerted on the lower hinge by the door has components $D_x = -27$ N and

$D_y = -140$ N. From the Pythagorean theorem,

$$D = \sqrt{D_x^2 + D_y^2} = \sqrt{(-27 \text{ N})^2 + (-140 \text{ N})^2} = \boxed{143 \text{ N}}$$

The angle θ is given by

$$\theta = \tan^{-1}\left(\frac{-140 \text{ N}}{-27 \text{ N}}\right) = 79°$$

The force is directed $\boxed{79° \text{ below the horizontal}}$.

---

21. $\boxed{\text{WWW}}$ **REASONING AND SOLUTION** The figure below shows the massless board and the forces that act on the board due to the person and the scales.

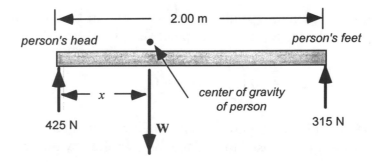

a. Applying Newton's second law to the vertical direction gives

$$315 \text{ N} + 425 \text{ N} - W = 0 \qquad \text{or} \qquad \boxed{W = 7.40 \times 10^2 \text{ N, downward}}$$

b. Let x be the position of the center of gravity relative to the scale at the person's head. Taking torques about an axis through the contact point between the scale at the person's head and the board gives

$$(315 \text{ N})(2.00 \text{ m}) - (7.40 \times 10^2 \text{ N})x = 0 \qquad \text{or} \qquad \boxed{x = 0.851 \text{ m}}$$

---

23. **REASONING** Since the man holds the ball motionless, the ball and the arm are in equilibrium. Therefore, the net force, as well as the net torque about any axis, must be zero.

    SOLUTION
    a. Using Equation 9.1, the net torque about an axis through the elbow joint is

    $$\Sigma\tau = M(0.0510 \text{ m}) - (22.0 \text{ N})(0.140 \text{ m}) - (178 \text{ N})(0.330 \text{ m}) = 0$$

    Solving this expression for M gives $\boxed{M = 1.21 \times 10^3 \text{ N}}$.

    b. The net torque about an axis through the center of gravity is

    $$\Sigma\tau = -(1210 \text{ N})(0.0890 \text{ m}) + F(0.140 \text{ m}) - (178 \text{ N})(0.190 \text{ m}) = 0$$

    Solving this expression for F gives $\boxed{F = 1.01 \times 10^3 \text{ N}}$. Since the forces must add to give a net force of zero, we know that the direction of **F** is $\boxed{\text{downward}}$.

---

29. **REASONING AND SOLUTION** For a rigid body rotating about a fixed axis, Newton's second law for rotational motion is given by Equation 9.7, $\Sigma\tau = I\alpha$, where I is the moment of inertia of the body and α is the angular acceleration expressed in rad/s^2. Equation 9.7 gives

    $$I = \frac{\Sigma\tau}{\alpha} = \frac{10.0 \text{ N·m}}{8.00 \text{ rad/s}^2} = \boxed{1.25 \text{ kg·m}^2}$$

---

33. **REASONING** The figure below shows eight particles, each one located at a different corner of an imaginary cube. As shown, if we consider an axis that lies along one edge of the cube, two of the particles lie on the axis, and for these particles $r = 0$. The next four particles closest to the axis have $r = \ell$, where ℓ is the length of one edge of the cube. The remaining

two particles have $r = d$, where d is the length of the diagonal along any one of the faces. From the Pythagorean theorem, $d = \sqrt{\ell^2 + \ell^2} = \ell\sqrt{2}$.

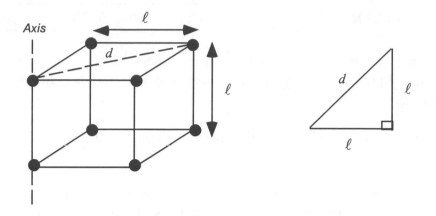

According to Equation 9.6, the moment of inertia of a system of particles is given by $I = \sum mr^2$.

SOLUTION   Direct application of Equation 9.6 gives

$$I = \sum mr^2 = 4(m\ell^2) + 2(md^2) = 4(m\ell^2) + 2(2m\ell^2) = 8m\ell^2$$

or

$$I = 8(0.12 \text{ kg})(0.25 \text{ m})^2 = \boxed{0.060 \text{ kg} \cdot \text{m}^2}$$

---

37.   **REASONING AND SOLUTION**
a.   From Table 9.1 in the text, the moment of inertia of the rod relative to an axis that is perpendicular to the rod at one end is given by

$$I_{\text{rod}} = \tfrac{1}{3} ML^2 = \tfrac{1}{3}(2.00 \text{ kg})(2.00 \text{ m})^2 = \boxed{2.67 \text{ kg} \cdot \text{m}^2}$$

b.   The moment of inertia of a point particle of mass m relative to a rotation axis located a perpendicular distance r from the particle is $I = mr^2$. Suppose that all the mass of the rod were located at a single point located a perpendicular distance R from the axis in part (a). If this point particle has the same moment of inertia as the rod in part (a), then the distance R, the radius of gyration, is given by

$$R = \sqrt{\frac{I}{m}} = \sqrt{\frac{2.67 \text{ kg} \cdot \text{m}^2}{2.00 \text{ kg}}} = \boxed{1.16 \text{ m}}$$

---

39. ***REASONING***   The angular acceleration of the bicycle wheel can be calculated from Equation 8.4. Once the angular acceleration is known, Equation 9.7 can be used to find the net torque caused by the brake pads. The normal force can be calculated from the torque using Equation 9.1.

SOLUTION   The angular acceleration of the wheel is, according to Equation 8.4,

$$\alpha = \frac{\omega - \omega_0}{t} = \frac{3.7 \text{ rad/s} - 13.1 \text{ rad/s}}{3.0 \text{ s}} = -3.1 \text{ rad/s}^2$$

If we assume that all the mass of the wheel is concentrated in the rim, we may treat the wheel as a hollow cylinder. From Table 9.1, we know that the moment of inertia of a hollow cylinder of mass m and radius r about an axis through its center is $I = mr^2$. The net torque that acts on the wheel due to the brake pads is, therefore,

$$\sum \tau = I\alpha = (mr^2)\alpha \tag{1}$$

From Equation 9.1, the net torque that acts on the wheel due to the action of the two brake pads is

$$\sum \tau = -2 f_k \ell \tag{2}$$

where f_k is the kinetic frictional force applied to the wheel by each brake pad, and $\ell = 0.33$ m is the lever arm between the axle of the wheel and the brake pad (see the drawing in the text). The factor of 2 accounts for the fact that there are two brake pads. The minus sign arises because the net torque must have the same sign as the angular acceleration. The kinetic frictional force can be written as (see Equation 4.8)

$$f_k = \mu_k F_N \tag{3}$$

where μ_k is the coefficient of kinetic friction and F_N is the magnitude of the normal force applied to the wheel by each brake pad.

Combining Equations (1), (2), and (3) gives

$$-2(\mu_k F_N)\ell = (mr^2)\alpha$$

$$F_N = \frac{-mr^2\alpha}{2\mu_k \ell} = \frac{-(1.3 \text{ kg})(0.33 \text{ m})^2(-3.1 \text{ rad/s}^2)}{2(0.85)(0.33 \text{ m})} = \boxed{0.78 \text{ N}}$$

43. ***REASONING***   The kinetic energy of the flywheel is given by Equation 9.9. The moment of inertia of the flywheel is the same as that of a solid disk, and, according to Table 9.1 in the text, is given by $I = \frac{1}{2}MR^2$. Once the moment of inertia of the flywheel is known,

Equation 9.9 can be solved for the angular speed ω in rad/s. This quantity can then be converted to rev/min.

SOLUTION Solving Equation 9.9 for ω, we obtain,

$$\omega = \sqrt{\frac{2(KE_R)}{I}} = \sqrt{\frac{2(KE_R)}{\frac{1}{2}MR^2}} = \sqrt{\frac{4(1.2 \times 10^9 \text{ J})}{(13 \text{ kg})(0.30 \text{ m})^2}} = 6.4 \times 10^4 \text{ rad/s}$$

Converting this answer into rev/min, we find that

$$\omega = \left(6.4 \times 10^4 \text{ rad/s}\right) \left(\frac{1 \text{ rev}}{2\pi \text{ rad}}\right) \left(\frac{60 \text{ s}}{1 \text{ min}}\right) = \boxed{6.1 \times 10^5 \text{ rev/min}}$$

---

45. ***REASONING AND SOLUTION***
     a. The tangential speed of each object is given by Equation 8.9, $v_T = r\omega$. Therefore,

For object 1: $\quad v_{T1} = (2.00 \text{ m})(6.00 \text{ rad/s}) = \boxed{12.0 \text{ m/s}}$

For object 2: $\quad v_{T2} = (1.50 \text{ m})(6.00 \text{ rad/s}) = \boxed{9.00 \text{ m/s}}$

For object 3: $\quad v_{T3} = (3.00 \text{ m})(6.00 \text{ rad/s}) = \boxed{18.0 \text{ m/s}}$

b. The total kinetic energy of this system can be calculated by computing the sum of the kinetic energies of each object in the system. Therefore,

$$KE = \tfrac{1}{2}m_1 v_1^2 + \tfrac{1}{2}m_2 v_2^2 + \tfrac{1}{2}m_3 v_3^2$$

$$KE = \tfrac{1}{2}\left[(6.00 \text{ kg})(12.0 \text{ m/s})^2 + (4.00 \text{ kg})(9.00 \text{ m/s})^2 + (3.00 \text{ kg})(18.0 \text{ m/s})^2\right] = \boxed{1.08 \times 10^3 \text{ J}}$$

c. The total moment of inertia of this system can be calculated by computing the sum of the moments of inertia of each object in the system. Therefore,

$$I = \sum mr^2 = m_1 r_1^2 + m_2 r_2^2 + m_3 r_3^2$$

$$I = (6.00 \text{ kg})(2.00 \text{ m})^2 + (4.00 \text{ kg})(1.50 \text{ m})^2 + (3.00 \text{ kg})(3.00 \text{ m})^2 = \boxed{60.0 \text{ kg} \cdot \text{m}^2}$$

d. The rotational kinetic energy of the system is, according to Equation 9.9,

$$KE_R = \tfrac{1}{2}I\omega^2 = \tfrac{1}{2}(60.0 \text{ kg} \cdot \text{m}^2)(6.00 \text{ rad/s})^2 = \boxed{1.08 \times 10^3 \text{ J}}$$

This agrees, as it should, with the result for part (b).

49. ***REASONING AND SOLUTION***   The only force that does work on the cylinders as they move up the incline is the conservative force of gravity; hence, the total mechanical energy is conserved as the cylinders ascend the incline. We will let $h = 0$ on the horizontal plane at the bottom of the incline. Applying the principle of conservation of mechanical energy to the solid cylinder, we have

$$mgh_s = \tfrac{1}{2}mv_0^2 + \tfrac{1}{2}I_s\omega_0^2 \tag{1}$$

where, from Table 9.1, $I_s = \tfrac{1}{2}mr^2$. In this expression, v_0 and ω_0 are the initial translational and rotational speeds, and h_s is the final height attained by the solid cylinder. Since the cylinder rolls without slipping, the rotational speed ω_0 and the translational speed v_0 are related according to Equation 8.12, $\omega_0 = v_0/r$. Then, solving Equation (1) for h_s, we obtain

$$h_s = \frac{3v_0^2}{4g}$$

Repeating the above for the hollow cylinder and using $I_h = mr^2$ we have

$$h_h = \frac{v_0^2}{g}$$

The height h attained by each cylinder is related to the distance s traveled along the incline and the angle θ of the incline by

$$s_s = \frac{h_s}{\sin\theta} \qquad \text{and} \qquad s_h = \frac{h_h}{\sin\theta}$$

Dividing these gives

$$\frac{s_s}{s_h} = \boxed{3/4}$$

---

53. **WWW**   ***REASONING***   Let the two disks constitute the system. Since there are no external torques acting on the system, the principle of conservation of angular momentum applies. Therefore we have $L_{\text{initial}} = L_{\text{final}}$, or

$$I_A\omega_A + I_B\omega_B = (I_A + I_B)\omega_{\text{final}}$$

This expression can be solved for the moment of inertia of disk B.

SOLUTION   Solving the above expression for I_B, we obtain

$$I_B = I_A\left(\frac{\omega_{\text{final}} - \omega_A}{\omega_B - \omega_{\text{final}}}\right) = (3.4 \text{ kg}\cdot\text{m}^2)\left[\frac{-2.4 \text{ rad/s} - 7.2 \text{ rad/s}}{-9.8 \text{ rad/s} - (-2.4 \text{ rad/s})}\right] = \boxed{4.4 \text{ kg}\cdot\text{m}^2}$$

57.   ***REASONING***   Let the space station and the people within it constitute the system. Then as the people move radially from the outer surface of the cylinder toward the axis, any torques that occur are internal torques. Since there are no external torques acting on the system, the principle of conservation of angular momentum can be employed.

SOLUTION   Since angular momentum is conserved,

$$I_{final}\omega_{final} = I_0\omega_0$$

Before the people move from the outer rim, the moment of inertia is

$$I_0 = I_{station} + 500m_{person}r^2_{person}$$

or

$$I_0 = 3.00\times10^9 \text{ kg}\cdot\text{m}^2 + (500)(70.0 \text{ kg})(82.5 \text{ m})^2 = 3.24\times10^9 \text{ kg}\cdot\text{m}^2$$

If the people all move to the center of the space station, the total moment of inertia is

$$I_{final} = I_{station} = 3.00\times10^9 \text{ kg}\cdot\text{m}^2$$

Therefore,

$$\frac{\omega_{final}}{\omega_0} = \frac{I_0}{I_{final}} = \frac{3.24\times10^9 \text{ kg}\cdot\text{m}^2}{3.00\times10^9 \text{ kg}\cdot\text{m}^2} = 1.08$$

This fraction represents a percentage increase of $\boxed{\text{8 percent}}$.

61. $\boxed{\text{WWW}}$ ***REASONING***   The maximum torque will occur when the force is applied perpendicular to the diagonal of the square as shown. The lever arm ℓ is half the length of the diagonal. From the Pythagorean theorem, the lever arm is, therefore,

$$\ell = \tfrac{1}{2}\sqrt{(0.40 \text{ m})^2 + (0.40 \text{ m})^2} = 0.28 \text{ m}$$

Since the lever arm is now known, we can use Equation 9.1 to obtain the desired result directly.

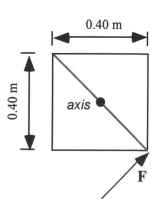

SOLUTION   Equation 9.1 gives

$$\tau = F\ell = (15 \text{ N})(0.28 \text{ m}) = \boxed{4.2 \text{ N}\cdot\text{m}}$$

65. ***REASONING AND SOLUTION***
a. The rim of the bicycle wheel can be treated as a hoop. Using the expression given in Table 9.1 in the text, we have

$$I_{hoop} = MR^2 = (1.20 \text{ kg})(0.330 \text{ m})^2 = \boxed{0.131 \text{ kg} \cdot \text{m}^2}$$

b. Any one of the spokes may be treated as a long, thin rod that can rotate about one end. The expression in Table 9.1 gives

$$I_{rod} = \tfrac{1}{3} ML^2 = \tfrac{1}{3}(0.010 \text{ kg})(0.330 \text{ m})^2 = \boxed{3.6 \times 10^{-4} \text{ kg} \cdot \text{m}^2}$$

c. The total moment of inertia of the bicycle wheel is the sum of the moments of inertia of each constituent part. Therefore, we have

$$I = I_{hoop} + 50 I_{rod} = 0.131 \text{ kg} \cdot \text{m}^2 + 50(3.6 \times 10^{-4} \text{ kg} \cdot \text{m}^2) = \boxed{0.149 \text{ kg} \cdot \text{m}^2}$$

---

67. ***REASONING*** The drawing shows the forces acting on the board, which has a length L. The ground exerts the vertical normal force **V** on the lower end of the board. The maximum force of static friction has a magnitude of $\mu_s V$ and acts horizontally on the lower end of the board. The weight **W** acts downward at the board's center. The vertical wall applies a force **P** to the upper end of the board, this force being perpendicular to the wall since the wall is smooth (i.e., there is no friction along the wall). We take upward and to the right as our positive directions. Then, since the horizontal forces balance to zero, we have

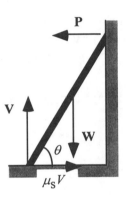

$$\mu_s V - P = 0 \qquad\qquad (1)$$

The vertical forces also balance to zero giving

$$V - W = 0 \qquad\qquad (2)$$

Using an axis through the lower end of the board, we express the fact that the torques balance to zero as

$$PL \sin \theta - W\left(\frac{L}{2}\right)\cos \theta = 0 \qquad\qquad (3)$$

Equations (1), (2), and (3) may then be combined to yield an expression for θ.

SOLUTION Rearranging Equation (3) gives

$$\tan \theta = \frac{W}{2P} \tag{4}$$

But, $P = \mu_s V$ according to Equation (1), and $W = V$ according to Equation (2). Substituting these results into Equation (4) gives

$$\tan \theta = \frac{V}{2\mu_s V} = \frac{1}{2\mu_s}$$

Therefore,

$$\theta = \tan^{-1}\left(\frac{1}{2\mu_s}\right) = \tan^{-1}\left[\frac{1}{2(0.650)}\right] = \boxed{37.6°}$$

---

71. $\boxed{\text{WWW}}$ *REASONING AND SOLUTION* Consider the left board, which has a length L and a weight of $(356 \text{ N})/2 = 178 \text{ N}$. Let $\mathbf{F_V}$ be the upward normal force exerted by the ground on the board. This force balances the weight, so $F_V = 178 \text{ N}$. Let $\mathbf{f_s}$ be the force of static friction, which acts horizontally on the end of the board in contact with the ground. $\mathbf{f_s}$ points to the right. Since the board is in equilibrium, the net torque acting on the board through any axis must be zero. Measuring the torques with respect to an axis through the apex of the triangle formed by the boards, we have

$$+ (178 \text{ N})(\sin 30.0°)\left(\frac{L}{2}\right) + f_s(L \cos 30.0°) - F_V (L \sin 30.0°) = 0$$

or

$$44.5 \text{ N} + f_s \cos 30.0° - F_V \sin 30.0° = 0$$

so that

$$f_s = \frac{(178 \text{ N})(\sin 30.0°) - 44.5 \text{ N}}{\cos 30.0°} = \boxed{51.4 \text{ N}}$$

CHAPTER 10 | SIMPLE HARMONIC MOTION AND ELASTICITY

1. **REASONING AND SOLUTION** Using Equation 10.1, we first determine the spring constant:

$$k = \frac{F_{Applied}}{x} = \frac{89.0 \text{ N}}{0.0191 \text{ m}} = 4660 \text{ N/m}$$

Again using Equation 10.1, we find that the force needed to compress the spring by 0.0508 m is

$$F_{Applied} = kx = (4660 \text{ N/m})(0.0508 \text{ m}) = \boxed{237 \text{ N}}$$

5. **REASONING AND SOLUTION** According to Newton's second law, the force required to accelerate the trailer is $F = ma$. The displacement of the spring is given by Equation 10.2; solving Equation 10.2 for x and using $F = ma$, we obtain

$$x = -\frac{F}{k} = -\frac{ma}{k} = -\frac{(92 \text{ kg})(0.30 \text{ m/s}^2)}{2300 \text{ N/m}} = -0.012 \text{ m}$$

The amount that the spring stretches is $\boxed{0.012 \text{ m}}$.

11. **REASONING** When the ball is whirled in a horizontal circle of radius r at speed v, the centripetal force is provided by the restoring force of the spring. From Equation 5.3, the magnitude of the centripetal force is mv^2/r, while the magnitude of the restoring force is kx (see Equation 10.2). Thus,

$$\frac{mv^2}{r} = kx \qquad\qquad (1)$$

The radius of the circle is equal to $L_0 + \Delta L$, where L_0 is the unstretched length of the spring and ΔL is the amount that the spring stretches. Equation (1) becomes

$$\frac{mv^2}{L_0 + \Delta L} = k\Delta L \qquad\qquad (1')$$

If the spring were attached to the ceiling and the ball were allowed to hang straight down, motionless, the net force must be zero: $mg - kx = 0$, where $-kx$ is the restoring force of the spring. If we let Δy be the displacement of the spring in the vertical direction, then

$$mg = k\Delta y$$

Solving for Δy, we obtain

$$\Delta y = \frac{mg}{k} \tag{2}$$

SOLUTION According to equation (1') above, the spring constant k is given by

$$k = \frac{mv^2}{\Delta L(L_0 + \Delta L)}$$

Substituting this expression for k into equation (2) gives

$$\Delta y = \frac{mg\Delta L(L_0 + \Delta L)}{mv^2} = \frac{g\Delta L(L_0 + \Delta L)}{v^2}$$

or

$$\Delta y = \frac{(9.80\,\text{m/s}^2)(0.010\,\text{m})(0.200\,\text{m} + 0.010\,\text{m})}{(3.00\,\text{m/s})^2} = \boxed{2.29\times10^{-3}\ \text{m}}$$

---

15. **REASONING AND SOLUTION** According to Equations 10.6 and 10.11, $2\pi f = \sqrt{k/m}$. According to the data in the problem, the frequency of vibration of either spring is

$$f = \frac{5.0\ \text{cycles}}{3.0\ \text{s}} = 1.67\ \text{Hz}$$

Squaring both sides of the equation $2\pi f = \sqrt{k/m}$ and solving for k, we obtain

$$k = 4\pi^2 f^2 m = 4\pi^2(1.67\ \text{Hz})^2(320\ \text{kg}) = \boxed{3.5\times10^4\ \text{N/m}}$$

---

19. $\boxed{\text{WWW}}$ **REASONING AND SOLUTION** According to Equation 10.6, $\omega = 2\pi f$. The maximum speed and maximum acceleration of the atoms may be calculated from Equations 10.8 and 10.10, respectively.

a. Combining Equations 10.6 and 10.8, we obtain

$$v_{\text{max}} = \omega A = (2\pi f)A = 2\pi(2.0\times10^{12}\ \text{Hz})(1.1\times10^{-11}\ \text{m}) = \boxed{140\ \text{m/s}}$$

b. Combining Equations 10.6 and 10.10, we obtain

$$a_{max} = \omega^2 A = (4\pi^2 f^2)A = 4\pi^2 (2.0 \times 10^{12} \text{ Hz})^2 (1.1 \times 10^{-11} \text{ m}) = \boxed{1.7 \times 10^{15} \text{ m/s}^2}$$

---

21. **REASONING** The frequency of vibration of the spring is related to the added mass m by Equations 10.6 and 10.11:

$$f = \frac{1}{2\pi}\sqrt{\frac{k}{m}} \qquad (1)$$

The spring constant can be determined from Equation 10.1.

SOLUTION Since the spring stretches by 0.018 m when a 2.8-kg object is suspended from its end, the spring constant is, according to Equation 10.1,

$$k = \frac{F_{Applied}}{x} = \frac{mg}{x} = \frac{(2.8 \text{ kg})(9.80 \text{ m/s}^2)}{0.018 \text{ m}} = 1.52 \times 10^3 \text{ N/m}$$

Solving Equation (1) for m, we find that the mass required to make the spring vibrate at 3.0 Hz is

$$m = \frac{k}{4\pi^2 f^2} = \frac{1.52 \times 10^3 \text{ N/m}}{4\pi^2 (3.0 \text{ Hz})^2} = \boxed{4.3 \text{ kg}}$$

---

25. **REASONING** The only force that acts on the block along the line of motion is the force due to the spring. Since the force due to the spring is a conservative force, the principle of conservation of mechanical energy applies. Initially, when the spring is unstrained, all of the mechanical energy is kinetic energy, $(1/2)mv_0^2$. When the spring is fully compressed, all of the mechanical energy is in the form of elastic potential energy, $(1/2)kx_{max}^2$, where x_{max}, the maximum compression of the spring, is the amplitude A. Therefore, the statement of energy conservation can be written as

$$\tfrac{1}{2}mv_0^2 = \tfrac{1}{2}kA^2$$

This expression may be solved for the amplitude A.

SOLUTION Solving for the amplitude A, we obtain

$$A = \sqrt{\frac{mv_0^2}{k}} = \sqrt{\frac{(1.00 \times 10^{-2} \text{ kg})(8.00 \text{ m/s})^2}{124 \text{ N/m}}} = \boxed{7.18 \times 10^{-2} \text{ m}}$$

29. ***REASONING AND SOLUTION*** If we neglect air resistance, only the conservative forces of the spring and gravity act on the ball. Therefore, the principle of conservation of mechanical energy applies.

When the 2.00 kg object is hung on the end of the vertical spring, it stretches the spring by an amount x, where

$$x = \frac{F}{k} = \frac{mg}{k} = \frac{(2.00 \text{ kg})(9.80 \text{ m/s}^2)}{50.0 \text{ N/m}} = 0.392 \text{ m} \qquad (10.1)$$

This position represents the equilibrium position of the system with the 2.00-kg object suspended from the spring. The object is then pulled down another 0.200 m and released from rest ($v_0 = 0$ m/s). At this point the spring is stretched by an amount of $0.392 \text{ m} + 0.200 \text{ m} = 0.592 \text{ m}$. This point represents the zero reference level ($h = 0$ m) for the gravitational potential energy.

h **= 0 m:**   The kinetic energy, the gravitational potential energy, and the elastic potential energy at the point of release are:

$$\text{KE} = \tfrac{1}{2}mv_0^2 = \tfrac{1}{2}m(0 \text{ m/s})^2 = \boxed{0 \text{ J}}$$

$$\text{PE}_{\text{gravity}} = mgh = mg(0 \text{ m}) = \boxed{0 \text{ J}}$$

$$\text{PE}_{\text{elastic}} = \tfrac{1}{2}kx_0^2 = \tfrac{1}{2}(50.0 \text{ N/m})(0.592 \text{ m})^2 = \boxed{8.76 \text{ J}}$$

The total mechanical energy E_0 at the point of release is the sum of the three energies above: $E_0 = \boxed{8.76 \text{ J}}$.

h **= 0.200 m:**   When the object has risen a distance of $h = 0.200$ m above the release point, the spring is stretched by an amount of $0.592 \text{ m} - 0.200 \text{ m} = 0.392 \text{ m}$. Since the total mechanical energy is conserved, its value at this point is still $E = \boxed{8.76 \text{ J}}$. The gravitational and elastic potential energies are:

$$\text{PE}_{\text{gravity}} = mgh = (2.00 \text{ kg})(9.80 \text{ m/s}^2)(0.200 \text{ m}) = \boxed{3.92 \text{ J}}$$

$$\text{PE}_{\text{elastic}} = \tfrac{1}{2}kx^2 = \tfrac{1}{2}(50.0 \text{ N/m})(0.392 \text{ m})^2 = \boxed{3.84 \text{ J}}$$

Since $\text{KE} + \text{PE}_{\text{gravity}} + \text{PE}_{\text{elastic}} = E$,

$$\text{KE} = E - \text{PE}_{\text{gravity}} - \text{PE}_{\text{elastic}} = 8.76 \text{ J} - 3.92 \text{ J} - 3.84 \text{ J} = \boxed{1.00 \text{ J}}$$

h = 0.400 m: When the object has risen a distance of h = 0.400 m above the release point, the spring is stretched by an amount of 0.592 m − 0.400 m = 0.192 m. At this point, the total mechanical energy is still E = $\boxed{8.76 \text{ J}}$. The gravitational and elastic potential energies are:

$$PE_{gravity} = mgh = (2.00 \text{ kg})(9.80 \text{ m/s}^2)(0.400 \text{ m}) = \boxed{7.84 \text{ J}}$$

$$PE_{elastic} = \frac{1}{2}kx^2 = \frac{1}{2}(50.0 \text{ N/m})(0.192 \text{ m})^2 = \boxed{0.92 \text{ J}}$$

The kinetic energy is

$$KE = E - PE_{gravity} - PE_{elastic} = 8.76 \text{ J} - 7.84 \text{ J} - 0.92 \text{ J} = \boxed{0 \text{ J}}$$

The results are summarized in the table below:

h	KE	PE_{grav}	$PE_{elastic}$	E
0.000 m	0.00 J	0.00 J	8.76 J	8.76 J
0.200 m	1.00 J	3.92 J	3.84 J	8.76 J
0.400 m	0.00 J	7.84 J	0.92 J	8.76 J

33. ***REASONING AND SOLUTION*** The amount by which the spring stretches due to the weight of the 1.1-kg object can be calculated using Equation 10.1, where the force F is equal to the weight of the object. The position of the object when the spring is stretched is the equilibrium position for the vertical harmonic motion of the object-spring system.

a. Solving Equation 10.1 for x with F equal to the weight of the object gives

$$x = \frac{F}{k} = \frac{mg}{k} = \frac{(1.1 \text{ kg})(9.80 \text{ m/s}^2)}{120 \text{ N/m}} = \boxed{9.0 \times 10^{-2} \text{ m}}$$

b. The object is then pulled down another 0.20 m and released from rest (v_0 = 0 m/s). At this point the spring is stretched by an amount of 0.090 m + 0.20 m = 0.29 m. We will let this point be the zero reference level (h = 0 m) for the gravitational potential energy.

The kinetic energy, the gravitational potential energy, and the elastic potential energy at the point of release are:

$$KE = \frac{1}{2}mv_0^2 = \frac{1}{2}m(0 \text{ m/s})^2 = 0 \text{ J}$$

$$PE_{gravity} = mgh = mg(0 \text{ m}) = 0 \text{ J}$$

$$PE_{elastic} = \frac{1}{2}kx_0^2 = \frac{1}{2}(120 \text{ N/m})(0.29 \text{ m})^2 = 5.0 \text{ J}$$

The total mechanical energy E_0 is the sum of these three energies, so $E_0 = 5.0$ J. When the object has risen a distance of $h = 0.20$ m above the release point, the spring is stretched by an amount of $0.29 \text{ m} - 0.20 \text{ m} = 0.090 \text{ m}$. Since the total mechanical energy is conserved, its value at this point is still 5.0 J. Thus,

$$E = KE + PE_{gravity} + PE_{elastic}$$

$$E = \frac{1}{2}mv^2 + mgh + \frac{1}{2}kx^2$$

$$5.0 \text{ J} = \frac{1}{2}(1.1 \text{ kg})v^2 + (1.1 \text{ kg})(9.80 \text{ m/s}^2)(0.20 \text{ m}) + \frac{1}{2}(120 \text{ N/m})(0.090 \text{ m})^2$$

Solving for v yields $v = \boxed{2.1 \text{ m/s}}$.

---

37. ***REASONING*** Using the principle of conservation of mechanical energy, the initial elastic potential energy stored in the elastic bands must be equal to the sum of the kinetic energy and the gravitational potential energy of the performer at the point of ejection:

$$\frac{1}{2}kx^2 = \frac{1}{2}mv_0^2 + mgh$$

where v_0 is the speed of the performer at the point of ejection and, from the figure at the right, $h = x \sin\theta$.

Thus,

$$\frac{1}{2}kx^2 = \frac{1}{2}mv_0^2 + mgx\sin\theta \qquad (1)$$

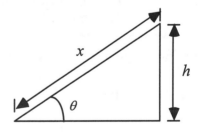

From the horizontal motion of the performer

$$v_{0x} = v_0 \cos\theta \qquad (2)$$

where

$$v_{0x} = \frac{s_x}{t} \qquad (3)$$

and $s_x = 50.0$ m. Combining equations (2) and (3) gives

$$v_0 = \frac{s_x}{t(\cos\theta)}$$

Equation (1) becomes:

$$\frac{1}{2}kx^2 = \frac{1}{2}m\frac{s_x^2}{t^2\cos^2\theta} + mgx\sin\theta$$

This expression can be solved for k, the spring constant of the firing mechanism.

SOLUTION Solving for k yields:

$$k = m\left(\frac{s_x}{xt\cos\theta}\right)^2 + \frac{2mg(\sin\theta)}{x}$$

$$k = (70.0\text{ kg})\left(\frac{50.0\text{ m}}{(3.00\text{ m})(4.00\text{ s})(\cos 40.0°)}\right)^2$$

$$+ \frac{2(70.0\text{ kg})(9.80\text{ m/s}^2)(\sin 40.0°)}{3.00\text{ m}} = \boxed{2.36\times10^3\text{ N/m}}$$

---

41. **REASONING AND SOLUTION** Recall that the relationship between frequency f and period T is $f = 1/T$. Then, according to Equations 10.6 and 10.16, the period of the simple pendulum is given by

$$T = 2\pi\sqrt{\frac{L}{g}}$$

where L is the length of the pendulum. Solving for g, we obtain

$$g = \frac{4\pi^2 L}{T^2} = \frac{4\pi^2(1.2\text{ m})}{(2.8\text{ s})^2} = \boxed{6.0\text{ m/s}^2}$$

---

43. **REASONING** For small-angle displacements, the frequency of simple harmonic motion for a physical pendulum is determined by $2\pi f = \sqrt{mgL/I}$, where L is the distance between the axis of rotation and the center of gravity of the rigid body of moment of inertia I. Since the frequency f and the period T are related by $f = 1/T$, the period of pendulum A is given by

$$T_A = 2\pi\sqrt{\frac{I}{mgL}}$$

Since the pendulum is made from a thin, rigid, uniform rod, its moment of inertia is given by $I = (1/3)md^2$, where d is the length of the rod. Since the rod is uniform, its center of gravity lies at its geometric center, and $L = d/2$ Therefore, the period of pendulum A is given by

$$T_A = 2\pi \sqrt{\frac{2d}{3g}}$$

For the simple pendulum we have

$$T_B = 2\pi \sqrt{\frac{d}{g}}$$

SOLUTION The ratio of the periods is, therefore,

$$\frac{T_A}{T_B} = \frac{2\pi\sqrt{2d/(3g)}}{2\pi\sqrt{d/g}} = \sqrt{\frac{2}{3}} = \boxed{0.816}$$

---

47. $\boxed{\text{WWW}}$ *REASONING AND SOLUTION* The amount of compression can be obtained from Equation 10.17,

$$F = Y\left(\frac{\Delta L}{L_0}\right)A$$

where F is the magnitude of the force on the stand due to the weight of the statue. Solving for ΔL gives

$$\Delta L = \frac{FL_0}{YA} = \frac{mgL_0}{YA} = \frac{(3500\text{ kg})(9.80\text{ m/s}^2)(1.8\text{ m})}{(2.3\times10^{10}\text{ N/m}^2)(7.3\times10^{-2}\text{ m}^2)} = \boxed{3.7\times10^{-5}\text{ m}}$$

---

49. *REASONING AND SOLUTION* If we assume that each block is subjected to one-fourth of the applied force F, then, according to Equation 10.18, the force on each block is given by

$$\frac{F}{4} = S\left(\frac{\Delta X}{L_0}\right)A$$

Solving for ΔX, we obtain

$$\Delta X = \frac{FL_0}{4SA} = \frac{(32\text{ N})(0.030\text{ m})}{4(2.6\times10^6\text{ N/m}^2)(1.2\times10^{-3}\text{ m}^2)} = \boxed{7.7\times10^{-5}\text{ m}}$$

53. ***REASONING AND SOLUTION***   Equation 10.20 gives the desired result. Solving for $\Delta V / V_0$, we obtain

$$\frac{\Delta V}{V_0} = -\frac{\Delta P}{B} = -\frac{-1.01 \times 10^5 \text{ Pa}}{7.1 \times 10^{10} \text{ N/m}^2} = \boxed{1.4 \times 10^{-6}}$$

59. ***REASONING AND SOLUTION***
a. When the jeep is suspended motionless, the tension T in the cable must be equal in magnitude and opposite in direction to the weight mg of the jeep. Therefore, from Equation 10.17,

$$mg = Y\left(\frac{\Delta L}{L_0}\right) A$$

Solving for ΔL, we obtain

$$\Delta L = \frac{mgL_0}{YA} = \frac{mgL_0}{Y(\pi r^2)} = \frac{(2100 \text{ kg})(9.80 \text{ m/s}^2)(48 \text{ m})}{(2.0 \times 10^{11} \text{ Pa})\pi(5.0 \times 10^{-3} \text{ m})^2} = \boxed{6.3 \times 10^{-2} \text{ m}}$$

b. When the jeep is hoisted upward with an acceleration a, Newton's second law gives

$$T - mg = ma \qquad \text{or} \qquad T = m(a + g)$$

Therefore, Equation 10.17 becomes

$$m(a + g) = Y\left(\frac{\Delta L}{L_0}\right) A$$

and solving for ΔL gives

$$\Delta L = \frac{m(a + g)L_0}{Y(\pi r^2)} = \frac{(2100 \text{ kg})(1.5 \text{ m/s}^2 + 9.80 \text{ m/s}^2)(48 \text{ m})}{(2.0 \times 10^{11} \text{ Pa})\pi(5.0 \times 10^{-3} \text{ m})^2} = \boxed{7.3 \times 10^{-2} \text{ m}}$$

61. $\boxed{\text{WWW}}$ ***REASONING*** The strain in the wire is given by $\Delta L / L_0$. From Equation 10.17, the strain is therefore given by

$$\frac{\Delta L}{L_0} = \frac{F}{YA} \tag{1}$$

where F must be equal to the magnitude of the centripetal force that keeps the stone moving in the circular path of radius R.

SOLUTION   Combining Equation (1) with Equation 5.3 for the magnitude of the centripetal force, we obtain

$$\frac{\Delta L}{L_0} = \frac{F}{Y(\pi r^2)} = \frac{(mv^2/R)}{Y(\pi r^2)} = \frac{(8.0 \text{ kg})(12 \text{ m/s})^2/(4.0 \text{ m})}{(2.0 \times 10^{11} \text{ Pa})\pi(1.0 \times 10^{-3} \text{ m})^2} = \boxed{4.6 \times 10^{-4}}$$

---

65. $\boxed{\text{WWW}}$ *REASONING* Equation 10.20 can be used to find the fractional change in volume of the brass sphere when it is exposed to the Venusian atmosphere. Once the fractional change in volume is known, it can be used to calculate the fractional change in radius.

SOLUTION According to Equation 10.20, the fractional change in volume is

$$\frac{\Delta V}{V_0} = -\frac{\Delta P}{B} = -\frac{8.9 \times 10^6 \text{ Pa}}{6.7 \times 10^{10} \text{ Pa}} = -1.33 \times 10^{-4}$$

The initial volume of the sphere is $V_0 = (4/3)\pi r^3$. If we assume that the change in the radius of the sphere is very small relative to the initial radius, we can think of the sphere's change in volume as the addition or subtraction of a spherical shell of volume ΔV, whose radius is r and whose thickness is Δr. Then, the change in volume of the sphere is equal to the volume of the shell and is given by $\Delta V = (4\pi r^2)\Delta r$. Combining the expressions for V_0 and ΔV, and solving for $\Delta r/r$, we have

$$\frac{\Delta r}{r} = \frac{1}{3}\frac{\Delta V}{V_0}$$

Therefore,

$$\frac{\Delta r}{r} = \tfrac{1}{3}(-1.33 \times 10^{-4}) = \boxed{-4.4 \times 10^{-5}}$$

---

71. *REASONING AND SOLUTION* The shearing stress is equal to the force per unit area applied to the rivet. Thus, when a shearing stress of 5.0×10^8 Pa is applied to each rivet, the force experienced by each rivet is

$$F = (\text{Stress})A = (\text{Stress})(\pi r^2) = (5.0 \times 10^8 \text{ Pa})\left[\pi (5.0 \times 10^{-3} \text{ m})^2\right] = 3.9 \times 10^4 \text{ N}$$

Therefore, the maximum tension T that can be applied to each beam, assuming that each rivet carries one-fourth of the total load, is $\boxed{4F = 1.6 \times 10^5 \text{ N}}$.

75. **REASONING AND SOLUTION** The force that acts on the block is given by Newton's Second law, $F = ma$. Since the block has a constant acceleration, the acceleration is given by Equation 2.8 with $v_0 = 0$ m/s; that is, $a = 2d/t^2$, where d is the distance through which the block is pulled. Therefore, the force that acts on the block is given by

$$F = ma = \frac{2md}{t^2}$$

The force acting on the block is the restoring force of the spring. Thus, according to Equation 10.2, $F = -kx$, where k is the spring constant and x is the displacement. Solving Equation 10.2 for x and using the expression above for F, we obtain

$$x = -\frac{F}{k} = -\frac{2md}{kt^2} = -\frac{2(7.00 \text{ kg})(4.00 \text{ m})}{(415 \text{ N/m})(0.750 \text{ s})^2} = -0.240 \text{ m}$$

The amount that the spring stretches is $\boxed{0.240 \text{ m}}$.

---

79. **REASONING AND SOLUTION** If we assume that the wire has a circular cross-section, then Equation 10.17 becomes

$$F = Y\left(\frac{\Delta L}{L_0}\right)\left(\pi r^2\right)$$

where F is the tension in the wire and r is the radius of the wire. Solving for r gives

$$r = \sqrt{\frac{FL_0}{\pi Y \Delta L}} \qquad (1)$$

Young's modulus for steel is given in Table 10.1 as $Y = 2.0 \times 10^{11}$ N/m^2. The change in length ΔL can be determined from the figure at the right. If L_0 represents the length of the wire when it is horizontal, and L its length when stretched, then the figure shows that

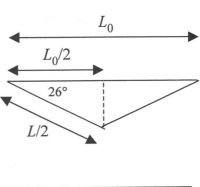

$$\cos 26° = \frac{L_0/2}{L/2} \quad \text{or} \quad L = \frac{L_0}{\cos 26°} = 1.11L_0 \qquad (2)$$

Therefore,
$$\Delta L = L - L_0 = 0.11L_0$$

The tension in the wire can be determined from the free-body diagram at the right. Applying Newton's second law to the forces in the vertical direction yields

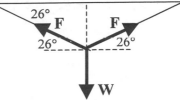

$$2F \sin 26° - W = 0 \quad \text{or} \quad F = \frac{W}{2 \sin 26°} \tag{3}$$

Substituting Equations (2) and (3) into Equation (1) yields

$$r = \sqrt{\frac{W L_0}{2\pi (\sin 26°) Y (0.11 L_0)}} = \sqrt{\frac{(96 \text{ N}) L_0}{2\pi (\sin 26°)(2.0 \times 10^{11} \text{ N/m}^2)(0.11 L_0)}} = \boxed{4.0 \times 10^{-5} \text{ m}}$$

---

83. **REASONING**   The angular frequency for simple harmonic motion is given by Equation 10.11 as $\omega = \sqrt{k/m}$. Since the frequency f is related to the angular frequency ω by $f = \omega/2\pi$, and f is related to the period T by $f = 1/T$, the period of the motion is given by $T = 2\pi/\omega = 2\pi\sqrt{m/k}$.

SOLUTION
a. When $m_1 = m_2 = 3.0$ kg, we have that

$$T_1 = T_2 = 2\pi \sqrt{\frac{3.0 \text{ kg}}{120 \text{ N/m}}} = 0.99 \text{ s}$$

Both particles will pass through the position $x = 0$ m for the first time one-quarter of the way through one cycle, or

$$\Delta t = \tfrac{1}{4}T_1 = \tfrac{1}{4}T_2 = \frac{0.99 \text{ s}}{4} = \boxed{0.25 \text{ s}}$$

b. $T_1 = 0.99$ s, as in part (a) above, while $T_2 = 2\pi\sqrt{(27.0 \text{ kg})/(120 \text{ N/m})} = 3.0$ s. Each particle will pass through the position $x = 0$ m every odd-quarter of a cycle, $\tfrac{1}{4}T, \tfrac{3}{4}T, \tfrac{5}{4}T, \ldots$ Thus, the two particles will pass through $x = 0$ m when

3.0-kg particle $t = \tfrac{1}{4}T_1, \tfrac{3}{4}T_1, \tfrac{5}{4}T_1, \ldots$

27.0-kg particle $t = \tfrac{1}{4}T_2, \tfrac{3}{4}T_2, \tfrac{5}{4}T_2, \ldots$

Since $T_2 = 3T_1$, we see that both particles will be at $x = 0$ m simultaneously when $t = \tfrac{3}{4}T_1$, or $t = \tfrac{1}{4}T_2 = \tfrac{3}{4}T_1$. Thus,

$$t = \tfrac{3}{4}T_1 = \tfrac{3}{4}(0.99 \text{ s}) = \boxed{0.75 \text{ s}}$$

CHAPTER 11 | FLUIDS

PROBLEMS

---

1. **REASONING AND SOLUTION** Equation 11.1 can be solved for the mass m. In order to use Equation 11.1, however, we must know the volume of the ice. Since the lake is circular, the volume of the ice is equal to the circular area of the lake ($A = \pi r^2$) multiplied by the thickness d of the ice.

$$m = \rho V = \rho(\pi r^2)d = (917 \text{ kg/m}^3)\pi(480 \text{ m})^2(0.010 \text{ m}) = \boxed{6.6 \times 10^6 \text{ kg}}$$

---

5. **REASONING** Equation 11.1 can be used to find the volume occupied by 1.00 kg of silver. Once the volume is known, the area of a sheet of silver of thickness d can be found from the fact that the volume is equal to the area of the sheet times its thickness.

SOLUTION Solving Equation 11.1 for V, the volume of 1.00 kg of silver is

$$V = \frac{m}{\rho} = \frac{1.00 \text{ kg}}{10\ 500 \text{ kg/m}^3} = 9.52 \times 10^{-5} \text{ m}^3$$

The area of the silver, is, therefore,

$$A = \frac{V}{d} = \frac{9.52 \times 10^{-5} \text{ m}^3}{3.00 \times 10^{-7} \text{ m}} = \boxed{317 \text{ m}^2}$$

---

9. **WWW** **REASONING** The total mass of the solution is the sum of the masses of its constituents. Therefore,

$$\rho_s V_s = \rho_w V_w + \rho_g V_g \qquad (1)$$

where the subscripts s, w, and g refer to the solution, the water, and the ethylene glycol, respectively. The volume of the water can be written as $V_w = V_s - V_g$. Making this substitution for V_w, Equation (1) above can be rearranged to give

$$\frac{V_g}{V_s} = \frac{\rho_s - \rho_w}{\rho_g - \rho_w} \qquad (2)$$

Equation (2) can be used to calculate the relative volume of ethylene glycol in the solution.

SOLUTION The density of ethylene glycol is given in the problem. The density of water is given in Table 11.1 as 1.000×10^3 kg/m^3. The specific gravity of the solution is given as 1.0730. Therefore, the density of the solution is

$$\rho_s = (\text{specific gravity of solution}) \times \rho_w$$

$$= (1.0730)(1.000 \times 10^3 \text{ kg/m}^3) = 1.0730 \times 10^3 \text{ kg/m}^3$$

Substituting the values for the densities into Equation (2), we obtain

$$\frac{V_g}{V_s} = \frac{\rho_s - \rho_w}{\rho_g - \rho_w} = \frac{1.0730 \times 10^3 \text{ kg/m}^3 - 1.000 \times 10^3 \text{ kg/m}^3}{1116 \text{ kg/m}^3 - 1.000 \times 10^3 \text{ kg/m}^3} = 0.63$$

Therefore, the volume percentage of ethylene glycol is $\boxed{63\%}$.

---

13. *REASONING AND SOLUTION* Equation 11.3 gives the desired result, where F is equal to the weight of the woman. Assuming that the heel has a circular cross section, its area is $A = \pi r^2$, where r is the radius.

$$P = \frac{F}{A} = \frac{mg}{\pi r^2} = \frac{(50.0 \text{ kg})(9.80 \text{ m/s}^2)}{\pi (6.00 \times 10^{-3} \text{ m})^2} = \boxed{4.33 \times 10^6 \text{ Pa}}$$

---

17. *REASONING AND SOLUTION* Both the cylinder and the hemisphere have circular cross sections. According to Equation 11.3, the pressure exerted on the ground by the hemisphere is

$$P = \frac{W_h}{A_h} = \frac{W_h}{\pi r_h^2}$$

where W_h and r_h are the weight and radius of the hemisphere. Similarly, the pressure exerted on the ground by the cylinder is

$$P = \frac{W_c}{A_c} = \frac{W_c}{\pi r_c^2}$$

where W_c and r_c are the weight and radius of the cylinder. Since each object exerts the same pressure on the ground, we can equate the right-hand sides of the expressions above to obtain

$$\frac{W_h}{\pi r_h^2} = \frac{W_c}{\pi r_c^2}$$

Solving for $r_h^{\,2}$, we obtain

$$r_h^2 = r_c^2 \frac{W_h}{W_c} \tag{1}$$

The weight of the hemisphere is

$$W_h = \rho g V_h = \rho g \left[\tfrac{1}{2} \left(\tfrac{4}{3} \pi r_h^3 \right) \right] = \tfrac{2}{3} \rho g \pi r_h^3$$

where ρ and V_h are the density and volume of the hemisphere, respectively. The weight of the cylinder is

$$W_c = \rho g V_c = \rho g \pi r_c^2 h$$

where ρ and V_c are the density and volume of the cylinder, respectively, and h is the height of the cylinder. Substituting these expressions for the weights into Equation (1) gives

$$r_h^2 = r_c^2 \frac{W_h}{W_c} = r_c^2 \frac{\tfrac{2}{3} \rho g \pi r_h^3}{\rho g \pi r_c^2 h}$$

Solving for r_h gives

$$r_h = \tfrac{3}{2} h = \tfrac{3}{2} (0.500 \text{ m}) = \boxed{0.750 \text{ m}}$$

---

19. ***REASONING AND SOLUTION*** Using Equation 11.4, we have

$$P_{\text{heart}} - P_{\text{brain}} = \rho g h = (1.000 \times 10^3 \text{ kg/m}^3)(9.80 \text{ m/s}^2)(12 \text{ m}) = \boxed{1.2 \times 10^5 \text{ Pa}}$$

---

23. ***REASONING*** Since the faucet is closed, the water in the pipe may be treated as a static fluid. The gauge pressure P_2 at the faucet on the first floor is related to the gauge pressure P_1 at the faucet on the second floor by Equation 11.4, $P_2 = P_1 + \rho g h$.

SOLUTION
a. Solving Equation 11.4 for P_1, we find the gauge pressure at the second-floor faucet is

$$P_1 = P_2 - \rho g h = 1.90 \times 10^5 \text{ Pa} - (1.00 \times 10^3 \text{ kg} / \text{m}^3)(9.80 \text{ m} / \text{s}^2)(6.50 \text{ m}) = \boxed{1.26 \times 10^5 \text{ Pa}}$$

b. If the second faucet were placed at a height h above the first-floor faucet so that the gauge pressure P_1 at the second faucet were zero, then no water would flow from the second faucet, even if it were open. Solving Equation 11.4 for h when P_1 equals zero, we obtain

$$h = \frac{P_2 - P_1}{\rho g} = \frac{1.90 \times 10^5 \text{ Pa} - 0}{(1.00 \times 10^3 \text{ kg} / \text{m}^3)(9.80 \text{ m} / \text{s}^2)} = \boxed{19.4 \text{ m}}$$

27. **REASONING** Let the length of the tube be denoted by L, and let the length of the liquid be denoted by ℓ. When the tube is whirled in a circle at a constant angular speed about an axis through one end, the liquid collects at the other end and experiences a centripetal force given by (see Equation 8.11, and use $F = ma$) $F = mr\omega^2 = mL\omega^2$.

Since there is no air in the tube, this is the only radial force experienced by the liquid, and it results in a pressure of

$$P = \frac{F}{A} = \frac{mL\omega^2}{A}$$

where A is the cross-sectional area of the tube. The mass of the liquid can be expressed in terms of its density ρ and volume V: $m = \rho V = \rho A\ell$. The pressure may then be written as

$$P = \frac{\rho A\ell L\omega^2}{A} = \rho\ell L\omega^2 \tag{1}$$

If the tube were completely filled with liquid and allowed to hang vertically, the pressure at the bottom of the tube (that is, where $h = L$) would be given by

$$P = \rho g L \tag{2}$$

SOLUTION   According to the statement of the problem, the quantities calculated by Equations (1) and (2) are equal, so that $\rho\ell L\omega^2 = \rho g L$. Solving for ω gives

$$\omega = \sqrt{\frac{g}{\ell}} = \sqrt{\frac{9.80 \text{ m/s}^2}{0.0100 \text{ m}}} = \boxed{31.3 \text{ rad/s}}$$

31. **REASONING** According to Equation 11.4, the initial pressure at the bottom of the pool is $P_0 = \left(P_{\text{atm}}\right)_0 + \rho g h$, while the final pressure is $P_{\text{f}} = \left(P_{\text{atm}}\right)_{\text{f}} + \rho g h$. Therefore, the change in pressure at the bottom of the pool is

$$\Delta P = P_{\text{f}} - P_0 = \left[\left(P_{\text{atm}}\right)_{\text{f}} + \rho g h\right] - \left[\left(P_{\text{atm}}\right)_0 + \rho g h\right] = \left(P_{\text{atm}}\right)_{\text{f}} - \left(P_{\text{atm}}\right)_0$$

According to Equation 11.3, $F = PA$, the change in the force at the bottom of the pool is

$$\Delta F = (\Delta P)A = \left[\left(P_{\text{atm}}\right)_{\text{f}} - \left(P_{\text{atm}}\right)_0\right]A$$

SOLUTION Direct substitution of the data given in the problem into the expression above yields

$$\Delta F = (765 \text{ mm Hg} - 755 \text{ mm Hg})(12 \text{ m})(24 \text{ m})\left(\frac{133 \text{ Pa}}{1.0 \text{ mm Hg}}\right) = \boxed{3.8 \times 10^5 \text{ N}}$$

Note that the conversion factor 133 Pa = 1.0 mm Hg is used to convert mm Hg to Pa.

---

35. **_REASONING_**   The pressure P' exerted on the bed of the truck by the plunger is $P' = P - P_{atm}$. According to Equation 11.3, $F = P'A$, so the force exerted on the bed of the truck can be expressed as $F = (P - P_{atm})A$. If we assume that the plunger remains perpendicular to the floor of the load bed, the torque that the plunger creates about the axis shown in the figure in the text is

$$\tau = F\ell = (P - P_{atm})A\ell = (P - P_{atm})(\pi r^2)\ell$$

SOLUTION   Direct substitution of the numerical data into the expression above gives

$$\tau = (3.54 \times 10^6 \text{ Pa} - 1.01 \times 10^5 \text{ Pa})\pi(0.150 \text{ m})^2(3.50 \text{ m}) = \boxed{8.50 \times 10^5 \text{ N} \cdot \text{m}}$$

---

39. **_REASONING_**   According to Archimedes principle, the buoyant force that acts on the block is equal to the weight of the water that is displaced by the block. The block displaces an amount of water V, where V is the volume of the block. Therefore, the weight of the water displaced by the block is $W = mg = (\rho_{water}V)g$.

SOLUTION   The buoyant force that acts on the block is, therefore,

$$F = \rho_{water}Vg = (1.00 \times 10^3 \text{ kg/m}^3)(0.10 \text{ m} \times 0.20 \text{ m} \times 0.30 \text{ m})(9.80 \text{ m/s}^2) = \boxed{59 \text{ N}}$$

---

43. $\boxed{\text{WWW}}$ **_REASONING AND SOLUTION_**   Under water, the weight of the person with empty lungs is $W_{empty} = W - \rho_{water}gV_{empty}$, where W is the weight of the person in air and V_{empty} is the volume of the empty lungs. Similarly, when the person's lungs are partially full under water, the weight of the person is $W_{full} = W - \rho_{water}gV_{full}$. Subtracting the second equation from the first equation and rearranging gives

$$V_{full} - V_{empty} = \frac{W_{empty} - W_{full}}{\rho_{water}g} = \frac{40.0 \text{ N} - 20.0 \text{ N}}{(1.00 \times 10^3 \text{ kg/m}^3)(9.80 \text{ m/s}^2)} = \boxed{2.04 \times 10^{-3} \text{ m}^3}$$

47. **_REASONING_** The height of the cylinder that is in the oil is given by $h_{oil} = V_{oil} / (\pi r^2)$, where V_{oil} is the volume of oil displaced by the cylinder and r is the radius of the cylinder. We must, therefore, find the volume of oil displaced by the cylinder. After the oil is poured in, the buoyant force that acts on the cylinder is equal to the sum of the weight of the water displaced by the cylinder and the weight of the oil displaced by the cylinder. Therefore, the magnitude of the buoyant force is given by $F = \rho_{water} g V_{water} + \rho_{oil} g V_{oil}$. Since the cylinder floats in the fluid, the net force that acts on the cylinder must be zero. Therefore, the buoyant force that supports the cylinder must be equal to the weight of the cylinder, or

$$\rho_{water} g V_{water} + \rho_{oil} g V_{oil} = mg$$

where m is the mass of the cylinder. Substituting values into the expression above leads to

$$V_{water} + (0.725)V_{oil} = 7.00 \times 10^{-3} \text{ m}^3 \tag{1}$$

From the figure in the text, $V_{cylinder} = V_{water} + V_{oil}$. Substituting values into the expression for $V_{cylinder}$ gives

$$V_{water} + V_{oil} = 8.48 \times 10^{-3} \text{ m}^3 \tag{2}$$

Subtracting Equation (1) from Equation (2) yields $V_{oil} = 5.38 \times 10^{-3} \text{ m}^3$.

SOLUTION The height of the cylinder that is in the oil is, therefore,

$$h_{oil} = \frac{V_{oil}}{\pi r^2} = \frac{5.38 \times 10^{-3} \text{ m}^3}{\pi (0.150 \text{ m})^2} = \boxed{7.6 \times 10^{-2} \text{ m}}$$

---

51. **_REASONING AND SOLUTION_** The mass flow rate Q_{mass} is the amount of fluid mass that flows per unit time. Therefore,

$$Q_{mass} = \frac{m}{t} = \frac{\rho V}{t} = \frac{(1030 \text{ kg/m}^3)(9.5 \times 10^{-4} \text{ m}^3)}{6.0 \text{ h}} \left(\frac{1.0 \text{ h}}{3600 \text{ s}}\right) = \boxed{4.5 \times 10^{-5} \text{ kg/s}}$$

---

55. **_REASONING AND SOLUTION_**
a. The volume flow rate is given by Equation 11.10. Assuming that the line has a circular cross section, $A = \pi r^2$, we have

$$Q = Av = (\pi r^2)v = \pi (0.0065 \text{ m})^2 (1.2 \text{ m/s}) = \boxed{1.6 \times 10^{-4} \text{ m}^3/\text{s}}$$

b. The volume flow rate calculated in part (a) above is the flow rate for all twelve holes. Therefore, the volume flow rate through one of the twelve holes is

$$Q_{hole} = \frac{Q}{12} = (\pi r_{hole}^2)v_{hole}$$

Solving for v_{hole} we have

$$v_{hole} = \frac{Q}{12\pi r_{hole}^2} = \frac{1.6 \times 10^{-4} \ m^3/s}{12\pi(4.6 \times 10^{-4} \ m)^2} = \boxed{2.0 \times 10^1 \ m/s}$$

---

57. **REASONING AND SOLUTION**
   a. Using Equation 11.12, the form of Bernoulli's equation with $y_1 = y_2$, we have

$$P_1 - P_2 = \tfrac{1}{2}\rho\left(v_2^2 - v_1^2\right) = \frac{1.29 \ kg/m^3}{2}\left[(15 \ m/s)^2 - 0\right] = \boxed{150 \ Pa}$$

   b. The pressure inside the roof is greater than the pressure on the outside. Therefore, there is a net outward force on the roof. If the wind speed is sufficiently high, some roofs are "blown outward."

---

59. **REASONING AND SOLUTION** Let the speed of the air below and above the wing be given by v_1 and v_2, respectively. According to Equation 11.12, the form of Bernoulli's equation with $y_1 = y_2$, we have

$$P_1 - P_2 = \tfrac{1}{2}\rho\left(v_2^2 - v_1^2\right) = \frac{1.29 \ kg/m^3}{2}\left[(251 \ m/s)^2 - (225 \ m/s)^2\right] = 7.98 \times 10^3 \ Pa$$

From Equation 11.3, the lifting force is, therefore,

$$F = \left(P_1 - P_2\right)A = (7.98 \times 10^3 \ Pa)(24.0 \ m^2) = \boxed{1.92 \times 10^5 \ N}$$

---

65. **REASONING** Since the pressure difference is known, Bernoulli's equation can be used to find the speed v_2 of the gas in the pipe. Bernoulli's equation also contains the unknown speed v_1 of the gas in the Venturi meter; therefore, we must first express v_1 in terms of v_2. This can be done by using Equation 11.9, the equation of continuity.

SOLUTION
   a. From the equation of continuity (Equation 11.9) it follows that $v_1 = \left(A_2/A_1\right)v_2$. Therefore,

$$v_1 = \frac{0.0700 \text{ m}^2}{0.0500 \text{ m}^2} v_2 = (1.40) v_2$$

Substituting this expression into Bernoulli's equation (Equation 11.12), we have

$$P_1 + \tfrac{1}{2}\rho(1.40\ v_2)^2 = P_2 + \tfrac{1}{2}\rho v_2^2$$

Solving for v_2, we obtain

$$v_2 = \sqrt{\frac{2(P_2 - P_1)}{\rho\left[(1.40)^2 - 1\right]}} = \sqrt{\frac{2(120 \text{ Pa})}{(1.30 \text{ kg} / \text{m}^3)\left[(1.40)^2 - 1\right]}} = \boxed{14 \text{ m} / \text{s}}$$

b.   According to Equation 11.10, the volume flow rate is

$$Q = A_2 v_2 = (0.0700 \text{ m}^2)(14 \text{ m} / \text{s}) = \boxed{0.98 \text{ m}^3 / \text{s}}$$

---

69.   ***REASONING AND SOLUTION*** Bernoulli's equation (Equation 11.12) is

$$P_1 + \tfrac{1}{2}\rho v_1^2 = P_2 + \tfrac{1}{2}\rho v_2^2$$

If we let the right hand side refer to the air above the plate, and the left hand side refer to the air below the plate, then $v_1 = 0$, since the air below the plate is stationary.   We wish to find v_2 for the situation illustrated in part *b* of the figure shown in the text. Solving the equation above for v_2 (with $v_2 = v_{2b}$ and $v_1 = 0$) gives

$$v_{2b} = \sqrt{\frac{2(P_1 - P_2)}{\rho}} \qquad (1)$$

In Equation (1), P_1 is atmospheric pressure and P_2 must be determined. We must first consider the situation in part *a* of the text figure. The figure at the right shows the forces that act on the rectangular plate in part *a* of the text drawing. F_1 is the force exerted on the plate from the air below the plate, and F_2 is the force exerted on the plate from the air above the plate. Applying Newton's second law, we have (taking "up" to be the positive direction),

$$F_1 - F_2 - mg = 0$$

$$F_1 - F_2 = mg$$

Thus, the difference in pressures exerted by the air on the plate in part *a* of the drawing is

$$P_1 - P_2 = \frac{F_1 - F_2}{A} = \frac{mg}{A} \qquad (2)$$

where A is the area of the plate. From Bernoulli's equation (Equation 11.12) we have, with $v_2 = v_{2a}$ and $v_{1a} = 0$,

$$P_1 - P_2 = \tfrac{1}{2}\rho v_{2a}^2 \qquad (3)$$

where v_{2a} is the speed of the air along the top of the plate in part *a* of the text drawing. Combining Equations (2) and (3) we have

$$\frac{mg}{A} = \tfrac{1}{2}\rho v_{2a}^2 \qquad (4)$$

The figure below, on the left, shows the forces that act on the plate in part *b* of the text drawing. The notation is the same as that used when the plate was horizontal (part *a* of the text figure). The figure at the right below shows the same forces resolved into components along the plate and perpendicular to the plate.

Applying Newton's second law we have $F_1 - F_2 - mg\sin\theta = 0$, or

$$F_1 - F_2 = mg\sin\theta$$

Thus, the difference in pressures exerted by the air on the plate in part *b* of the text figure is

$$P_1 - P_2 = \frac{F_1 - F_2}{A} = \frac{mg\sin\theta}{A}$$

Using Equation (4) above,

$$P_1 - P_2 = \tfrac{1}{2}\rho v_{2a}^2 \sin\theta$$

Thus, Equation (1) becomes

$$v_{2b} = \sqrt{\frac{2\left(\frac{1}{2}\rho v_{2a}^2 \sin\theta\right)}{\rho}}$$

Therefore,

$$v_{2b} = \sqrt{v_{2a}^2 \sin\theta} = \sqrt{(11.0 \text{ m/s})^2 \sin 30.0°} = \boxed{7.78 \text{ m/s}}$$

---

71.   **WWW**   ***REASONING AND SOLUTION***

a.   If the water behaves as an ideal fluid, and since the pipe is horizontal and has the same radius throughout, the speed and pressure of the water are the same at all points in the pipe. Since the right end of the pipe is open to the atmosphere, the pressure at the right end is atmospheric pressure; therefore, the pressure at the left end is also atmospheric pressure, or $\boxed{1.01 \times 10^5 \text{ Pa}}$.

b.   If the water is treated as a viscous fluid, the volume flow rate Q is described by Poiseuille's law (Equation 11.14):

$$Q = \frac{\pi R^4 (P_2 - P_1)}{8\eta L}$$

Let P_1 represent the pressure at the right end of the pipe, and let P_2 represent the pressure at the left end of the pipe. Solving for P_2 (with P_1 equal to atmospheric pressure), we obtain

$$P_2 = \frac{8\eta L Q}{\pi R^4} + P_1$$

Therefore,

$$P_2 = \frac{8(1.00 \times 10^{-3} \text{ Pa·s})(1.3 \text{ m})(9.0 \times 10^{-3} \text{ m}^3/\text{s})}{\pi (6.4 \times 10^{-3} \text{ m})^4} + 1.013 \times 10^5 \text{ Pa} = \boxed{1.19 \times 10^5 \text{ Pa}}$$

---

75.   ***REASONING AND SOLUTION***

a.   Using Stoke's law, the viscous force is

$$F = 6\pi\eta R v = 6\pi (1.00 \times 10^{-3} \text{ Pa·s})(5.0 \times 10^{-4} \text{ m})(3.0 \text{ m/s}) = \boxed{2.8 \times 10^{-5} \text{ N}}$$

b.   When the sphere reaches its terminal speed, the net force on the sphere is zero, so that the magnitude of **F** must be equal to the magnitude of $m\mathbf{g}$, or $F = mg$. Therefore, $6\pi\eta R v_T = mg$, where v_T is the terminal speed of the sphere. Solving for v_T, we have

$$v_T = \frac{mg}{6\pi\eta R} = \frac{(1.0\times10^{-5}\ \text{kg})(9.80\ \text{m/s}^2)}{6\pi(1.00\times10^{-3}\ \text{Pa}\cdot\text{s})(5.0\times10^{-4}\ \text{m})} = \boxed{1.0\times10^1\ \text{m/s}}$$

79. **REASONING**   The buoyant force exerted on the balloon by the air must be equal in magnitude to the weight of the balloon and its contents (load and hydrogen).   The magnitude of the buoyant force is given by $\rho_{air} V g$. Therefore,

$$\rho_{air} V g = W_{load} + \rho_{hydrogen} V g$$

where, since the balloon is spherical, $V = (4/3)\pi r^3$. Making this substitution for V and solving for r, we obtain

$$r = \left[\frac{3W_{load}}{4\pi g(\rho_{air} - \rho_{hydrogen})}\right]^{1/3}$$

SOLUTION   Direct substitution of the data given in the problem yields

$$r = \left[\frac{3(5750\ \text{N})}{4\pi(9.80\ \text{m/s}^2)(1.29\ \text{kg/m}^3 - 0.0899\ \text{kg/m}^3)}\right]^{1/3} = \boxed{4.89\ \text{m}}$$

83. **REASONING AND SOLUTION**   Using Equation 11.10 for the volume flow rate, we have $Q = Av$, where A is the cross-sectional area of the pipe and v is the speed of the fluid. Since the pipe has a circular cross section, $A = \pi r^2$. Therefore, solving Equation 11.10 for v, we have

$$v = \frac{Q}{A} = \frac{Q}{\pi r^2} = \frac{1.50\ \text{m}^3/\text{s}}{\pi(0.500\ \text{m})^2} = \boxed{1.91\ \text{m/s}}$$

87. **REASONING AND SOLUTION**
a. The pressure at the level of house A is given by Equation 11.4 as $P = P_{atm} + \rho g h$. Now the height h consists of the 15.0 m and the diameter d of the tank. We first calculate the radius of the tank, from which we can infer d. Since the tank is spherical, its full mass is given by $M = \rho V = \rho[(4/3)\pi r^3]$. Therefore,

$$r^3 = \frac{3M}{4\pi\rho} \quad \text{or} \quad r = \left(\frac{3M}{4\pi\rho}\right)^{1/3} = \left[\frac{3(5.25\times10^5\ \text{kg})}{4\pi(1.000\times10^3\ \text{kg/m}^3)}\right]^{1/3} = 5.00\ \text{m}$$

Therefore, the diameter of the tank is 10.0 m, and the height h is given by

$$h = 10.0 \text{ m} + 15.0 \text{ m} = 25.0 \text{ m}$$

According to Equation 11.4, the gauge pressure in house A is, therefore,

$$P - P_{atm} = \rho g h = (1.000 \times 10^3 \text{ kg/m}^3)(9.80 \text{ m/s}^2)(25.0 \text{ m}) = \boxed{2.45 \times 10^5 \text{ Pa}}$$

b. The pressure at house B is $P = P_{atm} + \rho g h$, where

$$h = 15.0 \text{ m} + 10.0 \text{ m} - 7.30 \text{ m} = 17.7 \text{ m}$$

According to Equation 11.4, the gauge pressure in house B is

$$P - P_{atm} = \rho g h = (1.000 \times 10^3 \text{ kg/m}^3)(9.80 \text{ m/s}^2)(17.7 \text{ m}) = \boxed{1.73 \times 10^5 \text{ Pa}}$$

---

89. ***REASONING AND SOLUTION***   The weight of the coin in air is equal to the sum of the weights of the silver and copper that make up the coin:

$$W_{air} = m_{coin} g = \left(m_{silver} + m_{copper} \right) g = \left(\rho_{silver} V_{silver} + \rho_{copper} V_{copper} \right) g \qquad (1)$$

The weight of the coin in water is equal to its weight in air minus the buoyant force exerted on the coin by the water. Therefore,

$$W_{water} = W_{air} - \rho_{water} \left(V_{silver} + V_{copper} \right) g \qquad (2)$$

Solving Equation (2) for the sum of the volumes gives

$$V_{silver} + V_{copper} = \frac{W_{air} - W_{water}}{\rho_{water} \, g} = \frac{m_{coin} g - W_{water}}{\rho_{water} \, g}$$

or

$$V_{silver} + V_{copper} = \frac{(1.150 \times 10^{-2} \text{ kg})(9.80 \text{ m/s}^2) - 0.1011 \text{ N}}{(1.000 \times 10^3 \text{ kg/m}^3)(9.80 \text{ m/s}^2)} = 1.184 \times 10^{-6} \text{ m}^3$$

From Equation (1) we have

$$V_{silver} = \frac{m_{coin} - \rho_{copper} V_{copper}}{\rho_{silver}} = \frac{1.150 \times 10^{-2} \text{ kg} - (8890 \text{ kg/m}^3) V_{copper}}{10\,500 \text{ kg/m}^3}$$

or

$$V_{silver} = 1.095 \times 10^{-6} \text{ m}^3 - (0.8467) V_{copper} \qquad (3)$$

Substitution of Equation (3) into Equation (2) gives

$$W_{water} = W_{air} - \rho_{water}\left[1.095 \times 10^{-6}~m^3 - (0.8467)V_{copper} + V_{copper}\right]g$$

Solving for V_{copper} gives $V_{copper} = 5.806 \times 10^{-7}~m^3$. Substituting into Equation (3) gives

$$V_{silver} = 1.095 \times 10^{-6}~m^3 - (0.8467)(5.806 \times 10^{-7}~m^3) = 6.034 \times 10^{-7}~m^3$$

From Equation 11.1, the mass of the silver is

$$m_{silver} = \rho_{silver}V_{silver} = (10~500~kg/m^3)(6.034 \times 10^{-7}~m^3) = \boxed{6.3 \times 10^{-3}~kg}$$

---

91. **WWW** **REASONING** According to Equation 11.4, the pressure $P_{mercury}$ at a point 7.10 cm below the ethyl alcohol-mercury interface is

$$P_{mercury} = P_{interface} + \rho_{mercury}gh_{mercury} \qquad (1)$$

where $P_{interface}$ is the pressure at the alcohol-mercury interface, and $h_{mercury} = 0.0710~m$. The pressure at the interface is

$$P_{interface} = P_{atm} + \rho_{ethyl}gh_{ethyl} \qquad (2)$$

Equation (2) can be used to find the pressure at the interface. This value can then be used in Equation (1) to determine the pressure 7.10 cm below the interface.

SOLUTION Direct substitution of the numerical data into Equation (2) yields

$$P_{interface} = 1.01 \times 10^5~Pa + (806~kg/m^3)(9.80~m/s^2)(1.10~m) = 1.10 \times 10^5~Pa$$

Therefore, the pressure 7.10 cm below the ethyl alcohol-mercury interface is

$$P_{mercury} = 1.10 \times 10^5~Pa + (13~600~kg/m^3)(9.80~m/s^2)(0.0710~m) = \boxed{1.19 \times 10^5~Pa}$$

CHAPTER 12 | *TEMPERATURE AND HEAT*

1. **REASONING AND SOLUTION** The difference between these two averages, expressed in Fahrenheit degrees, is

$$98.6 \text{ °F} - 98.2 \text{ °F} = 0.4 \text{ F°}$$

Since 1 C° is equal to $\frac{9}{5}$ F°, we can make the following conversion

$$(0.4 \text{ F°}) \left(\frac{1 \text{ C°}}{(9/5) \text{ F°}} \right) = \boxed{0.2 \text{ C°}}$$

---

5. **REASONING AND SOLUTION**   The temperature of –273.15 °C is 273.15 Celsius degrees *below* the ice point of 0 °C. This number of Celsius degrees corresponds to

$$(273.15 \text{ C°}) \left(\frac{(9/5) \text{ F°}}{1 \text{ C°}} \right) = 491.67 \text{ F°}$$

Subtracting 491.67 F° from the ice point of 32.00 °F on the Fahrenheit scale gives a Fahrenheit temperature of $\boxed{-459.67 \text{ °F}}$

---

7. **REASONING** The temperature at any pressure can be determined from the equation of the graph in Figure 12.4. Since the gas pressure is 5.00×10^3 Pa when the temperature is 0.00 °C, and the pressure is zero when the temperature is –273.15 °C, the slope of the line is given by

$$\frac{\Delta P}{\Delta T} = \frac{5.00 \times 10^3 \text{ Pa} - 0.00 \text{ Pa}}{0.00 \text{ °C} - (-273.15 \text{ °C})} = 18.3 \text{ Pa/C°}$$

Therefore, the equation of the line is $\Delta P = (18.3 \text{ Pa/C°}) \Delta T$, or

$$P - P_0 = (18.3 \text{ Pa/C°})(T - T_0)$$

where $P_0 = 5.00 \times 10^3$ Pa when $T_0 = 0.00$ °C .

SOLUTION Solving the equation above for T, we obtain

$$T = \frac{P - P_0}{18.3 \text{ Pa/C°}} + T_0 = \frac{2.00 \times 10^3 \text{ Pa} - 5.00 \times 10^3 \text{ Pa}}{18.3 \text{ Pa/C°}} + 0 \text{ °C} = \boxed{-164 \text{ °C}}$$

11.   WWW   *REASONING AND SOLUTION* The steel in the bridge expands according to Equation 12.2, $\Delta L = \alpha L_0 \Delta T$. Solving for L_0, we find that the approximate length of the Golden Gate bridge is

$$L_0 = \frac{\Delta L}{\alpha \, \Delta T} = \frac{0.53 \text{ m}}{\left[12 \times 10^{-6} \text{ (C°)}^{-1}\right] (32 \text{ °C} - 2 \text{ °C})} = \boxed{1500 \text{ m}}$$

---

15.   *REASONING AND SOLUTION*   Assuming that the rod expands linearly with heat, we first calculate the quantity $\Delta L / \Delta T$ using the data given in the problem.

$$\frac{\Delta L}{\Delta T} = \frac{8.47 \times 10^{-4} \text{ m}}{100.0 \text{ °C} - 25.0 \text{ °C}} = 1.13 \times 10^{-5} \text{ m/C°}$$

Therefore, when the rod is cooled from 25.0 °C, it will shrink by

$$\Delta L = (1.13 \times 10^{-5} \text{ m/C°}) \, \Delta T$$

$$= (1.13 \times 10^{-5} \text{ m/C°}) \, (0.00 \text{ °C} - 25.0 \text{ °C}) = \boxed{-2.82 \times 10^{-4} \text{ m}}$$

---

19.   WWW   *REASONING AND SOLUTION*   Recall that $\omega = 2\pi / T$, Equation 10.6, where ω is the angular frequency of the pendulum and T is the period. Using this fact and Equation 10.16, we know that the period of the pendulum before the temperature rise is given by $T_1 = 2\pi \sqrt{L_0 / g}$, where L_0 is the length of the pendulum. After the temperature has risen, the period becomes (using Equation 12.2), $T_2 = 2\pi \sqrt{\left(L_0 + \alpha L_0 \Delta T\right) / g}$. Dividing these expressions and solving for T_2 we have

$$T_2 = T_1 \sqrt{1 + \alpha \Delta T} = (2.0000 \text{ s}) \sqrt{1 + (19 \times 10^{-6} / \text{C°}) (140 \text{ C°})} = \boxed{2.0027 \text{ s}}$$

---

23.   *REASONING* The change in length of the wire is the sum of the change in length of each of the two segments: $\Delta L = \Delta L_{al} + \Delta L_{st}$. Using Equation 12.2 to express the ΔLs

$$\alpha L_0 \Delta T = \alpha_{al} L_{0al} \Delta T + \alpha_{st} L_{0st} \Delta T$$

Dividing both sides by L_0 and algebraically canceling ΔT gives

$$\alpha = \alpha_{al}\left(\frac{L_{0al}}{L_0}\right) + \alpha_{st}\left(\frac{L_{0st}}{L_0}\right)$$

The length of the steel segment of the wire is given by $L_{0st} = L_0 - L_{0al}$. Making this substitution leads to

$$\alpha = \alpha_{al}\left(\frac{L_{0al}}{L_0}\right) + \alpha_{st}\left(\frac{L_0 - L_{0al}}{L_0}\right)$$

$$= \alpha_{al}\left(\frac{L_{0al}}{L_0}\right) + \alpha_{st}\left(\frac{L_0}{L_0}\right) - \alpha_{st}\left(\frac{L_{0al}}{L_0}\right)$$

This expression can be solved for the desired quantity, L_{0al}/L_0.

SOLUTION Solving for the ratio (L_{0al}/L_0) gives

$$\frac{L_{0al}}{L_0} = \frac{\alpha - \alpha_{st}}{\alpha_{al} - \alpha_{st}} = \frac{19 \times 10^{-6}(C°)^{-1} - 12 \times 10^{-6}(C°)^{-1}}{23 \times 10^{-6}(C°)^{-1} - 12 \times 10^{-6}(C°)^{-1}} = \boxed{0.6}$$

---

27. **REASONING AND SOLUTION** According to Equation 12.3, $\Delta V = \beta V_0 \Delta T$. Therefore,

$$\frac{\Delta V_{air}}{\Delta V_{water}} = \frac{\beta_{air} V_0 \Delta T}{\beta_{water} V_0 \Delta T} = \frac{\beta_{air}}{\beta_{water}} = \frac{3.7 \times 10^{-3}(C°)^{-1}}{207 \times 10^{-6}(C°)^{-1}} = \boxed{18}$$

---

31. **REASONING AND SOLUTION** The cider will expand according to Equation 12.3; therefore, the change in volume of the cider is

$$\Delta V = \beta V_0 \Delta T = \left[280 \times 10^{-6}\ (C°)^{-1}\right](1.0\ \text{gal})(22\ C°) = 6.2 \times 10^{-3}\ \text{gal}$$

At a cost of two dollars per gallon, this amounts to

$$\left(6.2 \times 10^{-3}\ \text{gal}\right)\left(\frac{\$\,2.00}{\text{gal}}\right) = \$\,0.01 \quad \text{or} \quad \boxed{\text{one penny}}$$

---

33. **REASONING** In order to keep the water from expanding as its temperature increases from 15 to 25 °C, the atmospheric pressure must be increased to compress the water as it tries to expand. The magnitude of the pressure change ΔP needed to compress a substance by an amount ΔV is, according to Equation 10.20, $\Delta P = B(\Delta V/V_0)$. The ratio $\Delta V/V_0$ is,

according to Equation 12.3, $\Delta V / V_0 = \beta \Delta T$. Combining these two equations yields $\Delta P = B \beta \Delta T$.

SOLUTION The change in atmospheric pressure that is required to keep the water from expanding is, therefore,

$$\Delta P = (2.2 \times 10^9 \text{ N/m}^2)\left[207 \times 10^{-6} \text{ (C}^\circ)^{-1} \right](25 \text{ °C} - 15 \text{ °C})$$

$$= \left(4.6 \times 10^6 \text{ Pa}\right) \left(\frac{1 \text{ atm}}{1.01 \times 10^5 \text{ Pa}} \right) = \boxed{45 \text{ atm}}$$

---

37. $\boxed{\text{WWW}}$ **REASONING** The cavity that contains the liquid in either Pyrex thermometer expands according to Equation 12.3, $\Delta V_g = \beta_g V_0 \Delta T$. On the other hand, the volume of mercury expands by an amount $\Delta V_m = \beta_m V_0 \Delta T$, while the volume of alcohol expands by an amount $\Delta V_a = \beta_a V_0 \Delta T$. Therefore, the net change in volume for the mercury thermometer is

$$\Delta V_m - \Delta V_g = (\beta_m - \beta_g) V_0 \Delta T$$

while the net change in volume for the alcohol thermometer is

$$\Delta V_a - \Delta V_g = (\beta_a - \beta_g) V_0 \Delta T$$

In each case, this volume change is related to a movement of the liquid into a cylindrical region of the thermometer with volume $\pi r^2 h$, where r is the radius of the region and h is the height of the region. For the mercury thermometer, therefore,

$$h_m = \frac{(\beta_m - \beta_g) V_0 \Delta T}{\pi r^2}$$

Similarly, for the alcohol thermometer

$$h_a = \frac{(\beta_a - \beta_g) V_0 \Delta T}{\pi r^2}$$

These two expressions can be combined to give the ratio of the heights, h_a / h_m.

SOLUTION Division of the two expressions for the heights of the liquids in the thermometers gives

$$\frac{h_a}{h_m} = \frac{\beta_a - \beta_g}{\beta_m - \beta_g} = \frac{1200 \times 10^{-6} \text{ (C}^\circ)^{-1} - 9.9 \times 10^{-6} \text{ (C}^\circ)^{-1}}{182 \times 10^{-6} \text{ (C}^\circ)^{-1} - 9.9 \times 10^{-6} \text{ (C}^\circ)^{-1}} = 6.9$$

Therefore, the degree marks are $\boxed{\text{6.9 times further apart}}$ on the alcohol thermometer than on the mercury thermometer.

---

41. **REASONING** Let the system be comprised only of the metal forging and the oil. Then, according to the principle of energy conservation, the heat lost by the forging equals the heat gained by the oil, or $Q_{metal} = Q_{oil}$. According to Equation 12.4, the heat lost by the forging is $Q_{metal} = c_{metal} m_{metal} (T_{0metal} - T_{eq})$, where T_{eq} is the final temperature of the system at thermal equilibrium. Similarly, the heat gained by the oil is given by $Q_{oil} = c_{oil} m_{oil} (T_{eq} - T_{0oil})$.

SOLUTION

$$Q_{metal} = Q_{oil}$$

$$c_{metal} m_{metal} (T_{0metal} - T_{eq}) = c_{oil} m_{oil} (T_{eq} - T_{0oil})$$

Solving for T_{0metal}, we have

$$T_{0metal} = \frac{c_{oil} m_{oil} (T_{eq} - T_{0oil})}{c_{metal} m_{metal}} + T_{eq}$$

or

$$T_{0metal} = \frac{[2700 \text{ J/(kg} \cdot \text{C°)}](710 \text{ kg})(47 \text{ °C} - 32 \text{ °C})}{[430 \text{ J/(kg} \cdot \text{C°)}](75 \text{ kg})} + 47 \text{ °C} = \boxed{940 \text{ °C}}$$

---

45. **REASONING** According to Equation 12.4, the heat required to warm the pool can be calculated from $Q = cm\Delta T$. In order to use Equation 12.4, we must first determine the mass of the water in the pool. Equation 11.1 indicates that the mass can be calculated from $m = \rho V$, where ρ is the density of water and V is the volume of water in the pool.

SOLUTION Combining these two expressions, we have $Q = c\rho V \Delta T$, or

$$Q = \left[4186 \text{ J/(kg} \cdot \text{C°)}\right](1.00 \times 10^3 \text{ kg/m}^3)(12.0 \text{ m} \times 9.00 \text{ m} \times 1.5 \text{ m}) (27 \text{ °C} - 15 \text{ °C}) = 8.14 \times 10^9 \text{ J}$$

Using the fact that $1 \text{ kWh} = 3.6 \times 10^6 \text{ J}$, the cost of using electrical energy to heat the water in the pool at a cost of $0.10 per kWh is

$$(8.14 \times 10^9 \text{ J}) \left(\frac{\$0.10}{3.6 \times 10^6 \text{ J}} \right) = \boxed{\$230.}$$

---

49. ***REASONING AND SOLUTION*** According to Equation 12.4, the total heat per kilogram required to raise the temperature of the water is

$$\frac{Q}{m} = c\Delta T = [4186 \text{ J/(kg} \cdot \text{C}^\circ)]\ (32.0\,^\circ\text{C}) = 1.34 \times 10^5 \text{ J/kg}$$

The mass flow rate, $\Delta m/\Delta t$, is given by Equations 11.7 and 11.10 as $\Delta m/\Delta t = \rho A v = \rho Q_v$, where ρ and Q_v are the density and volume flow rate, respectively. We have, therefore,

$$\frac{\Delta m}{\Delta t} = (1.000 \times 10^3 \text{ kg/m}^3)(5.0 \times 10^{-6} \text{ m}^3/\text{s}) = 5.0 \times 10^{-3} \text{ kg/s}$$

Therefore, the minimum power rating of the heater must be

$$(1.34 \times 10^5 \text{ J/kg})(5.0 \times 10^{-3} \text{ kg/s}) = 6.7 \times 10^2 \text{ J/s} = \boxed{6.7 \times 10^2 \text{ W}}$$

---

51. ***REASONING*** Heat Q_1 must be added to raise the temperature of the aluminum in its solid phase from 130 °C to its melting point at 660 °C. According to Equation 12.4, $Q_1 = cm\Delta T$. Once the solid aluminum is at its melting point, additional heat Q_2 must be supplied to change its phase from solid to liquid. The additional heat required to melt or liquefy the aluminum is $Q_2 = mL_f$, where L_f is the latent heat of fusion of aluminum. Therefore, the total amount of heat which must be added to the aluminum in its solid phase to liquefy it is

$$Q_{\text{total}} = Q_1 + Q_2 = m(c\Delta T + L_f)$$

SOLUTION Substituting values, we obtain

$$Q_{\text{total}} = (0.45 \text{ kg}) \left([9.00 \times 10^2 \text{ J/(kg} \cdot \text{C}^\circ)](660\,^\circ\text{C} - 130\,^\circ\text{C}) + 4.0 \times 10^5 \text{ J/kg}\right) = \boxed{3.9 \times 10^5 \text{ J}}$$

---

55. ***REASONING*** From the principle of conservation of energy, the heat lost by the coin must be equal to the heat gained by the liquid nitrogen. The heat lost by the coin is, from Equation 12.4, $Q = c_{\text{coin}} m_{\text{coin}} \Delta T_{\text{coin}}$. If the liquid nitrogen is at its boiling point, –195.8 °C, then the heat gained by the nitrogen will cause it to change phase from a liquid to a vapor. The heat gained by the liquid nitrogen is $Q = m_{\text{nitrogen}} L_v$, where m_{nitrogen} is the mass of liquid nitrogen that vaporizes, and L_v is the latent heat of vaporization for nitrogen.

SOLUTION

$$\underset{\substack{\text{lost by} \\ \text{coin}}}{Q} = \underset{\substack{\text{gained by} \\ \text{nitrogen}}}{Q}$$

$$c_{\text{coin}} m_{\text{coin}} \Delta T_{\text{coin}} = m_{\text{nitrogen}} L_v$$

Solving for the mass of the nitrogen that vaporizes

$$m_{nitrogen} = \frac{c_{coin}m_{coin}\Delta T_{coin}}{L_v}$$

$$= \frac{[235 \text{ J/(kg} \cdot \text{C}°)](1.5 \times 10^{-2} \text{ kg})[25 °C - (-195.8 °C)]}{2.00 \times 10^5 \text{ J/kg}} = \boxed{3.9 \times 10^{-3} \text{ kg}}$$

---

61. ***REASONING*** According to the statement of the problem, the initial state of the system is comprised of the ice and the steam. From the principle of energy conservation, the heat lost by the steam equals the heat gained by the ice, or $Q_{steam} = Q_{ice}$. When the ice and the steam are brought together, the steam immediately begins losing heat to the ice. An amount $Q_{1(lost)}$ is released as the temperature of the steam drops from 130 °C to 100 °C, the boiling point of water. Then an amount of heat $Q_{2(lost)}$ is released as the steam condenses into liquid water at 100 °C. The remainder of the heat lost by the "steam" $Q_{3(lost)}$ is the heat that is released as the water at 100 °C cools to the equilibrium temperature of $T_{eq} = 50.0 °C$. According to Equation 12.4, $Q_{1(lost)}$ and $Q_{3(lost)}$ are given by

$$Q_{1(lost)} = c_{steam}m_{steam}(T_{steam} - 100.0 °C) \quad \text{and} \quad Q_{3(lost)} = c_{water}m_{steam}(100.0 °C - T_{eq})$$

$Q_{2(lost)}$ is given by $Q_{2(lost)} = m_{steam}L_v$, where L_v is the latent heat of vaporization of water. The total heat lost by the steam has three effects on the ice. First, a portion of this heat $Q_{1(gained)}$ is used to raise the temperature of the ice to its melting point at 0.00 °C. Then, an amount of heat $Q_{2(gained)}$ is used to melt the ice completely (we know this because the problem states that after thermal equilibrium is reached the liquid phase is present at 50.0 °C). The remainder of the heat $Q_{3(gained)}$ gained by the "ice" is used to raise the temperature of the resulting liquid at 0.0 °C to the final equilibrium temperature. According to Equation 12.4, $Q_{1(gained)}$ and $Q_{3(gained)}$ are given by

$$Q_{1(gained)} = c_{ice}m_{ice}(0.00 °C - T_{ice}) \quad \text{and} \quad Q_{3(gained)} = c_{water}m_{ice}(T_{eq} - 0.00 °C)$$

$Q_{2(gained)}$ is given by $Q_{2(gained)} = m_{ice}L_f$, where L_f is the latent heat of fusion of ice.

SOLUTION

$$Q_{steam} = Q_{ice}$$

$$Q_{1(lost)} + Q_{2(lost)} + Q_{3(lost)} = Q_{1(gained)} + Q_{2(gained)} + Q_{3(gained)}$$

or

$$c_{steam}m_{steam}(T_{steam} - 100.0 °C) + m_{steam}L_v + c_{water}m_{steam}(100.0 °C - T_{eq})$$

$$= c_{ice}m_{ice}(0.00\ °C - T_{ice}) + m_{ice}L_f + c_{water}m_{ice}(T_{eq} - 0.00\ °C)$$

Solving for the ratio of the masses gives

$$\frac{m_{steam}}{m_{ice}} = \frac{c_{ice}(0.00\ °C - T_{ice}) + L_f + c_{water}(T_{eq} - 0.00\ °C)}{c_{steam}(T_{steam} - 100.0\ °C) + L_v + c_{water}(100.0\ °C - T_{eq})}$$

$$= \frac{\left[2.00\times10^3\,\text{J/(kg}\cdot\text{C°)}\right]\left[0.0\ °C - (-10.0\ °C)\right] + 33.5\times10^4\ \text{J/kg} + \left[4186\ \text{J/(kg}\cdot\text{C°)}\right](50.0\ °C - 0.0\ °C)}{\left[2020\ \text{J/(kg}\cdot\text{C°)}\right](130\ °C - 100.0\ °C) + 22.6\times10^5\ \text{J/kg} + \left[4186\ \text{J/(kg}\cdot\text{C°)}\right](100.0\ °C - 50.0\ °C)}$$

or

$$\frac{m_{steam}}{m_{ice}} = \boxed{0.223}$$

---

65. **REASONING** In order to melt, the bullet must first heat up to 327.3 °C (its melting point) and then undergo a phase change. According to Equation 12.4, the amount of heat necessary to raise the temperature of the bullet to 327.3 °C is $Q = cm(327.3\ °C - 30.0\ °C)$, where m is the mass of the bullet. The amount of heat required to melt the bullet is given by $Q_{melt} = mL_f$, where L_f is the latent heat of fusion of lead.

The lead bullet melts completely when it comes to a sudden halt; all of the kinetic energy of the bullet is converted into heat; therefore,

$$KE = Q + Q_{melt}$$

$$\tfrac{1}{2}mv^2 = cm(327.3\ °C - 30.0\ °C) + mL_f$$

This expression can be solved for v, the minimum speed of the bullet for such an event to occur.

SOLUTION Solving for v, we find that the minimum speed of the lead bullet is

$$v = \sqrt{2L_f + 2c\ (327.3\ °C - 30.0\ °C)}$$

$$v = \sqrt{2(2.32\times10^4\ \text{J/kg}) + 2\left[128\ \text{J/(kg}\cdot\text{C°)}\right](327.3\ °C - 30.0\ °C)} = \boxed{3.50\times10^2\ \text{m/s}}$$

67. ***REASONING AND SOLUTION***   From inspection of the graph that accompanies this problem, a pressure of 3.5×10^6 Pa corresponds to a temperature of $0 \, °C$. Therefore, liquid carbon dioxide exists in equilibrium with its vapor phase at $\boxed{0 \, °C}$ when the pressure is 3.5×10^6 Pa.

---

73. ***REASONING AND SOLUTION*** From the vapor pressure curve that accompanies problem 71, it is seen that the partial pressure of water vapor in the atmosphere at 10 °C is about 1400 Pa, and that the equilibrium vapor pressure at 30 °C is about 4200 Pa. The relative humidity is, from Equation 12.6,

$$\begin{matrix} \text{Percent} \\ \text{relative} \\ \text{humidity} \end{matrix} = \left(\frac{1400 \text{ Pa}}{4200 \text{ Pa}} \right) \times 100 = \boxed{33\%}$$

---

75. ***REASONING***   We must first find the equilibrium temperature T_{eq} of the iced tea. Once this is known, we can use the vapor pressure curve that accompanies problem 71 to find the partial pressure of water vapor at that temperature and then estimate the relative humidity using Equation 12.6.

According to the principle of energy conservation, when the ice is mixed with the tea, the heat lost by the tea is gained by the ice, or $Q_{tea} = Q_{ice}$. The heat gained by the ice is used to melt the ice at 0.0 °C; the remainder of the heat is used to bring the water at 0.0 °C up to the final equilibrium temperature T_{eq}.

SOLUTION

$$Q_{tea} = Q_{ice}$$

$$c_{water} \, m_{tea} \, (30.0 \, °C - T_{eq}) = m_{ice} L_f + c_{water} \, m_{ice} (T_{eq} - 0.00 \, °C)$$

Solving for T_{eq}, we have

$$T_{eq} = \frac{c_{water} \, m_{tea} \, (30.0 \, °C) - m_{ice} L_f}{c_{water} \, (m_{tea} + m_{ice})}$$

$$= \frac{\left[4186 \text{ J/(kg} \cdot \text{C°)} \right] (0.300 \text{ kg})(30.0 \, °C) - (0.0670 \text{ kg})(33.5 \times 10^4 \text{ J/kg})}{\left[4186 \text{ J/(kg} \cdot \text{C°)} \right] (0.300 \text{ kg} + 0.0670 \text{ kg})} = 9.91 \, °C$$

According to the vapor pressure curve that accompanies problem 71, at a temperature of 9.91 °C, the equilibrium vapor pressure is approximately 1250 Pa. At 30 °C, the equilibrium vapor pressure is approximately 4400 Pa. Therefore, according to Equation 12.6, the percent relative humidity is approximately

$$\text{Percent relative humidity} = \left(\frac{1250 \text{ Pa}}{4400 \text{ Pa}}\right) \times 100 = \boxed{28\%}$$

---

79. **REASONING** From the conservation of energy, the heat lost by the mercury is equal to the heat gained by the water. As the mercury loses heat, its temperature decreases; as the water gains heat, its temperature rises to its boiling point. Any remaining heat gained by the water will then be used to vaporize the water.

According to Equation 12.4, the heat lost by the mercury is $Q_{mercury} = (cm\Delta T)_{mercury}$. The heat required to vaporize the water is, from Equation 12.5, $Q_{vap} = (m_{vap}L_v)_{water}$. Thus, the total amount of heat gained by the water is $Q_{water} = (cm\Delta T)_{water} + (m_{vap}L_v)_{water}$.

SOLUTION

$$\underset{\substack{\text{lost by}\\\text{mercury}}}{Q} = \underset{\substack{\text{gained by}\\\text{water}}}{Q}$$

$$(cm\Delta T)_{mercury} = (cm\Delta T)_{water} + (m_{vap}L_v)_{water}$$

where $\Delta T_{mercury} = (205\ °C - 100.0\ °C)$ and $\Delta T_{water} = (100.0\ °C - 80.0\ °C)$. Solving for the mass of the water that vaporizes gives

$$m_{vap} = \frac{c_{mercury}m_{mercury}\Delta T_{mercury} - c_{water}m_{water}\Delta T_{water}}{(L_v)_{water}}$$

$$= \frac{[139 \text{ J/(kg} \cdot \text{C°)}](2.10 \text{ kg})(105 \text{ C°}) - [4186 \text{ J/(kg} \cdot \text{C°)}](0.110 \text{ kg})(20.0 \text{ C°})}{22.6 \times 10^5 \text{ J/kg}}$$

$$= \boxed{9.49 \times 10^{-3} \text{ kg}}$$

---

83. **REASONING AND SOLUTION** Both the gasoline and the tank expand as the temperature increases. According to Equation 12.3, the volume expansion of the gasoline is

$$\Delta V_g = \beta_g V_0 \Delta T = \left[950 \times 10^{-6}\ (\text{C°})^{-1}\right](20.0 \text{ gal})(18 \text{ C°}) = 0.34 \text{ gal}$$

while the volume of the steel tank expands by an amount

$$\Delta V_s = \beta_s V_0 \Delta T = \left[36 \times 10^{-6} \, (C^\circ)^{-1}\right](20.0 \text{ gal})(18 \text{ C}^\circ) = 0.013 \text{ gal}$$

The amount of gasoline which spills out is

$$\Delta V_g - \Delta V_s = \boxed{0.33 \text{ gal}}$$

---

87. $\boxed{\text{WWW}}$ *REASONING* The system is comprised of the unknown material, the glycerin, and the aluminum calorimeter. From the principle of energy conservation, the heat gained by the unknown material is equal to the heat lost by the glycerin and the calorimeter. The heat gained by the unknown material is used to melt the material and then raise its temperature from the initial value of –25.0 °C to the final equilibrium temperature of $T_{eq} = 20.0 \, ^\circ\text{C}$.

SOLUTION

$$\underset{\substack{\text{gained by} \\ \text{unknown}}}{Q} = \underset{\substack{\text{lost by} \\ \text{glycerine}}}{Q} + \underset{\substack{\text{lost by} \\ \text{calorimeter}}}{Q}$$

$$m_u L_f + c_u m_u \Delta T_u = c_{gl} m_{gl} \Delta T_{gl} + c_{al} m_{al} \Delta T_{al}$$

or

$$(0.10 \text{ kg})L_f + [160 \text{ J/(kg} \cdot C^\circ)](0.10 \text{ kg})(45.0 \text{ C}^\circ) = [2410 \text{ J/(kg} \cdot C^\circ)](0.100 \text{ kg})(7.0 \text{ C}^\circ)$$
$$+ [9.0 \times 10^2 \text{ J/(kg} \cdot C^\circ)](0.150 \text{ kg})(7.0 \text{ C}^\circ)$$

Solving for L_f yields,

$$L_f = \boxed{1.9 \times 10^4 \text{ J/kg}}$$

---

91. *REASONING AND SOLUTION* As the rock falls through a distance h, its initial potential energy $m_{rock} g h$ is converted into kinetic energy. This kinetic energy is then converted into heat when the rock is brought to rest in the pail. If we ignore the heat absorbed by the pail, then

$$m_{rock} g h = c_{rock} m_{rock} \Delta T + c_{water} m_{water} \Delta T$$

Solving for ΔT yields

$$\Delta T = \frac{m_{rock} g h}{c_{rock} m_{rock} + c_{water} m_{water}}$$

Substituting values yields

$$\Delta T = \frac{(0.20 \text{ kg})(9.80 \text{ m/s}^2)(15 \text{ m})}{[1840 \text{ J/(kg} \cdot C^\circ)](0.20 \text{ kg}) + [4186 \text{ J/(kg} \cdot C^\circ)](0.35 \text{ kg})} = \boxed{0.016 \text{ C}^\circ}$$

CHAPTER 13 | *THE TRANSFER OF HEAT*

PROBLEMS

1. ***REASONING***   The heat conducted through the iron poker is given by Equation 13.1, $Q = (kA \, \Delta T)t \, / \, L$. If we assume that the poker has a circular cross-section, then its cross-sectional area is $A = \pi r^2$. Table 13.1 gives the thermal conductivity of iron as $79 \, \text{J} \, / \, (\text{s} \cdot \text{m} \cdot \text{C}°)$.

   SOLUTION   The amount of heat conducted from one end of the poker to the other in 5.0 s is, therefore,

   $$Q = \frac{(k \, A \, \Delta T)t}{L} = \frac{\left[79 \, \text{J} \, / \, (\text{s} \cdot \text{m} \cdot \text{C}°)\right] \pi \left(5.0 \times 10^{-3} \, \text{m}\right)^2 \left(502 \, °\text{C} - 26 \, °\text{C}\right)\left(5.0 \, \text{s}\right)}{1.2 \, \text{m}} = \boxed{12 \, \text{J}}$$

5. ***REASONING***   The heat transferred in a time t is given by Equation 13.1, $Q = (k A \, \Delta T)t \, / \, L$. If the same amount of heat per second is conducted through the two plates, then $(Q/t)_{\text{al}} = (Q/t)_{\text{st}}$. Using Equation 13.1, this becomes

   $$\frac{k_{\text{al}} A \, \Delta T}{L_{\text{al}}} = \frac{k_{\text{st}} A \, \Delta T}{L_{\text{st}}}$$

   This expression can be solved for L_{st}.

   SOLUTION   Solving for L_{st} gives

   $$L_{\text{st}} = \frac{k_{\text{st}}}{k_{\text{al}}} L_{\text{al}} = \frac{14 \, \text{J} \, / \, (\text{s} \cdot \text{m} \cdot \text{C}°)}{240 \, \text{J} \, / \, (\text{s} \cdot \text{m} \cdot \text{C}°)} (0.035 \, \text{m}) = \boxed{2.0 \times 10^{-3} \, \text{m}}$$

7. $\boxed{\text{WWW}}$ ***REASONING AND SOLUTION***   The conductance of an 0.080 mm thick sample of Styrofoam of cross-sectional area A is

   $$\frac{k_{\text{s}} A}{L_{\text{s}}} = \frac{\left[0.010 \, \text{J} \, / \, (\text{s} \cdot \text{m} \cdot \text{C}°)\right] A}{0.080 \times 10^{-3} \, \text{m}} = \left[125 \, \text{J} \, / \, (\text{s} \cdot \text{m}^2 \cdot \text{C}°)\right] A$$

   The conductance of a 3.5 mm thick sample of air of cross-sectional area A is

$$\frac{k_a A}{L_a} = \frac{[0.0256 \text{ J}/(\text{s}\cdot\text{m}\cdot\text{C}°)]\, A}{3.5\times 10^{-3}\text{ m}} = [7.3 \text{ J}/(\text{s}\cdot\text{m}^2\cdot\text{C}°)]\, A$$

Dividing the conductance of Styrofoam by the conductance of air for samples of the same cross-sectional area A, gives

$$\frac{[125 \text{ J}/(\text{s}\cdot\text{m}^2\cdot\text{C}°)]\, A}{[7.3 \text{ J}/(\text{s}\cdot\text{m}^2\cdot\text{C}°)]\, A} = 17$$

Therefore, the body can adjust the conductance of the tissues beneath the skin by $\boxed{\text{a factor of 17}}$.

---

11. **REASONING AND SOLUTION**

a. If we ignore the loss of heat through the sides of the rod and assume that heat does not accumulate at any point, then, from energy conservation, the rate at which heat is conducted through the two-rod combination must be the same at all points. In particular, at the interface of the two rods, $Q_a/t = Q_c/t$. Let T represent the temperature at the interface. Using Equation 13.1, we have

$$\frac{k_a A (T_a - T)}{L} = \frac{k_c A (T - T_c)}{L}$$

Solving for T, we find

$$T = \frac{k_a T_a + k_c T_c}{k_a + k_c} = \frac{[240 \text{ J}/(\text{s}\cdot\text{m}\cdot\text{C}°)](302 \text{ °C}) + [390 \text{ J}/(\text{s}\cdot\text{m}\cdot\text{C}°)](25 \text{ °C})}{240 \text{ J}/(\text{s}\cdot\text{m}\cdot\text{C}°) + 390 \text{ J}/(\text{s}\cdot\text{m}\cdot\text{C}°)} = \boxed{130 \text{ °C}}$$

b. Now that the temperature of the interface is known, Equation 13.1 can be used to calculate the amount of heat Q that flows through either section of the unit in a time t. Since the rate of heat flow is the same at all points throughout the unit as discussed above, we need only calculate Q for one of the rods that make up the unit. Using the aluminum rod, and considering the heat flow from its free end to the interface, we have

$$Q = \frac{k_a A (T_a - T)}{L} t = \frac{[240 \text{ J}/(\text{s}\cdot\text{m}\cdot\text{C}°)](4.0\times 10^{-4}\text{ m}^2)(302 \text{ °C} - 130 \text{ °C})}{4.0\times 10^{-2}\text{ m}}(2.0 \text{ s}) = \boxed{830 \text{ J}}$$

c. According to Equation 13.1, the temperature T_d at a distance d from the hot end of the aluminum rod is

$$T_a - T_d = \frac{Qd}{k_a A t} = \frac{(Q/t)d}{k_a A}$$

where, from the answer to part (b), we know that $Q/t = (830 \text{ J})/(2.0 \text{ s}) = 415 \text{ J/s}$. Solving for T_d, we have

$$T_d = T_a - \frac{(Q/t)d}{k_a A} = 302 \text{ °C} - \frac{(415 \text{ J/s})(1.5 \times 10^{-2} \text{ m})}{[240 \text{ J/(s·m·C°)}](4.0 \times 10^{-4} \text{ m}^2)} = \boxed{237 \text{ °C}}$$

---

15.   **WWW**   *REASONING*   If the cylindrical rod were made of solid copper, the amount of heat it would conduct in a time t is, according to Equation 13.1, $Q_{copper} = (k_{copper} A_2 \Delta T / L)t$. Similarly, the amount of heat conducted by the lead-copper combination is the sum of the heat conducted through the copper portion of the rod and the heat conducted through the lead portion: $Q_{combination} = \left[k_{copper}(A_2 - A_1)\Delta T / L + k_{lead} A_1 \Delta T / L \right] t$. Since the lead-copper combination conducts one-half the amount of heat than does the solid copper rod, $Q_{combination} = \frac{1}{2} Q_{copper}$, or

$$\frac{k_{copper}(A_2 - A_1)\Delta T}{L} + \frac{k_{lead} A_1 \Delta T}{L} = \frac{1}{2} \frac{k_{copper} A_2 \Delta T}{L}$$

This expression can be solved for A_1 / A_2, the ratio of the cross-sectional areas. Since the cross-sectional area of a cylinder is circular, $A = \pi r^2$. Thus, once the ratio of the areas is known, the ratio of the radii can be determined.

SOLUTION   Solving for the ratio of the areas, we have

$$\frac{A_1}{A_2} = \frac{k_{copper}}{2(k_{copper} - k_{lead})}$$

The cross-sectional areas are circular so that $A_1 / A_2 = (\pi r_1^2)/(\pi r_2^2) = (r_1/r_2)^2$; therefore,

$$\frac{r_1}{r_2} = \sqrt{\frac{k_{copper}}{2(k_{copper} - k_{lead})}} = \sqrt{\frac{390 \text{ J/(s·m·C°)}}{2[390 \text{ J/(s·m·C°)} - 35 \text{ J/(s·m·C°)}]}} = \boxed{0.74}$$

---

19.   **WWW**   *REASONING AND SOLUTION*   Solving the Stefan-Boltzmann law, Equation 13.2, for the time t, and using the fact that $Q_{blackbody} = Q_{bulb}$, we have

$$t_{blackbody} = \frac{Q_{blackbody}}{\sigma T^4 A} = \frac{Q_{bulb}}{\sigma T^4 A} = \frac{P_{bulb} t_{bulb}}{\sigma T^4 A}$$

where P_{bulb} is the power rating of the light bulb. Therefore,

$$t_{blackbody} = \frac{(100.0 \text{ J}/\text{s}) (3600 \text{ s})}{\left[5.67 \times 10^{-8} \text{ J}/(\text{s} \cdot \text{m}^2 \cdot \text{K}^4)\right] (303 \text{ K})^4 \left[(6 \text{ sides})(0.0100 \text{ m})^2 / \text{side}\right]}$$

$$\times \left(\frac{1 \text{ h}}{3600 \text{ s}}\right)\left(\frac{1 \text{ da}}{24 \text{ h}}\right) = \boxed{14.5 \text{ da}}$$

---

21. ***REASONING AND SOLUTION*** The net power generated by the stove is given by Equation 13.3, $P_{net} = e\sigma A(T^4 - T_0^4)$. Solving for T gives

$$T = \left(\frac{P_{net}}{e\sigma A} + T_0^4\right)^{1/4}$$

$$= \left\{\frac{7300 \text{ W}}{(0.900)[5.67 \times 10^{-8} \text{ J}/(\text{s} \cdot \text{m}^2 \cdot \text{K}^4)](2.00 \text{ m}^2)} + (302 \text{ K})^4\right\}^{1/4} = \boxed{532 \text{ K}}$$

---

25. ***REASONING*** The total radiant power emitted by an object that has a Kelvin temperature T, surface area A, and emissivity e can be found by rearranging Equation 13.2, the Stefan-Boltzmann law: $Q = e\sigma T^4 At$. The emitted power is $P = Q/t = e\sigma T^4 A$. Therefore, when the original cylinder is cut perpendicular to its axis into N smaller cylinders, the ratio of the power radiated by the pieces to that radiated by the original cylinder is

$$\frac{P_{pieces}}{P_{original}} = \frac{e\sigma T^4 A_2}{e\sigma T^4 A_1} \tag{1}$$

where A_1 is the surface area of the original cylinder, and A_2 is the sum of the surface areas of all N smaller cylinders. The surface area of the original cylinder is the sum of the surface area of the ends and the surface area of the cylinder body; therefore, if L and r represent the length and cross-sectional radius of the original cylinder, with $L = 10r$,

$$A_1 = (\text{area of ends}) + (\text{area of cylinder body})$$

$$= 2(\pi r^2) + (2\pi r)L = 2(\pi r^2) + (2\pi r)(10r) = 22\pi r^2$$

When the original cylinder is cut perpendicular to its axis into N smaller cylinders, the total surface area A_2 is

$$A_2 = N2(\pi r^2) + (2\pi r)L = N2(\pi r^2) + (2\pi r)(10r) = (2N + 20)\pi r^2$$

Substituting the expressions for A_1 and A_2 into Equation (1), we obtain the following expression for the ratio of the power radiated by the N pieces to that radiated by the original cylinder

$$\frac{P_{\text{pieces}}}{P_{\text{original}}} = \frac{e\sigma T^4 A_2}{e\sigma T^4 A_1} = \frac{(2N+20)\pi r^2}{22\pi r^2} = \frac{N+10}{11}$$

SOLUTION Since the total radiant power emitted by the N pieces is twice that emitted by the original cylinder, $P_{\text{pieces}} / P_{\text{original}} = 2$, we have $(N + 10)/11 = 2$. Solving this expression for N gives $N = 12$. Therefore, there are 12 smaller cylinders .

---

27. **REASONING AND SOLUTION** According to Equation 13.1, the heat per second lost is

$$\frac{Q}{t} = \frac{kA\,\Delta T}{L} = \frac{[0.040 \text{ J/(s·m·C°)}]\,(1.6 \text{ m}^2)(25 \text{ C°})}{2.0 \times 10^{-3} \text{ m}} = \boxed{8.0 \times 10^2 \text{ J/s}}$$

---

31. **REASONING AND SOLUTION** The power radiated per square meter by the car when it has reached a temperature T is given by the Stefan-Boltzmann law, Equation 13.2, $P_{\text{radiated}} / A = e\sigma T^4$, where $P_{\text{radiated}} = Q/t$. Solving for T we have

$$T = \left[\frac{(P_{\text{radiated}} / A)}{e\sigma}\right]^{1/4} = \left\{\frac{560 \text{ W/m}^2}{(1.00)\left[5.67 \times 10^{-8} \text{ J/(s·m}^2 \cdot \text{K}^4)\right]}\right\}^{1/4} = \boxed{320 \text{ K}}$$

---

35. $\boxed{\text{WWW}}$ **REASONING** The rate at which heat is conducted along either rod is given by Equation 13.1, $Q/t = (kA\,\Delta T)/L$. Since both rods conduct the same amount of heat per second, we have

$$\frac{k_s A_s\,\Delta T}{L_s} = \frac{k_i A_i\,\Delta T}{L_i} \tag{1}$$

Since the same temperature difference is maintained across both rods, we can algebraically cancel the ΔTs. Because both rods have the same mass, $m_s = m_i$; in terms of the densities of silver and iron, the statement about the equality of the masses becomes $\rho_s(L_s A_s) = \rho_i(L_i A_i)$, or

$$\frac{A_s}{A_i} = \frac{\rho_i L_i}{\rho_s L_s} \tag{2}$$

Equations (1) and (2) may be combined to find the ratio of the lengths of the rods. Once the ratio of the lengths is known, Equation (2) can be used to find the ratio of the cross-sectional areas of the rods. If we assume that the rods have circular cross sections, then each has an area of $A = \pi r^2$. Hence, the ratio of the cross-sectional areas can be used to find the ratio of the radii of the rods.

SOLUTION

a.   Solving Equation (1) for the ratio of the lengths and substituting the right hand side of Equation (2) for the ratio of the areas, we have

$$\frac{L_s}{L_i} = \frac{k_s A_s}{k_i A_i} = \frac{k_s \left(\rho_i L_i \right)}{k_i \left(\rho_s L_s \right)} \quad \text{or} \quad \left(\frac{L_s}{L_i} \right)^2 = \frac{k_s \rho_i}{k_i \rho_s}$$

Solving for the ratio of the lengths, we have

$$\frac{L_s}{L_i} = \sqrt{\frac{k_s \rho_i}{k_i \rho_s}} = \sqrt{\frac{[420\ \text{J}/(\text{s}\cdot\text{m}\cdot\text{C}°)](7860\ \text{kg}/\text{m}^3)}{[79\ \text{J}/(\text{s}\cdot\text{m}\cdot\text{C}°)](10\ 500\ \text{kg}/\text{m}^3)}} = \boxed{2.0}$$

b.   From Equation (2) we have

$$\frac{\pi r_s^2}{\pi r_i^2} = \frac{\rho_i L_i}{\rho_s L_s} \quad \text{or} \quad \left(\frac{r_s}{r_i} \right)^2 = \frac{\rho_i L_i}{\rho_s L_s}$$

Solving for the ratio of the radii, we have

$$\frac{r_s}{r_i} = \sqrt{\frac{\rho_i}{\rho_s}\left(\frac{L_i}{L_s} \right)} = \sqrt{\frac{7860\ \text{kg}/\text{m}^3}{10\ 500\ \text{kg}/\text{m}^3}\left(\frac{1}{2.0} \right)} = \boxed{0.61}$$

CHAPTER 14 | *THE IDEAL GAS LAW AND KINETIC THEORY*

1. ***REASONING AND SOLUTION*** Since hemoglobin has a molecular mass of 64 500 u, the mass per mole of hemoglobin is 64 500 g/mol. The number of hemoglobin molecules per mol is Avogadro's number, or 6.022×10^{23} mol^{-1}. Therefore, one molecule of hemoglobin has a mass (in kg) of

$$\left(\frac{64\ 500\ \text{g/mol}}{6.022 \times 10^{23}\ \text{mol}^{-1}} \right) \left(\frac{1\ \text{kg}}{1000\ \text{g}} \right) = \boxed{1.07 \times 10^{-22}\ \text{kg}}$$

5. ***REASONING AND SOLUTION*** The number n of moles contained in a sample is equal to the number N of atoms in the sample divided by the number N_A of atoms per mole (Avogadro's number):

$$n = \frac{N}{N_A} = \frac{30.1 \times 10^{23}}{6.022 \times 10^{23}\ \text{mol}^{-1}} = 5.00\ \text{mol}$$

Since the sample has a mass of 135 g, the mass per mole is

$$\frac{135\ \text{g}}{5.00\ \text{mol}} = 27.0\ \text{g/mol}$$

The mass per mole (in g/mol) of a substance has the same numerical value as the atomic mass of the substance. Therefore, the atomic mass is 27.0 u. The periodic table of the elements reveals that the unknown element is $\boxed{\text{aluminum}}$.

7. $\boxed{\text{WWW}}$ ***REASONING AND SOLUTION*** For a given mass of the liquid phase and the vapor phase, $m_{\text{liquid}} = m_{\text{vapor}}$, or using Equation 11.1, $\rho_{\text{liquid}} V_{\text{liquid}} = \rho_{\text{vapor}} V_{\text{vapor}}$. In this result, ρ and V stand for density and volume, respectively. The ratio of the volumes, is therefore,

$$\frac{V_{\text{vapor}}}{V_{\text{liquid}}} = \frac{\rho_{\text{liquid}}}{\rho_{\text{vapor}}}$$

If we assume that the volume of each phase is filled with N cubes, with one molecule at the center of each cube, then the volume is $V = Nd^3$, where d is the length of one side of a cube. The number N is the same for each phase, since each has the same mass. Since each cube contains one molecule at its center, d is also the distance between neighboring molecules. It follows that, for either phase, $d = \sqrt[3]{V/N}$. Therefore,

$$\frac{d_{\text{vapor}}}{d_{\text{liquid}}} = \left(\frac{V_{\text{vapor}}/N}{V_{\text{liquid}}/N}\right)^{1/3} = \left(\frac{\rho_{\text{liquid}}}{\rho_{\text{vapor}}}\right)^{1/3} = \left(\frac{958 \text{ kg/m}^3}{0.598 \text{ kg/m}^3}\right)^{1/3} = \boxed{11.7}$$

---

11. **REASONING** According to the ideal gas law (Equation 14.1), $PV = nRT$. Since n, the number of moles, is constant, $n_1 R = n_2 R$. Thus, according to Equation 14.1, we have

$$\frac{P_1 V_1}{T_1} = \frac{P_2 V_2}{T_2}$$

SOLUTION Solving for T_2, we have

$$T_2 = \left(\frac{P_2}{P_1}\right)\left(\frac{V_2}{V_1}\right) T_1 = \left(\frac{48.5\, P_1}{P_1}\right)\left[\frac{V_1/16}{V_1}\right](305 \text{ K}) = \boxed{925 \text{ K}}$$

---

15. **REASONING AND SOLUTION** According to the ideal gas law (Equation 14.1), the total number of moles of fresh air in the sample is

$$n = \frac{PV}{RT} = \frac{(1.0 \times 10^5 \text{ Pa})(5.0 \times 10^{-4} \text{ m}^3)}{[8.31 \text{ J/(mole} \cdot \text{K)}](310 \text{ K})} = 1.94 \times 10^{-2} \text{ mol}$$

The total number of molecules in the sample is nN_A, where N_A is Avogadro's number. Since the sample contains approximately 21% oxygen, the total number of oxygen molecules in the sample is $(0.21)nN_A$ or

$$(0.21)(1.94 \times 10^{-2} \text{ mol})(6.022 \times 10^{23} \text{ mol}^{-1}) = \boxed{2.5 \times 10^{21}}$$

---

19. **REASONING** The graph that accompanies problem 71 in Chapter 12 can be used to determine the equilibrium vapor pressure of water in the air when the temperature is 30.0 °C (303 K). Equation 12.6 can then be used to find the partial pressure of water in the air at this temperature. Using this pressure, the ideal gas law can then be used to find the number of moles of water vapor per cubic meter.

SOLUTION According to the graph that accompanies problem 71 in Chapter 12, the equilibrium vapor pressure of water vapor at 30.0 °C is approximately 4250 Pa. According to Equation 12.6,

$$\begin{pmatrix} \text{Partial} \\ \text{pressure of} \\ \text{water vapor} \end{pmatrix} = \begin{pmatrix} \text{Percent relative} \\ \text{humidity} \end{pmatrix} \begin{pmatrix} \text{Equilibrium vapor pressure of} \\ \text{water at the existing temperature} \end{pmatrix} \times \frac{1}{100\%}$$

$$= \frac{(55\%)(4250 \text{ Pa})}{100\%} = 2.34 \times 10^3 \text{ Pa}$$

The ideal gas law then gives the number of moles of water vapor per cubic meter of air as

$$\frac{n}{V} = \frac{P}{RT} = \frac{(2.34 \times 10^3 \text{ Pa})}{[8.31 \text{ J/(mol} \cdot \text{K})](303 \text{ K})} = \boxed{0.93 \text{ mol/m}^3} \tag{14.1}$$

---

23. **WWW** ***REASONING*** Since the temperature of the confined air is constant, Boyle's law applies, and $P_{\text{surface}} V_{\text{surface}} = P_h V_h$, where P_{surface} and V_{surface} are the pressure and volume of the air in the tank when the tank is at the surface of the water, and P_h and V_h are the pressure and volume of the trapped air after the tank has been lowered a distance h below the surface of the water. Since the tank is completely filled with air at the surface, V_{surface} is equal to the volume V_{tank} of the tank. Therefore, the fraction of the tank's volume that is filled with air when the tank is a distance h below the water's surface is

$$\frac{V_h}{V_{\text{tank}}} = \frac{V_h}{V_{\text{surface}}} = \frac{P_{\text{surface}}}{P_h}$$

We can find the absolute pressure at a depth h using Equation 11.4. Once the absolute pressure is known at a depth h, we can determine the ratio of the pressure at the surface to the pressure at the depth h.

SOLUTION According to Equation 11.4, the trapped air pressure at a depth $h = 40.0$ m is

$$P_{\text{h}} = P_{\text{surface}} + \rho g h = (1.01 \times 10^5 \text{ Pa}) + \left[(1.00 \times 10^3 \text{ kg/m}^3)(9.80 \text{ m/s}^2)(40.0 \text{ m}) \right]$$

$$= 4.93 \times 10^5 \text{ Pa}$$

where we have used a value of $\rho = 1.00 \times 10^3 \text{ kg/m}^3$ for the density of water. The desired volume fraction is

$$\frac{V_h}{V_{\text{tank}}} = \frac{P_{\text{surface}}}{P_h} = \frac{1.01 \times 10^5 \text{ Pa}}{4.93 \times 10^5 \text{ Pa}} = \boxed{0.205}$$

---

27. ***REASONING*** According to the ideal gas law (Equation 14.1), $PV = nRT$. Since n, the number of moles of the gas, is constant, $n_1 R = n_2 R$. Therefore, $P_1 V_1 / T_1 = P_2 V_2 / T_2$, where $T_1 = 273$ K and T_2 is the temperature we seek. Since the beaker is cylindrical, the volume V of the gas is equal to Ad, where A is the cross-sectional area of the cylindrical volume and

d is the height of the region occupied by the gas, as measured from the bottom of the beaker. With this substitution for the volume, the expression obtained from the ideal gas law becomes

$$\frac{P_1 d_1}{T_1} = \frac{P_2 d_2}{T_2} \tag{1}$$

where the pressures P_1 and P_2 are equal to the sum of the atmospheric pressure and the pressure caused by the mercury in each case. These pressures can be determined using Equation 11.4. Once the pressures are known, Equation (1) can be solved for T_2.

SOLUTION Using Equation 11.4, we obtain the following values for the pressures P_1 and P_2. Note that the initial height of the mercury is $h_1 = (1.520 \text{ m})/2 = 0.760 \text{ m}$, while the final height of the mercury is $h_2 = (1.520 \text{ m})/4 = 0.380 \text{ m}$.

$$P_1 = P_0 + \rho g h_1 = 1.01 \times 10^5 \text{ Pa} + (1.36 \times 10^4 \text{ kg/m}^3)(9.80 \text{ m/s}^2)(0.760 \text{ m}) = 2.02 \times 10^5 \text{ Pa}$$

$$P_2 = P_0 + \rho g h_2 = 1.01 \times 10^5 \text{ Pa} + (1.36 \times 10^4 \text{ kg/m}^3)(9.80 \text{ m/s}^2)(0.380 \text{ m}) = 1.52 \times 10^5 \text{ Pa}$$

In these pressure calculations, the density of mercury is $\rho = 1.36 \times 10^4 \text{ kg/m}^3$. In Equation (1) we note that $d_1 = 0.760 \text{ m}$ and $d_2 = 1.14 \text{ m}$. Solving Equation (1) for T_2 and substituting values, we obtain

$$T_2 = \left(\frac{P_2 d_2}{P_1 d_1}\right) T_1 = \left[\frac{(1.52 \times 10^5 \text{ Pa})(1.14 \text{ m})}{(2.02 \times 10^5 \text{ Pa})(0.760 \text{ m})}\right] (273 \text{ K}) = \boxed{308 \text{ K}}$$

---

31. **REASONING AND SOLUTION** Using the expressions for $\overline{v^2}$ and $(\overline{v})^2$ given in the statement of the problem, we obtain:

a. $$\overline{v^2} = \frac{1}{3}(v_1^2 + v_2^2 + v_3^2) = \frac{1}{3}\left[(3.0 \text{ m/s})^2 + (7.0 \text{ m/s})^2 + (9.0 \text{ m/s})^2\right] = \boxed{46.3 \text{ m}^2/\text{s}^2}$$

b. $$(\overline{v})^2 = \left[\frac{1}{3}(v_1 + v_2 + v_3)\right]^2 = \left[\frac{1}{3}(3.0 \text{ m/s} + 7.0 \text{ m/s} + 9.0 \text{ m/s})\right]^2 = \boxed{40.1 \text{ m}^2/\text{s}^2}$$

$\overline{v^2}$ and $(\overline{v})^2$ are *not* equal, because they are two different physical quantities.

---

33. **REASONING** The internal energy of the neon at any Kelvin temperature T is given by Equation 14.7, $U = (3/2)nRT$. Therefore, when the temperature of the neon increases from an initial temperature T_i to a final temperature T_f, the internal energy of the neon increases by an amount

$$\Delta U = U_f - U_i = \frac{3}{2} nR \ (T_f - T_i)$$

In order to use this equation, we must first determine n. Since the neon is confined to a tank, the number of moles n is constant, and we can use the information given concerning the initial conditions to determine an expression for the quantity nR. According to the ideal gas law (Equation 14.1),

$$nR = \frac{P_i V_i}{T_i}$$

These two expressions can be combined to obtain an equation in terms of the variables that correspond to the data given in the problem statement.

SOLUTION   Combining the two expressions and substituting the given values yields

$$\Delta U = \frac{3}{2} \left(\frac{P_i V_i}{T_i} \right) (T_f - T_i)$$

$$= \frac{3}{2} \left[\frac{(1.01 \times 10^5 \text{ Pa})(680 \text{ m}^3)}{293.2 \text{ K}} \right] (294.3 \text{ K} - 293.2 \text{ K}) = \boxed{3.9 \times 10^5 \text{ J}}$$

---

37. $\boxed{\text{WWW}}$ **REASONING AND SOLUTION**   The average force exerted by one electron on the screen is, from the impulse-momentum theorem (Equation 7.4), $\overline{F} = \Delta p / \Delta t = m \, \Delta v / \Delta t$. Therefore, in a time Δt, N electrons exert an average force $\overline{F} = N m \, \Delta v / \Delta t = (N / \Delta t) m \, \Delta v$. Since the pressure on the screen is the average force per unit area (Equation 10.19), we have

$$P = \frac{\overline{F}}{A} = \frac{(N / \Delta t) m \Delta v}{A}$$

$$= \frac{(6.2 \times 10^{16} \text{ electrons/s})(9.11 \times 10^{-31} \text{ kg})(8.4 \times 10^7 \text{ m/s} - 0 \text{ m/s})}{1.2 \times 10^{-7} \text{ m}^2} = \boxed{4.0 \times 10^1 \text{ Pa}}$$

---

41.   **REASONING AND SOLUTION**   According to Fick's law of diffusion (Equation 14.8), the mass of ethanol that diffuses through the cylinder in one hour (3600 s) is

$$m = \frac{(DA \, \Delta C)t}{L} = \frac{(12.4 \times 10^{-10} \text{ m}^2/\text{s})(4.00 \times 10^{-4} \text{ m}^2)(1.50 \text{ kg/m}^3)(3600 \text{ s})}{0.0200 \text{ m}} = \boxed{1.34 \times 10^{-7} \text{ kg}}$$

---

47. ***REASONING AND SOLUTION*** The number of moles of air that is pumped into the tire is $\Delta n = n_f - n_i$. According to the ideal gas law (Equation 14.1), $n = PV/(RT)$; therefore,

$$\Delta n = n_f - n_i = \frac{P_f V_f}{R T_f} - \frac{P_i V_i}{R T_i} = \frac{V}{RT}\left(P_f - P_i\right)$$

Substitution of the data given in the problem leads to

$$\Delta n = \frac{4.1 \times 10^{-4} \text{ m}^3}{\left[8.31 \text{ J/(mol} \cdot \text{K)}\right]\left(296 \text{ K}\right)}\left[\left(6.2 \times 10^5 \text{ Pa}\right) - \left(4.8 \times 10^5 \text{ Pa}\right)\right] = \boxed{2.3 \times 10^{-2} \text{ mol}}$$

---

49. WWW ***REASONING*** Since mass is conserved, the mass flow rate is the same at all points, as described by the equation of continuity (Equation 11.8). Therefore, the mass flow rate at which CCl_4 enters the tube is the same as that at point A. The concentration difference of CCl_4 between point A and the left end of the tube, ΔC, can be calculated by using Fick's law of diffusion (Equation 14.8). The concentration of CCl_4 at point A can be found from $C_A = C_{\text{left end}} - \Delta C$.

SOLUTION
a. As discussed above in the reasoning, the mass flow rate of CCl_4 as it passes point A is the same as the mass flow rate at which CCl_4 enters the left end of the tube; therefore, the mass flow rate of CCl_4 at point A is $\boxed{5.00 \times 10^{-13} \text{ kg/s}}$.

b. Solving Fick's law for ΔC, we obtain

$$\Delta C = \frac{mL}{DAt} = \frac{(m/t)L}{DA}$$

$$= \frac{(5.00 \times 10^{-13} \text{ kg/s})(5.00 \times 10^{-3} \text{ m})}{(20.0 \times 10^{-10} \text{ m}^2/\text{s})(3.00 \times 10^{-4} \text{ m}^2)} = 4.2 \times 10^{-3} \text{ kg/m}^3$$

Then,

$$C_A = C_{\text{left end}} - \Delta C = (1.00 \times 10^{-2} \text{ kg/m}^3) - (4.2 \times 10^{-3} \text{ kg/m}^3) = \boxed{5.8 \times 10^{-3} \text{ kg/m}^3}$$

---

53. ***REASONING AND SOLUTION*** The number of moles of water in the drop is $n = m/M$, where m is the mass of the drop and M is the mass per mole of water. The mass of a drop of volume V and density ρ can be obtained from Equation 11.1 as $m = \rho V$. Thus, the number of molecules in the drop is

$$N = N_A n = \frac{N_A \rho V}{M}$$

where $N_A = 6.022 \times 10^{23}$ mol^{-1} is Avogadro's number. Since the drop is spherical, its volume is given by $V = \frac{4}{3}\pi r^3$, where r is the radius of the drop. Substituting the values for the data given in the problem, using a density of $\rho = 1.00 \times 10^3$ kg/m^3 for water, and using $M = 18.015 \times 10^{-3}$ kg/mol for water yields

$$N = \frac{\left(6.022 \times 10^{23} \text{ mol}^{-1}\right)\left(1.00 \times 10^3 \, \frac{\text{kg}}{\text{m}^3}\right)\frac{4}{3}\pi (9.00 \times 10^{-4}\text{m})^3}{18.015 \times 10^{-3} \, \frac{\text{kg}}{\text{mol}}} = \boxed{1.02 \times 10^{20}}$$

---

55. ***REASONING AND SOLUTION***   Since the density of aluminum is 2700 kg/m^3, the number of atoms of aluminum per cubic meter is, using the given data,

$$N/V = \left(\frac{2700 \times 10^3 \text{ g/m}^3}{26.9815 \text{ g/mol}}\right)\left(6.022 \times 10^{23} \text{ mol}^{-1}\right) = 6.0 \times 10^{28} \text{ atoms/m}^3$$

Assuming that the volume of the solid contains many small cubes, with one atom at the center of each, then there are $(6.0 \times 10^{28})^{1/3}$ atoms/m $= 3.9 \times 10^9$ atoms/m along each edge of a 1.00-m^3 cube. Therefore, the spacing between the centers of neighboring atoms is

$$d = \frac{1}{3.9 \times 10^9 \text{/m}} = \boxed{2.6 \times 10^{-10} \text{ m}}$$

CHAPTER 15 | *THERMODYNAMICS*

PROBLEMS

1. **REASONING**   Since the change in the internal energy and the heat released in the process are given, the first law of thermodynamics (Equation 15.1) can be used to find the work done. Since we are told how much work is required to make the car go one mile, we can determine how far the car can travel. When the gasoline burns, its internal energy decreases and heat flows into the surroundings; therefore, both ΔU and Q are negative.

   SOLUTION   According to the first law of thermodynamics, the work that is done when one gallon of gasoline is burned in the engine is

$$W = Q - \Delta U = -1.00 \times 10^8 \text{ J} - (-1.19 \times 10^8 \text{ J}) = 0.19 \times 10^8 \text{ J}$$

Since 6.0×10^5 J of work is required to make the car go one mile, the car can travel

$$0.19 \times 10^8 \text{ J} \left(\frac{1 \text{ mile}}{6.0 \times 10^5 \text{ J}} \right) = \boxed{32 \text{ miles}}$$

---

5. **REASONING AND SOLUTION**   In going from the initial state to the intermediate state, the internal energy of the system changes according to Equation 15.1:

$$\Delta U_1 = U_f - U_i = Q - W = +165 \text{ J} - (+312 \text{ J}) = -147 \text{ J}$$

Since internal energy is a function of state, the change in the internal energy in going from the initial state to the intermediate state and back to the initial state is zero. That is, $\Delta U_{\text{total}} = \Delta U_1 + \Delta U_2 = 0$. Therefore, in returning to the initial state, the change in the internal energy is $\Delta U_2 = -\Delta U_1 = +147 \text{ J}$.

   a. Thus, the work involved is
$$W = Q - \Delta U_2 = -114 \text{ J} - (+147 \text{ J}) = \boxed{-261 \text{ J}}$$

   b. Since the work is negative, $\boxed{\text{work is done on the system}}$.

---

9. **REASONING**   The work done in the process is equal to the "area" under the curved line between A and B in the drawing. From the graph, we find that there are about 78 "squares" under the curve. Each square has an "area" of

$$(2.0 \times 10^4 \text{ Pa})(2.0 \times 10^{-3} \text{ m}^3) = 4.0 \times 10^1 \text{ J}$$

SOLUTION
a. The work done in the process has a magnitude of

$$W = (78)(4.0 \times 10^1 \text{ J}) = \boxed{3100 \text{ J}}$$

b. The final volume is smaller than the initial volume, so the gas is compressed. Therefore, work is done on the gas so the work is $\boxed{\text{negative}}$.

---

13. **REASONING**   According to Equation 15.2, $W = P\Delta V$, the average pressure $\overline{P}$ of the expanding gas is equal to $\overline{P} = W / \Delta V$, where the work W done by the gas on the bullet can be found from the work-energy theorem (Equation 6.3). Assuming that the barrel of the gun is cylindrical with radius r, the volume of the barrel is equal to its length L multiplied by the area (πr^2) of its cross section. Thus, the change in volume of the expanding gas is $\Delta V = L\pi r^2$.

SOLUTION   The work done by the gas on the bullet is given by Equation 6.3 as

$$W = \tfrac{1}{2} m(v_{\text{final}}^2 - v_{\text{initial}}^2) = \tfrac{1}{2}(2.6 \times 10^{-3} \text{ kg})[(370 \text{ m/s})^2 - 0] = 180 \text{ J}$$

The average pressure of the expanding gas is, therefore,

$$\overline{P} = \frac{W}{\Delta V} = \frac{180 \text{ J}}{(0.61 \text{ m})\pi(2.8 \times 10^{-3} \text{ m})^2} = \boxed{1.2 \times 10^7 \text{ Pa}}$$

---

15. $\boxed{\text{WWW}}$ **REASONING AND SOLUTION**   The first law of thermodynamics states that $\Delta U = Q - W$. The work W involved in an isobaric process is, according to Equation 15.2, $W = P\Delta V$. Combining these two expressions leads to $\Delta U = Q - P\Delta V$. Solving for Q gives

$$Q = \Delta U + P\Delta V \tag{1}$$

Since this is an expansion, $\Delta V > 0$, so $P\Delta V > 0$. From the ideal gas law, $PV = nRT$, we have $P\Delta V = nR\Delta T$. Since $P\Delta V > 0$, it follows that $nR\Delta T > 0$. The internal energy of an ideal gas is directly proportional to its Kelvin temperature T. Therefore, since $nR\Delta T > 0$, it follows that $\Delta U > 0$. Since both terms on the right hand side of Equation (1) are positive, the left hand side of Equation (1) must also be positive. Thus, Q is positive. By the convention described in the text, this means that

$$\boxed{\text{heat can only flow into an ideal gas during an isobaric expansion}}$$

19. ***REASONING***   According to the first law of thermodynamics (Equation 15.1), $\Delta U = Q - W$. For a monatomic ideal gas (Equation 14.7), $U = \frac{3}{2}nRT$. Therefore, for the process in question, the change in the internal energy is $\Delta U = \frac{3}{2}nR\Delta T$. Combining the last expression for ΔU with Equation 15.1 yields

$$\tfrac{3}{2}nR\Delta T = Q - W$$

This expression can be solved for ΔT.

SOLUTION
a.   The heat is $Q = +1200 \text{ J}$, since it is absorbed by the system. The work is $W = +2500 \text{ J}$, since it is done *by* the system. Solving the above expression for ΔT and substituting the values for the data given in the problem statement, we have

$$\Delta T = \frac{Q - W}{\frac{3}{2}nR} = \frac{1200 \text{ J} - 2500 \text{ J}}{\frac{3}{2}(0.50 \text{ mol})[8.31 \text{ J}/(\text{mol}\cdot\text{K})]} = \boxed{-2.1 \times 10^2 \text{ K}}$$

b.   Since $\Delta T = T_{\text{final}} - T_{\text{initial}}$ is negative, T_{initial} must be greater than T_{final}; this change represents a $\boxed{\text{decrease}}$ in temperature.

Alternatively, one could deduce that the temperature decreases from the following physical argument. Since the system loses more energy in doing work than it gains in the form of heat, the internal energy of the system decreases. Since the internal energy of an ideal gas depends only on the temperature, a decrease in the internal energy must correspond to a decrease in the temperature.

---

23. ***REASONING***   When the expansion is isothermal, the work done can be calculated from Equation (15.3): $W = nRT \ln(V_f / V_i)$. When the expansion is adiabatic, the work done can be calculated from Equation 15.4: $W = \frac{3}{2}nR(T_i - T_f)$.

Since the gas does the same amount of work whether it expands adiabatically or isothermally, we can equate the right hand sides of these two equations. We also note that since the initial temperature is the same for both cases, the temperature T in the isothermal expansion is the same as the initial temperature T_i for the adiabatic expansion.   We then have

$$nRT_i \ln\left(\frac{V_f}{V_i}\right) = \tfrac{3}{2}nR(T_i - T_f)$$

or

$$\ln\left(\frac{V_f}{V_i}\right) = \frac{\frac{3}{2}(T_i - T_f)}{T_i}$$

SOLUTION Solving for the ratio of the volumes gives

$$\frac{V_f}{V_i} = e^{\frac{3}{2}(T_i - T_f)/T_i} = e^{\frac{3}{2}(405 \text{ K} - 245 \text{ K})/(405 \text{ K})} = \boxed{1.81}$$

---

29. **REASONING AND SOLUTION**
    a. The final temperature of the adiabatic process is given by solving Equation 15.4 for T_f.

$$T_f = T_i - \frac{W}{\frac{3}{2}nR} = 393 \text{ K} - \frac{825 \text{ J}}{\frac{3}{2}(1.00 \text{ mol})[8.31 \text{ J}/(\text{mol} \cdot \text{K})]} = \boxed{327 \text{ K}}$$

b. According to Equation 15.5 for the adiabatic expansion of an ideal gas, $P_i V_i^\gamma = P_f V_f^\gamma$.
Therefore,

$$V_f^\gamma = V_i^\gamma \left(\frac{P_i}{P_f}\right)$$

From the ideal gas law, $PV = nRT$; therefore, the ratio of the pressures is given by

$$\frac{P_i}{P_f} = \left(\frac{T_i}{T_f}\right)\left(\frac{V_f}{V_i}\right)$$

Combining the previous two equations gives

$$V_f^\gamma = V_i^\gamma \left(\frac{T_i}{T_f}\right)\left(\frac{V_f}{V_i}\right)$$

Solving for V_f we obtain

$$\frac{V_f^\gamma}{V_f} = \left(\frac{T_i}{T_f}\right)\left(\frac{V_i^\gamma}{V_i}\right) \qquad \text{or} \qquad V_f^{(\gamma - 1)} = \left(\frac{T_i}{T_f}\right) V_i^{(\gamma - 1)}$$

$$V_f = \left[\left(\frac{T_i}{T_f}\right) V_i^{(\gamma - 1)}\right]^{1/(\gamma - 1)} = V_i \left(\frac{T_i}{T_f}\right)^{1/(\gamma - 1)}$$

Therefore,

$$V_f = V_i \left(\frac{T_i}{T_f}\right)^{1/(\gamma - 1)} = (0.100 \text{ m}^3)\left(\frac{393 \text{ K}}{327 \text{ K}}\right)^{1/(2/3)} = (0.100 \text{ m}^3)\left(\frac{393 \text{ K}}{327 \text{ K}}\right)^{3/2} = \boxed{0.132 \text{ m}^3}$$

31. ***REASONING AND SOLUTION*** The amount of heat required to change the temperature of the gas is given by Equation 15.6, where C_P is given by Equation 15.7.

$$Q = C_p n\Delta T = \tfrac{5}{2} Rn\Delta T = \tfrac{5}{2}[8.31 \text{ J} / (\text{mol} \cdot \text{K})] \, (1.5 \text{ mol}) \, (77 \text{ K}) = \boxed{2400 \text{ J}}$$

---

33. ***REASONING AND SOLUTION*** According to the first law of thermodynamics (Equation 15.1), $\Delta U = U_f - U_i = Q - W$. Since the internal energy of this gas is doubled by the addition of heat, the initial and final internal energies are U and $2U$, respectively. Therefore,

$$\Delta U = U_f - U_i = 2U - U = U$$

Equation 15.1 for this situation then becomes $U = Q - W$. Solving for Q gives

$$Q = U + W \tag{1}$$

The initial internal energy of the gas can be calculated from Equation 14.7:

$$U = \frac{3}{2} nRT = \frac{3}{2}(2.5 \text{ mol})[8.31 \text{ J}/(\text{mol} \cdot \text{K})](350 \text{ K}) = 1.1 \times 10^4 \text{ J}$$

a. If the process is carried out isochorically (i.e., at constant volume), then $W = 0$, and the heat required to double the internal energy is

$$Q = U + W = U + 0 = \boxed{1.1 \times 10^4 \text{ J}}$$

b. If the process is carried out isobarically (i.e., at constant pressure), then $W = P\Delta V$, and Equation (1) above becomes

$$Q = U + W = U + P\Delta V \tag{2}$$

From the ideal gas law, $PV = nRT$, we have that $P\Delta V = nR\Delta T$, and Equation (2) becomes

$$Q = U + nR\Delta T \tag{3}$$

The internal energy of an ideal gas is directly proportional to its Kelvin temperature. Since the internal energy of the gas is doubled, the final Kelvin temperature will be twice the initial Kelvin temperature, or $\Delta T = 350$ K. Substituting values into Equation (3) gives

$$Q = 1.1 \times 10^4 \text{ J} + (2.5 \text{ mol})[8.31 \text{ J}/(\text{mol} \cdot \text{K})](350 \text{ K}) = \boxed{1.8 \times 10^4 \text{ J}}$$

---

37. ***REASONING*** According to Equations 15.6 and 15.7, the heat supplied to a monatomic ideal gas at constant pressure is $Q = C_P n \Delta T$, with $C_P = \tfrac{5}{2} R$. Thus, $Q = \tfrac{5}{2} nR\Delta T$. The percentage of this heat used to increase the internal energy by an amount ΔU is

$$\text{Percentage} = \left(\frac{\Delta U}{Q}\right) \times 100\ \% = \left(\frac{\Delta U}{\tfrac{5}{2}nR\,\Delta T}\right) \times 100\ \% \qquad (1)$$

But according to the first law of thermodynamics, $\Delta U = Q - W$. The work W is $W = P\Delta V$, and for an ideal gas $P\Delta V = nR\Delta T$. Therefore, the work W becomes $W = P\Delta V = nR\Delta T$ and the change in the internal energy is $\Delta U = Q - W = \tfrac{5}{2}nR\Delta T - nR\Delta T = \tfrac{3}{2}nR\Delta T$. Combining this expression for ΔU with Equation (1) above yields a numerical value for the percentage of heat being supplied to the gas that is used to increase its internal energy.

SOLUTION
a. The percentage is

$$\text{Percentage} = \left(\frac{\Delta U}{\tfrac{5}{2}nR\Delta T}\right) \times 100\ \% = \left(\frac{\tfrac{3}{2}nR\,\Delta T}{\tfrac{5}{2}nR\,\Delta T}\right) \times 100\ \% = \left(\frac{3}{5}\right) \times 100\ \% = \boxed{60.0\ \%}$$

b. The remainder of the heat, or $\boxed{40.0\%}$, is used for the work of expansion.

---

41. ***REASONING AND SOLUTION***   The efficiency of a heat engine is defined by Equation 15.11 as $e = W/Q_H$, where W is the work done and Q_H is the heat input. The principle of energy conservation requires that $Q_H = W + Q_C$, where Q_C is the heat rejected to the cold reservoir (Equation 15.12). Combining Equations 15.11 and 15.12 gives

$$e = \frac{W}{W + Q_C} = \frac{16\ 600\ \text{J}}{16\ 600\ \text{J} + 9700\ \text{J}} = \boxed{0.631}$$

---

45. $\boxed{\text{WWW}}$ ***REASONING***   The efficiency of either engine is given by Equation 15.13, $e = 1 - (Q_C/Q_H)$. Since engine A receives three times more input heat, produces five times more work, and rejects two times more heat than engine B, it follows that $Q_{HA} = 3Q_{HB}$, $W_A = 5W_B$, and $Q_{CA} = 2Q_{CB}$. As required by the principle of energy conservation for engine A (Equation 15.12),

$$\underbrace{Q_{HA}}_{3Q_{HB}} = \underbrace{Q_{CA}}_{2Q_{CB}} + \underbrace{W_A}_{5W_B}$$

Thus,

$$3Q_{HB} = 2Q_{CB} + 5W_B \qquad (1)$$

Since engine B also obeys the principle of energy conservation (Equation 15.12),

$$Q_{HB} = Q_{CB} + W_B \qquad (2)$$

Substituting Q_{HB} from Equation (2) into Equation (1) yields

$$3(Q_{CB} + W_B) = 2Q_{CB} + 5W_B$$

Solving for W_B gives

$$W_B = \frac{1}{2}Q_{CB}$$

Therefore, Equation (2) predicts for engine B that

$$Q_{HB} = Q_{CB} + W_B = \frac{3}{2}Q_{CB}$$

SOLUTION

a. Substituting $Q_{CA} = 2Q_{CB}$ and $Q_{HA} = 3Q_{HB}$ into Equation 15.13 for engine A, we have

$$e_A = 1 - \frac{Q_{CA}}{Q_{HA}} = 1 - \frac{2Q_{CB}}{3Q_{HB}} = 1 - \frac{2Q_{CB}}{3(\frac{3}{2}Q_{CB})} = 1 - \frac{4}{9} = \boxed{\frac{5}{9}}$$

b. Substituting $Q_{HB} = \frac{3}{2}Q_{CB}$ into Equation 15.13 for engine B, we have

$$e_B = 1 - \frac{Q_{CB}}{Q_{HB}} = 1 - \frac{Q_{CB}}{\frac{3}{2}Q_{CB}} = 1 - \frac{2}{3} = \boxed{\frac{1}{3}}$$

---

49. **REASONING**   The efficiency e of a Carnot engine is given by Equation 15.15, $e = 1 - (T_C / T_H)$, where, according to Equation 15.14, $(Q_C / Q_H) = (T_C / T_H)$.   Since the efficiency is given along with T_C and Q_C, Equation 15.15 can be used to calculate T_H. Once T_H is known, the ratio T_C / T_H is thus known, and Equation 15.14 can be used to calculate Q_H.

SOLUTION

a. Solving Equation 15.15 for T_H gives

$$T_H = \frac{T_C}{1 - e} = \frac{378 \text{ K}}{1 - 0.700} = \boxed{1260 \text{ K}}$$

b. Solving Equation 15.14 for Q_H gives

$$Q_H = Q_C \left(\frac{T_H}{T_C} \right) = (5230 \text{ J}) \left(\frac{1260 \text{ K}}{378 \text{ K}} \right) = \boxed{1.74 \times 10^4 \text{ J}}$$

---

53. **_REASONING_** The maximum efficiency e at which the power plant can operate is given by Equation 15.15, $e = 1 - (T_C / T_H)$. The power output is given; it can be used to find the work output W for a 24 hour period. With the efficiency and W known, Equation 15.11, $e = W / Q_H$, can be used to find Q_H. Q_C can then be found from Equation 15.12, $Q_H = W + Q_C$.

SOLUTION
a. The maximum efficiency is

$$e = 1 - \frac{T_C}{T_H} = 1 - \frac{323 \text{ K}}{505 \text{ K}} = \boxed{0.360}$$

b. Since the power output of the power plant is $P = 84\,000$ kW, the required heat input Q_H for a 24 hour period is

$$Q_H = \frac{W}{e} = \frac{Pt}{e} = \frac{(8.4 \times 10^7 \text{ J/s})(24 \text{ h})}{0.360} \left(\frac{3600 \text{ s}}{1 \text{ h}} \right) = 2.02 \times 10^{13} \text{ J}$$

Therefore, solving Equation 15.12 for Q_C, we have

$$Q_C = Q_H - W = 2.02 \times 10^{13} \text{ J} - 7.3 \times 10^{12} \text{ J} = \boxed{1.3 \times 10^{13} \text{ J}}$$

---

55. **_REASONING AND SOLUTION_** The efficiency e of the power plant is three-fourths its Carnot efficiency so, according to Equation 15.15,

$$e = 0.75 \left(1 - \frac{T_C}{T_H} \right) = 0.75 \left(1 - \frac{40 \text{ K} + 273 \text{ K}}{285 \text{ K} + 273 \text{ K}} \right) = 0.33$$

The power output of the plant is 1.2×10^9 watts. According to Equation 15.11, $e = W / Q_H = (\text{Power} \cdot t) / Q_H$. Therefore, at 33% efficiency, the heat input per unit time is

$$\frac{Q_H}{t} = \frac{\text{Power}}{e} = 3.6 \times 10^9 \text{ J/s}$$

From the principle of conservation of energy, the heat output per unit time must be

$$\frac{Q_C}{t} = \frac{Q_H}{t} - \text{Power} = 2.4 \times 10^9 \text{ J/s}$$

The rejected heat is carried away by the flowing water and, according to Equation 12.4, $Q_C = cm\Delta T$. Therefore,

$$\frac{Q_C}{t} = \frac{cm\Delta T}{t} \qquad \text{or} \qquad \frac{Q_C}{t} = c(m/t)\Delta T$$

Solving the last equation for ΔT, we have

$$\Delta T = \frac{Q_C}{tc(m/t)} = \frac{(Q_C/t)}{c(m/t)} = \frac{2.4 \times 10^9 \text{ J/s}}{[4186 \text{ J/(kg} \cdot \text{C}^\circ)](1.0 \times 10^5 \text{ kg/s})} = \boxed{5.7 \text{ C}^\circ}$$

59. **WWW**   ***REASONING AND SOLUTION***   Equation 15.14 holds for a Carnot air conditioner as well as a Carnot engine. Therefore, solving Equation 15.14 for Q_C, we have

$$Q_C = Q_H \left(\frac{T_C}{T_H} \right) = (6.12 \times 10^5 \text{ J}) \left(\frac{299 \text{ K}}{312 \text{ K}} \right) = \boxed{5.86 \times 10^5 \text{ J}}$$

65. **WWW**   ***REASONING***   Let the coefficient of performance be represented by the variable η. Then according to Equation 15.16, $\eta = Q_C / W$. From the statement of energy conservation for a Carnot refrigerator (Equation 15.12), $W = Q_H - Q_C$. Combining Equations 15.16 and 15.12 leads to

$$\eta = \frac{Q_C}{Q_H - Q_C} = \frac{Q_C / Q_C}{(Q_H - Q_C)/Q_C} = \frac{1}{(Q_H / Q_C) - 1}$$

Replacing the ratio of the heats with the ratio of the Kelvin temperatures, according to Equation 15.14, leads to

$$\eta = \frac{1}{(T_H / T_C) - 1} \tag{1}$$

The heat Q_C that must be removed from the refrigerator when the water is cooled can be calculated using Equation 12.4, $Q_C = cm\Delta T$; therefore,

$$W = \frac{Q_C}{\eta} = \frac{cm\Delta T}{\eta} \tag{2}$$

SOLUTION
a. Substituting values into Equation (1) gives

$$\eta = \frac{1}{\dfrac{T_H}{T_C} - 1} = \frac{1}{\dfrac{(20.0 + 273.15) \text{ K}}{(6.0 + 273.15) \text{ K}} - 1} = \boxed{2.0 \times 10^1}$$

b. Substituting values into Equation (2) gives

$$W = \frac{cm\Delta T}{\eta} = \frac{[4186 \text{ J/(kg} \cdot \text{C}^\circ)](5.00 \text{ kg})(14.0 \text{ C}^\circ)}{2.0 \times 10^1} = \boxed{1.5 \times 10^4 \text{ J}}$$

67. **REASONING**   The efficiency of the Carnot engine is, according to Equation 15.15,

$$e = 1 - \frac{T_C}{T_H} = 1 - \frac{842 \text{ K}}{1684 \text{ K}} = \frac{1}{2}$$

Therefore, the work delivered by the engine is, according to Equation 15.11,

$$W = e\, Q_H = \frac{1}{2} Q_H$$

The heat pump removes an amount of heat Q_H from the cold reservoir. Thus, the amount of heat Q' delivered to the hot reservoir of the heat pump is

$$Q' = Q_H + W = Q_H + \frac{1}{2} Q_H = \frac{3}{2} Q_H$$

Therefore, $Q'/Q_H = 3/2$. According to Equation 15.14, $Q'/Q_H = T'/T_C$, so $T'/T_C = 3/2$.

SOLUTION   Solving for T' gives

$$T' = \frac{3}{2} T_C = \frac{3}{2}(842 \text{ K}) = \boxed{1.26 \times 10^3 \text{ K}}$$

---

71. **REASONING AND SOLUTION**
   a.   Since the energy that becomes unavailable for doing work is zero for the process, we have from Equation 15.19, $W_{unavailable} = T_0 \Delta S_{universe} = 0$. Therefore, $\Delta S_{universe} = 0$ and according to the discussion in Section 15.11, the process is $\boxed{\text{reversible}}$.

   b.   Since the process is reversible, we have (see Section 15.11)

$$\Delta S_{universe} = \Delta S_{system} + \Delta S_{surroundings} = 0$$

Therefore,

$$\Delta S_{surroundings} = -\Delta S_{system} = \boxed{-125 \text{ J/K}}$$

---

75. **REASONING AND SOLUTION**
   a.   Since the temperature of the gas is kept constant at all times, the process is isothermal; therefore, the internal energy of an ideal gas does not change and $\boxed{\Delta U = 0}$.

   b.   From the first law of thermodynamics (Equation 15.1), $\Delta U = Q - W$. But $\Delta U = 0$, so that $Q = W$. Since work is done on the gas, the work is negative, and $\boxed{Q = -6.1 \times 10^3 \text{ J}}$.

   c.   The work done in an isothermal compression is given by Equation 15.3:

$$W = nRT \ln\left(\frac{V_f}{V_i}\right)$$

Therefore, the temperature of the gas is

$$T = \frac{W}{nR \ln\left(V_f / V_i\right)} = \frac{-6.1 \times 10^3 \text{ J}}{(3.0 \text{ mol})[8.31 \text{ J}/(\text{mol} \cdot \text{K})] \ln\left[(2.5 \times 10^{-2} \text{ m}^3)/(5.5 \times 10^{-2} \text{ m}^3)\right]} = \boxed{310 \text{ K}}$$

---

79. **REASONING AND SOLUTION**   The change in entropy ΔS of a system for a process in which heat Q enters or leaves the system reversibly at a constant temperature T is given by Equation 15.18, $\Delta S = (Q/T)_R$. For a phase change, $Q = mL$, where L is the latent heat (see Section 12.8).

a.   If we imagine a reversible process in which 3.00 kg of ice melts into water at 273 K, the change in entropy of the water molecules is

$$\Delta S = \left(\frac{Q}{T}\right)_R = \left(\frac{mL_f}{T}\right)_R = \frac{(3.00 \text{ kg})(3.35 \times 10^5 \text{ J/kg})}{273 \text{ K}} = \boxed{3.68 \times 10^3 \text{ J/K}}$$

b.   Similarly, if we imagine a reversible process in which 3.00 kg of water changes into steam at 373 K, the change in entropy of the water molecules is

$$\Delta S = \left(\frac{Q}{T}\right)_R = \left(\frac{mL_v}{T}\right)_R = \frac{(3.00 \text{ kg})(2.26 \times 10^6 \text{ J/kg})}{373 \text{ K}} = \boxed{1.82 \times 10^4 \text{ J/K}}$$

c.   Since the change in entropy is greater for vaporization than for fusion, the $\boxed{\text{vaporization process creates more disorder}}$ in the collection of water molecules.

---

83. **REASONING**   The work done in an isobaric process is given by Equation 15.2, $W = P\Delta V$; therefore, the pressure is equal to $P = W/\Delta V$. In order to use this expression, we must first determine a numerical value for the work done; this can be calculated using the first law of thermodynamics (Equation 15.1), $\Delta U = Q - W$.

SOLUTION   Solving Equation 15.1 for the work W, we find

$$W = Q - \Delta U = 1500 \text{ J} - (+4500 \text{ J}) = -3.0 \times 10^3 \text{ J}$$

Therefore, the pressure is

$$P = \frac{W}{\Delta V} = \frac{-3.0 \times 10^3 \text{ J}}{-0.010 \text{ m}^3} = \boxed{3.0 \times 10^5 \text{ Pa}}$$

The change in volume ΔV, which is the final volume minus the initial volume, is negative because the final volume is 0.010 m^3 *less* than the initial volume.

---

87. **REASONING**   According to Equation 15.5, $P_i V_i^{\gamma} = P_f V_f^{\gamma}$. The ideal gas law states that $V = nRT / P$ for both the initial and final conditions. Thus, we have

$$P_i \left(\frac{nRT_i}{P_i} \right)^{\gamma} = P_f \left(\frac{nRT_f}{P_f} \right)^{\gamma} \quad \text{or} \quad \frac{P_f}{P_i} = \left(\frac{T_i}{T_f} \right)^{\gamma/(1-\gamma)}$$

Since the ratio of the temperatures is known, the last expression can be solved for the final pressure P_f.

SOLUTION Since $T_i / T_f = 1/2$, and $\gamma = 5/3$, we find that

$$P_f = P_i \left(\frac{T_i}{T_f} \right)^{\gamma/(1-\gamma)} = \left(1.50 \times 10^5 \text{ Pa} \right) \left(\frac{1}{2} \right)^{(5/3)/[1-(5/3)]} = \boxed{8.49 \times 10^5 \text{ Pa}}$$

---

91. **REASONING AND SOLUTION**   We wish to find an expression for the overall efficiency e in terms of the efficiencies e_1 and e_2. From the problem statement, the overall efficiency of the two-engine device is

$$e = \frac{W_1 + W_2}{Q_H} \tag{1}$$

where Q_H is the input heat to engine 1. The efficiency of a heat engine is defined by Equation 15.11, $e = W / Q_H$, so we can write

$$W_1 = e_1 Q_H \tag{2}$$

and

$$W_2 = e_2 Q_{H2}$$

Since the heat rejected by engine 1 is used as input heat for the second engine, $Q_{H2} = Q_{C1}$, and the expression above for W_2 can be written as

$$W_2 = e_2 Q_{C1} \tag{3}$$

According to Equation 15.12, we have $Q_{C1} = Q_H - W_1$, so that Equation (3) becomes

$$W_2 = e_2(Q_H - W_1) \tag{4}$$

Substituting Equations (2) and (4) into Equation (1) gives

$$e = \frac{e_1 Q_H + e_2(Q_H - W_1)}{Q_H} = \frac{e_1 Q_H + e_2(Q_H - e_1 Q_H)}{Q_H}$$

Algebraically canceling the Q_H's in the right hand side of the last expression gives the desired result:

$$\boxed{e = e_1 + e_2 - e_1 e_2}$$

CHAPTER 16 | *WAVES AND SOUND*

1. ***REASONING AND SOLUTION*** The frequency f (in hertz) is the number of wave cycles that pass an observer and is the reciprocal of the period T (in seconds). The period T is the time required for one complete cycle of the wave to pass a fixed point. From the data given in the problem, $T = $ (time for n cycles)$/n = $ (120 s)$/10 = 12$ s. Therefore,

$$f = \frac{1}{T} = \frac{1}{12 \text{ s}} = \boxed{0.083 \text{ Hz}}$$

3. ***REASONING*** When the end of the Slinky is moved up and down continuously, a transverse wave is produced. The distance between two adjacent crests on the wave, is, by definition, one wavelength. The wavelength is related to the speed and frequency of a periodic wave by Equation 16.1, $\lambda = v/f$. In order to use Equation 16.1, we must first determine the frequency of the wave. The wave on the Slinky will have the same frequency as the simple harmonic motion of the hand. According to the data given in the problem statement, the frequency is $f = $ (2.00 cycles)$/$(1 s)$ = 2.00$ Hz.

   SOLUTION Substituting the values for λ and f, we find that the distance between crests is

$$\lambda = \frac{v}{f} = \frac{0.50 \text{ m/s}}{2.00 \text{ Hz}} = \boxed{0.25 \text{ m}}$$

7. ***REASONING*** As the transverse wave propagates, the colored dot moves up and down in simple harmonic motion with a frequency of 5.0 Hz. The amplitude (1.3 cm) is the magnitude of the maximum displacement of the dot from its equilibrium position.

   SOLUTION The period T of the simple harmonic motion of the dot is $T = 1/f = 1/(5.0 \text{ Hz}) = 0.20$ s. In one period the dot travels through one complete cycle of its motion, and covers a vertical distance of $4 \times (1.3 \text{ cm}) = 5.2$ cm. Therefore, in 3.0 s the dot will have traveled a *total vertical distance* of

$$\left(\frac{3.0 \text{ s}}{0.20 \text{ s}}\right) (5.2 \text{ cm}) = \boxed{78 \text{ cm}}$$

13. ***REASONING*** The tension F in the violin string can be found by solving Equation 16.2 for F to obtain $F = mv^2 / L$, where v is the speed of waves on the string and can be found from Equation 16.1 as $v = f\lambda$.

SOLUTION Combining Equations 16.2 and 16.1 and using the given data, we obtain

$$F = \frac{mv^2}{L} = (m/L)f^2\lambda^2 = (7.8 \times 10^{-4} \text{ kg/m})(440 \text{ Hz})^2 (65 \times 10^{-2} \text{ m})^2 = \boxed{64 \text{ N}}$$

---

15. $\boxed{\text{WWW}}$ *REASONING* According to Equation 16.2, the linear density of the string is given by $(m/L) = F/v^2$, where the speed v of waves on the middle C string is given by Equation 16.1, $v = \lambda f = \lambda / T$.

SOLUTION Combining Equations 16.2 and 16.1 and using the given data, we obtain

$$m/L = \frac{F}{v^2} = \frac{FT^2}{\lambda^2} = \frac{(944 \text{ N})(3.82 \times 10^{-3} \text{ s})^2}{(1.26 \text{ m})^2} = \boxed{8.68 \times 10^{-3} \text{ kg/m}}$$

---

19. *REASONING* Newton's second law can be used to analyze the motion of the blocks using the methods introduced and developed in Chapter 4. In this way we can determine an expression that relates the magnitude of the pulling force **P** to the magnitude of the tension **F** in the wire. Equation 16.2 $[v = \sqrt{F/(m/L)}]$ can then be used to find the tension in the wire.

SOLUTION The following figures show a schematic of the situation described in the problem and the free-body diagrams for each block, where $m_1 = 42.0$ kg and $m_2 = 19.0$ kg.

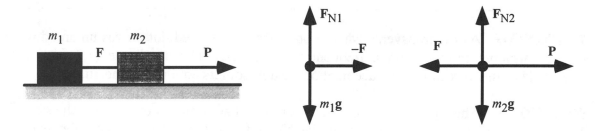

Newton's second law for block 1 is, taking forces that point to the right as positive, $F = m_1 a$, or $a = F/m_1$. For block 2, we obtain $P - F = m_2 a$. Using the expression for a obtained from the equation for block 1, we have

$$P - F = F\left(\frac{m_2}{m_1}\right) \qquad \text{or} \qquad P = F\left(\frac{m_2}{m_1}\right) + F = F\left(\frac{m_2}{m_1} + 1\right)$$

According to Equation 16.2, $F = v^2(m/L)$, where m/L is the mass per unit length of the wire. Combining this expression for F with the expression for P, we have

$$P = v^2(m/L)\left(\frac{m_2}{m_1}+1\right) = (352 \text{ m/s})^2(8.50\times 10^{-4} \text{ kg/m})\left(\frac{19.0 \text{ kg}}{42.0 \text{ kg}}+1\right) = \boxed{153 \text{ N}}$$

---

23. **REASONING** The mathematical form for the displacement of a wave traveling in the $-x$ direction is given by Equation 16.4: $y = A \sin\left(2\pi f t + \frac{2\pi x}{\lambda}\right)$. Using Equation 16.1 and the fact that $f = 1/T$, we obtain the following numerical values for f and λ: $f = 1/T = 1.3$ Hz, and $\lambda = v/f = 9.2$ m. Omitting units and substituting these values for f and λ into Equation 16.4 gives

$$\boxed{y = (0.37 \text{ m}) \sin (2.6\,\pi\, t + 0.22\,\pi\, x)}$$

---

27. **REASONING** According to Equation 16.2, the tension F in the string is given by $F = v^2(m/L)$. Since $v = \lambda f$ from Equation 16.1, the expression for F can be written

$$F = (\lambda f)^2 \left(\frac{m}{L}\right) \tag{1}$$

where the quantity m/L is the linear density of the string. In order to use Equation (1), we must first obtain values for f and λ; these quantities can be found by analyzing the expression for the displacement of a string particle.

SOLUTION The displacement is given by $y = (0.021 \text{ m})\sin(25t - 2.0x)$. Inspection of this equation and comparison with Equation 16.3, $y = A \sin\left(2\pi f t - \frac{2\pi x}{\lambda}\right)$, gives

$$2\pi f = 25 \text{ rad/s} \quad \text{or} \quad f = \frac{25}{2\pi} \text{ Hz}$$

and

$$\frac{2\pi}{\lambda} = 2.0 \text{ m}^{-1} \quad \text{or} \quad \lambda = \frac{2\pi}{2.0} \text{ m}$$

Substituting these values f and λ into Equation (1) gives

$$F = (\lambda f)^2\left(\frac{m}{L}\right) = \left[\left(\frac{2\pi}{2.0} \text{ m}\right)\left(\frac{25 \text{ Hz}}{2\pi}\right)\right]^2 (1.6\times 10^{-2} \text{ kg/m}) = \boxed{2.5 \text{ N}}$$

29. ***REASONING AND SOLUTION*** The speed of sound in an ideal gas is given by Equation 16.5, $v = \sqrt{\gamma kT/m}$. The ratio of the speed of sound v_2 in the container after the temperature is raised to the speed v_1 before the temperature change is

$$\frac{v_2}{v_1} = \sqrt{\frac{T_2}{T_1}}$$

Thus, the new speed is

$$v_2 = v_1 \sqrt{\frac{T_2}{T_1}} = (1220 \text{ m/s}) \sqrt{\frac{405 \text{ K}}{201 \text{ K}}} = \boxed{1730 \text{ m/s}}$$

---

33. ***REASONING*** If we treat the sample of argon atoms like an ideal monatomic gas ($\gamma = 1.67$) at 298 K, Equation 14.6 $\left(\frac{1}{2}mv_{rms}^2 = \frac{3}{2}kT\right)$ can be solved for the root-mean-square speed v_{rms} of the argon atoms. The speed of sound in argon can be found from Equation 16.5: $v = \sqrt{\gamma kT/m}$.

SOLUTION We first find the mass of an argon atom. Since the molecular mass of argon is 39.9 u, argon has a mass per mole of 39.9×10^{-3} kg/mol. Thus, the mass of a single argon atom is

$$m = \frac{39.9 \times 10^{-3} \text{ kg/mol}}{6.022 \times 10^{23} \text{ mol}^{-1}} = 6.63 \times 10^{-26} \text{ kg}$$

a. Solving Equation 14.6 for v_{rms} and substituting the data given in the problem statement, we find

$$v_{rms} = \sqrt{\frac{3kT}{m}} = \sqrt{\frac{3(1.38 \times 10^{-23} \text{ J/K})(298 \text{ K})}{6.63 \times 10^{-26} \text{ kg}}} = \boxed{431 \text{ m/s}}$$

b. The speed of sound in argon is, according to Equation 16.5,

$$v = \sqrt{\frac{\gamma kT}{m}} = \sqrt{\frac{(1.67)(1.38 \times 10^{-23} \text{ J/K})(298 \text{ K})}{6.63 \times 10^{-26} \text{ kg}}} = \boxed{322 \text{ m/s}}$$

---

39. ***REASONING*** Equation 16.7 relates the Young's modulus Y, the mass density ρ, and the speed of sound v in a long slender solid bar. According to Equation 16.7, the Young's modulus is given by $Y = \rho v^2$. The data given in the problem can be used to compute values for both ρ and v.

SOLUTION Using the values of the data given in the problem statement, we find that the speed of sound in the bar is

$$v = \frac{L}{t} = \frac{0.83 \text{ m}}{1.9 \times 10^{-4} \text{ s}} = 4.4 \times 10^3 \text{ m/s}$$

where L is the length of the rod and t is the time required for the wave to travel the length of the rod. The mass density of the bar is, from Equation 11.1, $\rho = m/V = m/(LA)$ where m, and A are, respectively, the mass and the cross-sectional area of the rod. The density of the rod is, therefore,

$$\rho = \frac{m}{LA} = \frac{2.1 \text{ kg}}{(0.83 \text{ m})(1.3 \times 10^{-4} \text{ m}^2)} = 1.9 \times 10^4 \text{ kg/m}^3$$

Using these values, we find that the bulk modulus for the rod is

$$Y = \rho v^2 = (1.9 \times 10^4 \text{ kg/m}^3)(4.4 \times 10^3 \text{ m/s})^2 = 3.7 \times 10^{11} \text{ N/m}^2$$

Comparing this value to those given in Table 10.1, we conclude that the bar is most likely made of $\boxed{\text{tungsten}}$.

---

41. *REASONING* The sound will spread out uniformly in all directions. For the purposes of counting the echoes, we will consider only the sound that travels in a straight line parallel to the ground and reflects from the vertical walls of the cliff. Let the distance between the hunter and the closer cliff be x_1 and the distance from the hunter to the further cliff be x_2.

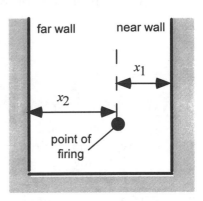

The first echo arrives at the location of the hunter after traveling a total distance $2x_1$ in a time t_1, so that, if v_s is the speed of sound, $t_1 = 2x_1/v_s$. Similarly, the second echo arrives after reflection from the far wall and in an amount of time t_2 after the firing of the gun. The quantity t_2 is related to the distance x_2 and the speed of sound v_s according to $t_2 = 2x_2/v_s$.

The time difference between the first and second echo is, therefore,

$$\Delta t = t_2 - t_1 = \frac{2}{v_s}(x_2 - x_1) \tag{1}$$

The third echo arrives in a time t_3 after the second echo. It arises from the sound of the second echo that is reflected from the closer cliff wall. Thus, $t_3 = 2x_1/v_s$, or, solving for x_1, we have

$$x_1 = \frac{v_s t_3}{2} \tag{2}$$

Combining Equations (1) and (2), we obtain

$$\Delta t = t_2 - t_1 = \frac{2}{v_s}\left(x_2 - \frac{v_s t_3}{2}\right)$$

Solving for x_2, we have

$$x_2 = \frac{v_s}{2}(\Delta t + t_3) \tag{3}$$

The distance between the cliffs can be found from $d = x_1 + x_2$, where x_1 and x_2 can be determined from Equations (2) and (3), respectively.

SOLUTION According to Equation (2), the distance x_1 is

$$x_1 = \frac{(343 \text{ m/s})(1.1 \text{ s})}{2} = 190 \text{ m}$$

According to Equation (3), the distance x_2 is

$$x_2 = \frac{(343 \text{ m/s})}{2}(1.6 \text{ s} + 1.1 \text{ s}) = 460 \text{ m}$$

Therefore, the distance between the cliffs is

$$d = x_1 + x_2 = 190 \text{ m} + 460 \text{ m} = \boxed{650 \text{ m}}$$

---

47. $\boxed{\text{WWW}}$ **REASONING** We must determine the time t for the warning to travel the vertical distance $h = 10.0$ m from the prankster to the ears of the man when he is just under the window. The desired distance above the man's ears is the distance that the balloon would travel in this time and can be found with the aid of the equations of kinematics.

SOLUTION Since sound travels with constant speed v_s, the distance h and the time t are related by $h = v_s t$. Therefore, the time t required for the warning to reach the ground is

$$t = \frac{h}{v_s} = \frac{10.0 \text{ m}}{343 \text{ m/s}} = 0.0292 \text{ s}$$

We now proceed to find the distance that the balloon travels in this time. To this end, we must find the balloon's final speed v_y after falling from rest for 10.0 m. Since the balloon is dropped from rest, we use Equation 3.6b ($v_y^2 = v_{0y}^2 + 2a_y y$) with $v_{0y} = 0$ m/s:

$$v_y = \sqrt{v_{0y}^2 + 2a_y y} = \sqrt{(0 \text{ m/s})^2 + 2(9.80 \text{ m/s}^2)(10.0 \text{ m})} = 14.0 \text{ m/s}$$

Using this result, we can find the balloon's speed 0.0292 seconds before it hits the man by solving Equation 3.3b ($v_y = v_{0y} + a_y t$) for v_{0y}:

$$v_{0y} = v_y - a_y t = 14.0 \text{ m/s} - (9.80 \text{ m/s}^2)(0.0292 \text{ s}) = 13.7 \text{ m/s}$$

Finally, we can find the desired distance y above the man's head from Equation 3.5b:

$$y = v_{0y} t + \tfrac{1}{2} a_y t^2 = (13.7 \text{ m/s})(0.0292 \text{ s}) + \tfrac{1}{2}(9.80 \text{ m/s}^2)(0.0292 \text{ s})^2 = \boxed{0.404 \text{ m}}$$

---

49. $\boxed{\text{WWW}}$ **REASONING AND SOLUTION** Since the sound radiates uniformly in all directions, at a distance r from the source, the energy of the sound wave is distributed over the area of a sphere of radius r. Therefore, according to Equation 16.9 $[I = P/(4\pi r^2)]$ with $r = 3.8$ m, the power radiated from the source is

$$P = 4\pi I r^2 = 4\pi(3.6 \times 10^{-2} \text{ W/m}^2)(3.8 \text{ m})^2 = \boxed{6.5 \text{ W}}$$

---

53. **REASONING AND SOLUTION** According to Equation 16.8, the power radiated by the speaker is $P = IA = I\pi r^2$, where r is the radius of the circular opening. Thus, the radiated power is

$$P = (17.5 \text{ W/m}^2)(\pi)(0.0950 \text{ m})^2 = 0.496 \text{ W}$$

As a percentage of the electrical power, this is

$$\frac{0.496 \text{ W}}{25.0 \text{ W}} \times 100 \% = \boxed{1.98 \%}$$

---

57. **REASONING** Since the sound is emitted from the rocket uniformly in all directions, the energy carried by the sound wave spreads out uniformly over concentric spheres of increasing radii $r_1, r_2, r_3, \ldots$ as the wave propagates. Let r_1 represent the radius when the measured intensity at the ground is I (position 1 for the rocket) and r_2 represent the radius when the measured intensity at the ground is $\frac{1}{3}I$ (position 2 of the rocket). The time required for the energy to spread out over the sphere of radius r_1 is $t_1 = r_1/v_s$, where v_s is the speed of sound. Similarly, the time required for the energy to spread out over a sphere of radius r_2 is $t_2 = r_2/v_s$. As the sound wave emitted at position 1 spreads out uniformly, the

rocket continues to accelerate upward to position 2 with acceleration a_y for a time t_{12}. Therefore, the time that elapses between the two intensity measurements is

$$t = t_2 - t_1 + t_{12} \qquad (1)$$

From Equation 3.3b, the time t_{12} is

$$t_{12} = \frac{v_2 - v_1}{a_y} \qquad (2)$$

where v_1 and v_2 are the speeds of the rocket at positions 1 and 2, respectively. Combining Equations (1) and (2), we obtain

$$t = t_2 - t_1 + \frac{v_2 - v_1}{a_y} \qquad (3)$$

The respective values of v_1 and v_2 can be found from Equation 3.6b ($v_y^2 = v_{0y}^2 + 2a_y y$) with $v_{0y} = 0$ m/s, and $y = r$. Once values for v_1 and v_2 are known, Equation (3) can be solved directly.

SOLUTION From the data given in the problem statement, $r_1 = 562$ m. We can find r_2 using the following reasoning: from the definition of intensity, $I_1 = P/(4\pi r_1^2)$ and $I_2 = P/(4\pi r_2^2)$. Since $I_1 = 3I_2$, we have

$$P/(4\pi r_1^2) = 3P/(4\pi r_2^2) \qquad \text{or} \qquad r_2 = r_1\sqrt{3} = 973 \text{ m}$$

The times t_1 and t_2 are

$$t_1 = \frac{r_1}{v_s} = \frac{562 \text{ m}}{343 \text{ m/s}} = 1.64 \text{ s}$$

and

$$t_2 = \frac{r_2}{v_s} = \frac{973 \text{ m}}{343 \text{ m/s}} = 2.84 \text{ s}$$

Then, taking up as the positive direction, we have from Equation 3.6b,

$$v_1 = \sqrt{2a_y r_1} = \sqrt{2(58.0 \text{ m/s}^2)(562 \text{ m})} = 255 \text{ m/s}$$

and

$$v_2 = \sqrt{2a_y r_2} = \sqrt{2(58.0 \text{ m/s}^2)(973 \text{ m})} = 336 \text{ m/s}$$

Substituting these values into Equation (3), we have

$$t = 2.84 \text{ s} - 1.64 \text{ s} + \frac{(336 \text{ m/s} - 255 \text{ m/s})}{58.0 \text{ m/s}^2} = \boxed{2.6 \text{ s}}$$

---

61. **REASONING AND SOLUTION**   The intensity level β in decibels (dB) is related to the sound intensity I according to Equation 16.10:

$$\beta = (10 \text{ dB}) \log \left(\frac{I}{I_0} \right)$$

where the quantity I_0 is the reference intensity. Therefore, we have

$$\beta_2 - \beta_1 = (10 \text{ dB}) \log \left(\frac{I_2}{I_0} \right) - (10 \text{ dB}) \log \left(\frac{I_1}{I_0} \right) = (10 \text{ dB}) \log \left(\frac{I_2 / I_0}{I_1 / I_0} \right) = (10 \text{ dB}) \log \left(\frac{I_2}{I_1} \right)$$

Solving for the ratio I_2 / I_1, we find

$$30.0 \text{ dB} = (10 \text{ dB}) \log \left(\frac{I_2}{I_1} \right) \qquad or \qquad \frac{I_2}{I_1} = 10^{3.0} = 1000$$

Thus, we conclude that the sound intensity $\boxed{\text{increases by a factor of } 1000}$.

---

63. **REASONING**   According to Equation 16.10, the sound intensity level β in decibels (dB) is related to the sound intensity I according to $\beta = (10 \text{ dB}) \log (I / I_0)$, where the quantity I_0 is the reference intensity. Since the sound is emitted uniformly in all directions, the intensity, or power per unit area, is given by $I = P / (4\pi r^2)$. Thus, the sound intensity at position 1 can be written as $I_1 = P / (4\pi r_1^2)$, while the sound intensity at position 2 can be written as $I_2 = P / (4\pi r_2^2)$. Therefore, the difference in the sound intensity level β_{21} between the two positions is

$$\beta_{21} = \beta_2 - \beta_1 = (10 \text{ dB}) \log \left(\frac{I_2}{I_0} \right) - (10 \text{ dB}) \log \left(\frac{I_1}{I_0} \right) = (10 \text{ dB}) \log \left(\frac{I_2 / I_0}{I_1 / I_0} \right) = (10 \text{ dB}) \log \left(\frac{I_2}{I_1} \right)$$

$$\beta_{21} = (10 \text{ dB}) \log \left[\frac{P/(4\pi r_2^2)}{P/(4\pi r_1^2)} \right] = (10 \text{ dB}) \log \left(\frac{r_1^2}{r_2^2} \right) = (10 \text{ dB}) \log \left(\frac{r_1}{r_2} \right)^2$$

$$= (20 \text{ dB}) \log \left(\frac{r_1}{r_2} \right) = (20 \text{ dB}) \log \left(\frac{r_1}{2r_1} \right) = (20 \text{ dB}) \log (1/2) = \boxed{-6.0 \text{ dB}}$$

The negative sign indicates that the sound intensity level decreases.

---

69. ***REASONING AND SOLUTION*** The sound intensity level β in decibels (dB) is related to the sound intensity I according to Equation 16.10, $\beta = (10 \text{ dB}) \log (I/I_0)$, where the quantity I_0 is the reference intensity. According to the problem statement, when the sound intensity level triples, the sound intensity also triples; therefore,

$$3\beta = (10 \text{ dB}) \log \left(\frac{3I}{I_0} \right)$$

Then,

$$3\beta - \beta = (10 \text{ dB}) \log \left(\frac{3I}{I_0} \right) - (10 \text{ dB}) \log \left(\frac{I}{I_0} \right) = (10 \text{ dB}) \log \left(\frac{3I/I_0}{I/I_0} \right)$$

Thus, $2\beta = (10 \text{ dB}) \log 3$ and

$$\beta = (5 \text{ dB}) \log 3 = \boxed{2.39 \text{ dB}}$$

---

73. ***REASONING*** Since you detect a frequency that is smaller than that emitted by the car when the car is stationary, the car must be moving away from you. Therefore, according to Equation 16.12, the frequency f_0 heard by a stationary observer from a source moving away from the observer is given by

$$f_0 = f_s \left(\frac{1}{1 + \dfrac{v_s}{v}} \right)$$

where f_s is the frequency emitted from the source when it is stationary with respect to the observer, v is the speed of sound, and v_s is the speed of the moving source. This expression can be solved for v_s.

SOLUTION We proceed to solve for v_s and substitute the data given in the problem statement. Rearrangement gives

$$\frac{v_s}{v} = \frac{f_s}{f_o} - 1$$

Solving for v_s and noting that $f_o / f_s = 0.86$ yields

$$v_s = v\left(\frac{f_s}{f_o} - 1\right) = (343 \text{ m/s})\left(\frac{1}{0.86} - 1\right) = \boxed{56 \text{ m/s}}$$

---

75. **REASONING AND SOLUTION**   In this situation, both the source and the observer are moving through the air. Therefore, the frequency of sound heard by the crew is given by Equation 16.15 using the plus signs in both the numerator and the denominator.

$$f_o = f_s \left(\frac{1 + \dfrac{v_o}{v}}{1 + \dfrac{v_s}{v}}\right) = (1550 \text{ Hz})\left(\frac{1 + \dfrac{13.0 \text{ m/s}}{343 \text{ m/s}}}{1 + \dfrac{67.0 \text{ m/s}}{343 \text{ m/s}}}\right) = \boxed{1350 \text{ Hz}}$$

---

79. **REASONING**   We can use the Doppler shift formula to determine the speed of the motorcycle. Once this speed is known, the equations of kinematics (specifically, Equation 2.9) can be used to determine the distance covered by the motorcycle.

SOLUTION   As the motorcycle accelerates, the frequency of the siren, as heard by the driver, decreases When the frequency of the siren heard by the driver is 90.0% of the value it has when the motorcycle is stationary, we have $f_o / f_s = 0.900$. Since the driver (the observer) is moving away from the source, we can use Equation 16.14 $\{f_o = f_s[1-(v_o/v)]\}$ to obtain the speed of the driver, and hence the speed of the motorcycle;

$$\frac{f_o}{f_s} = 1 - \frac{v_o}{v} = 0.900$$

Solving for v_o, we have
$$v_o = 0.100 \, v = 0.100 \, (343 \text{ m/s}) = 34.3 \text{ m/s}$$

According to Equation 2.9 ($v^2 = v_0^2 + 2ax$), since the motorcycle starts from rest so that $v_0 = 0$ m/s, the distance traveled by the motorcycle is

$$x = \frac{v^2}{2a} = \frac{(34.3 \text{ m/s})^2}{2(2.81 \text{ m/s}^2)} = \boxed{209 \text{ m}}$$

83. ***REASONING AND SOLUTION*** The speed of sound in a liquid is given by Equation 16.6, $v = \sqrt{B_{ad}/\rho}$, where B_{ad} is the adiabatic bulk modulus and ρ is the density of the liquid. Solving for B_{ad}, we obtain $B_{ad} = v^2\rho$. The ratio of the adiabatic bulk modulus of fresh water to that of ethanol at 20 °C is, therefore,

$$\frac{\left(B_{ad}\right)_{water}}{\left(B_{ad}\right)_{ethanol}} = \frac{v^2_{water}\rho_{water}}{v^2_{ethanol}\rho_{ethanol}} = \frac{(1482 \text{ m/s})^2(998 \text{ kg/m}^3)}{(1162 \text{ m/s})^2(789 \text{ kg/m}^3)} = \boxed{2.06}$$

87. ***REASONING AND SOLUTION*** The sound intensity level β in decibels (dB) is related to the sound intensity I according to Equation 16.10, $\beta = (10 \text{ dB})\log(I/I_0)$, where I_0 is the reference intensity. If β_1 and β_2 represent two different sound level intensities, then,

$$\beta_2 - \beta_1 = (10 \text{ dB})\log\left(\frac{I_2}{I_0}\right) - (10 \text{ dB})\log\left(\frac{I_1}{I_0}\right) = (10 \text{ dB})\log\left(\frac{I_2/I_0}{I_1/I_0}\right) = (10 \text{ dB})\log\left(\frac{I_2}{I_1}\right)$$

Therefore,

$$\frac{I_2}{I_1} = 10^{(\beta_2-\beta_1)/(10 \text{ dB})}$$

When the difference in sound intensity levels is $\beta_2 - \beta_1 = 1.0 \text{ dB}$, the ratio of the sound intensities is

$$\frac{I_2}{I_1} = 10^{(1 \text{ dB})/(10 \text{ dB})} = \boxed{1.3}$$

91. ***REASONING AND SOLUTION: METHOD 1*** Since the police car is matching the speed of the speeder, there is no relative motion between the police car and the listener (the speeder). Therefore, the frequency that the speeder hears when the siren is turned on is the same as if the police car and the listener are stationary. Thus, the speeder hears a frequency of $\boxed{860 \text{ Hz}}$.

REASONING AND SOLUTION: METHOD 2 In this situation, both the source (the siren) and the observer (the speeder) are moving through the air. Therefore, the frequency of the siren heard by the speeder is given by Equation 16.15 using the minus signs in both the numerator and the denominator.

$$f_o = f_s\left(\frac{1 - \dfrac{v_o}{v}}{1 - \dfrac{v_s}{v}}\right) = (860 \text{ Hz})\left(\frac{1 - \dfrac{38 \text{ m/s}}{343 \text{ m/s}}}{1 - \dfrac{38 \text{ m/s}}{343 \text{ m/s}}}\right) = \boxed{860 \text{ Hz}}$$

95. WWW   *REASONING* Using the procedures developed in Chapter 4 for using Newton's second law to analyze the motion of bodies and neglecting the weight of the wire relative to the tension in the wire lead to the following equations of motion for the two blocks:

$$\Sigma F_x = F - m_1 g \,(\sin 30.0°) = 0 \tag{1}$$

$$\Sigma F_y = F - m_2 g = 0 \tag{2}$$

where F is the tension in the wire. In Equation (1) we have taken the direction of the $+x$ axis for block 1 to be parallel to and up the incline. In Equation (2) we have taken the direction of the $+y$ axis to be upward for block 2. This set of equations consists of two equations in three unknowns, m_1, m_2, and F. Thus, a third equation is needed in order to solve for any of the unknowns. A useful third equation can be obtained by solving Equation 16.2 for F:

$$F = (m/L)v^2 \tag{3}$$

Combining Equation (3) with Equations (1) and (2) leads to

$$(m/L)v^2 - m_1 g \sin 30.0° = 0 \tag{4}$$

$$(m/L)v^2 - m_2 g = 0 \tag{5}$$

Equations (4) and (5) can be solved directly for the masses m_1 and m_2.

SOLUTION Substituting values into Equation (4), we obtain

$$m_1 = \frac{(m/L)v^2}{g \sin 30.0°} = \frac{(0.0250 \text{ kg/m})(75.0 \text{ m/s})^2}{(9.80 \text{ m/s}^2) \sin 30.0°} = \boxed{28.7 \text{ kg}}$$

Similarly, substituting values into Equation (5), we obtain

$$m_2 = \frac{(m/L)v^2}{g} = \frac{(0.0250 \text{ kg/m})(75.0 \text{ m/s})^2}{(9.80 \text{ m/s}^2)} = \boxed{14.3 \text{ kg}}$$

99. **REASONING** We must first find the intensities that correspond to the given sound intensity levels (in decibels). The total intensity is the sum of the two intensities. Once the total intensity is known, Equation 16.10 can be used to find the total sound intensity level in decibels.

SOLUTION Since, according to Equation 16.10, $\beta = (10 \text{ dB}) \log (I / I_0)$, where I_0 is the reference intensity corresponding to the threshold of hearing ($I_0 = 1.00 \times 10^{-12} \text{ W/m}^2$), it follows that $I = I_0 \ 10^{\beta/(10 \text{ dB})}$. Therefore, if $\beta_1 = 75.0$ dB and $\beta_2 = 72.0$ dB at the point in question, the corresponding intensities are

$$I_1 = I_0 \ 10^{\beta_1/(10 \text{ dB})} = (1.00 \times 10^{-12} \text{ W/m}^2) \ 10^{(75.0 \text{ dB})/(10 \text{ dB})} = 3.16 \times 10^{-5} \text{ W/m}^2$$

$$I_2 = I_0 \ 10^{\beta_2/(10 \text{ dB})} = (1.00 \times 10^{-12} \text{ W/m}^2) \ 10^{(72.0 \text{ dB})/(10 \text{ dB})} = 1.58 \times 10^{-5} \text{ W/m}^2$$

Therefore, the total intensity I_{total} at the point in question is

$$I_{\text{total}} = I_1 + I_2 = (3.16 \times 10^{-5} \text{ W/m}^2) + (1.58 \times 10^{-5} \text{ W/m}^2) = 4.74 \times 10^{-5} \text{ W/m}^2$$

and the corresponding intensity level β_{total} is

$$\beta_{\text{total}} = (10 \text{ dB}) \log \left(\frac{I_{\text{total}}}{I_0} \right) = (10 \text{ dB}) \log \left(\frac{4.74 \times 10^{-5} \text{ W/m}^2}{1.00 \times 10^{-12} \text{ W/m}^2} \right) = \boxed{76.8 \text{ dB}}$$

CHAPTER 17 | *THE PRINCIPLE OF LINEAR SUPERPOSITION AND INTERFERENCE PHENOMENA*

PROBLEMS

1. ***REASONING AND SOLUTION*** According to the principle of linear superposition, when two or more waves are present simultaneously at the same place, the resultant wave is the sum of the individual waves. Therefore, the shape of the string at the indicated times looks like the following:

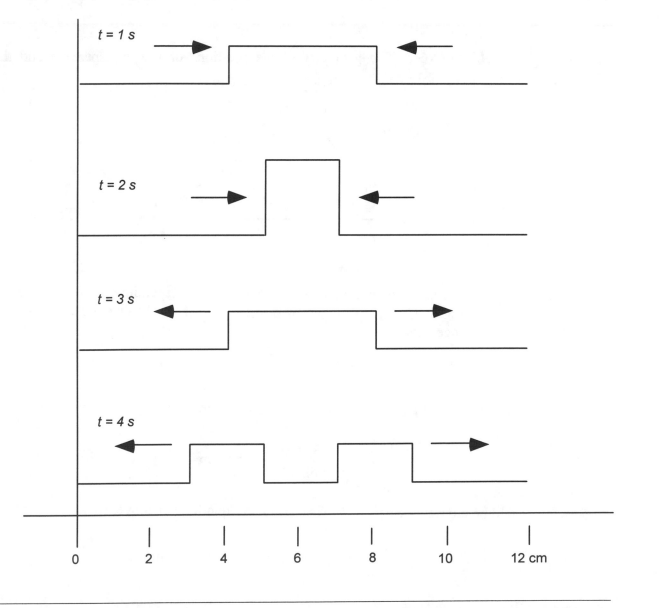

5. **REASONING** When constructive interference occurs again at point C, the path length difference is two wavelengths, or $\Delta s = 2\lambda = 3.20$ m. Therefore, we can write the expression for the path length difference as

$$s_{AC} - s_{BC} = \sqrt{s_{AB}^2 + s_{BC}^2} - s_{BC} = 3.20 \text{ m}$$

This expression can be solved for s_{AB}.

SOLUTION Solving for s_{AB}, we find that

$$s_{AB} = \sqrt{(3.20 \text{ m} + 2.40 \text{ m})^2 - (2.40 \text{ m})^2} = \boxed{5.06 \text{ m}}$$

---

7. **WWW** **REASONING** The geometry of the positions of the loudspeakers and the listener is shown in the following drawing.

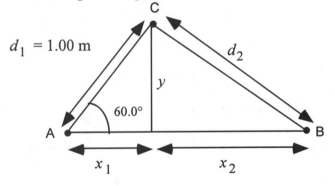

The listener at C will hear either a loud sound or no sound, depending upon whether the interference occurring at C is constructive or destructive. If the listener hears no sound, destructive interference occurs, so

$$d_2 - d_1 = \frac{n\lambda}{2} \qquad n = 1, 3, 5, \ldots \tag{1}$$

SOLUTION Since $v = \lambda f$, according to Equation 16.1, the wavelength of the tone is

$$\lambda = \frac{v}{f} = \frac{343 \text{ m/s}}{68.6 \text{ Hz}} = 5.00 \text{ m}$$

Speaker B will be closest to Speaker A when $n = 1$ in Equation (1) above, so

$$d_2 = \frac{n\lambda}{2} + d_1 = \frac{5.00 \text{ m}}{2} + 1.00 \text{ m} = 3.50 \text{ m}$$

From the figure above we have that,

$$x_1 = (1.00 \text{ m}) \cos 60.0° = 0.500 \text{ m}$$

$$y = (1.00 \text{ m}) \sin 60.0° = 0.866 \text{ m}$$

Then

$$x_2^2 + y^2 = d_2^2 = (3.50 \text{ m})^2 \qquad \text{or} \qquad x_2 = \sqrt{(3.50 \text{ m})^2 - (0.866 \text{ m})^2} = 3.39 \text{ m}$$

Therefore, the closest that speaker A can be to speaker B so that the listener hears no sound is $x_1 + x_2 = 0.500 \text{ m} + 3.39 \text{ m} = \boxed{3.89 \text{ m}}$.

---

11. **REASONING** The diffraction angle for the first minimum for a circular opening is given by Equation 17.2: $\sin \theta = 1.22 \lambda / D$, where D is the diameter of the opening.

SOLUTION
a. Using Equation 16.1, we must first find the wavelength of the 2.0-kHz tone:

$$\lambda = \frac{v}{f} = \frac{343 \text{ m/s}}{2.0 \times 10^3 \text{ Hz}} = 0.17 \text{ m}$$

The diffraction angle for a 2.0-kHz tone is, therefore,

$$\theta = \sin^{-1}\left(1.22 \times \frac{0.17 \text{ m}}{0.30 \text{ m}}\right) = \boxed{44°}$$

b. The wavelength of a 6.0-kHz tone is

$$\lambda = \frac{v}{f} = \frac{343 \text{ m/s}}{6.0 \times 10^3 \text{ Hz}} = 0.057 \text{ m}$$

Therefore, if we wish to generate a 6.0-kHz tone whose diffraction angle is as wide as that for the 2.0-kHz tone in part (a), we will need a speaker of diameter D, where

$$D = \frac{1.22 \lambda}{\sin \theta} = \frac{(1.22)(0.057 \text{ m})}{\sin 44°} = \boxed{0.10 \text{ m}}$$

---

17. **REASONING AND SOLUTION** Two ultrasonic sound waves combine and form a beat frequency that is in the range of human hearing. The frequency of one of the ultrasonic waves is 70 kHz. The beat frequency is the difference between the two sound frequencies. The smallest possible value for the ultrasonic frequency can be found by subtracting the upper limit of human hearing from the value of 70 kHz. The largest possible value for the ultrasonic frequency can be determined by adding the upper limit of human hearing to the

value of 70 kHz. We know that the frequency range of human hearing is from 20 Hz to 20 kHz.

a. The smallest possible frequency of the other ultrasonic wave is

$$f = 70 \text{ kHz} - 20 \text{ kHz} = \boxed{50 \text{ kHz}}$$

which results in a beat frequency of 70 kHz – 50 kHz = 20 kHz.

b. The largest possible frequency for the other wave is

$$f = 70 \text{ kHz} + 20 \text{ kHz} = \boxed{90 \text{ kHz}}$$

which results in a beat frequency of 90 kHz – 70 kHz = 20 kHz.

---

19. **REASONING**   The beat frequency of two sound waves is the difference between the two sound frequencies. From the graphs, we see that the period of the wave in the upper figure is 0.020 s, so its frequency is $f_1 = 1/T_1 = 1/(0.020 \text{ s}) = 5.0 \times 10^1$ Hz. The frequency of the wave in the lower figure is $f_2 = 1/(0.024 \text{ s}) = 4.2 \times 10^1$ Hz.

SOLUTION   The beat frequency of the two sound waves is

$$f_{\text{beat}} = f_1 - f_2 = 5.0 \times 10^1 \text{ Hz} - 4.2 \times 10^1 \text{ Hz} = \boxed{8 \text{ Hz}}$$

---

23. **REASONING**   For standing waves on a string that is clamped at both ends, Equations 17.3 and 16.2 indicate that the standing wave frequencies are

$$f_n = n\left(\frac{v}{2L}\right) \qquad \text{where} \qquad v = \sqrt{\frac{F}{m/L}}$$

Combining these two expressions, we have, with $n = 1$ for the fundamental frequency,

$$f_1 = \frac{1}{2L}\sqrt{\frac{F}{m/L}}$$

This expression can be used to find the ratio of the two fundamental frequencies.

SOLUTION   The ratio of the two fundamental frequencies is

$$\frac{f_{old}}{f_{new}} = \frac{\frac{1}{2L}\sqrt{\frac{F_{old}}{m/L}}}{\frac{1}{2L}\sqrt{\frac{F_{new}}{m/L}}} = \sqrt{\frac{F_{old}}{F_{new}}}$$

Since $F_{new} = 4F_{old}$, we have

$$f_{new} = f_{old}\sqrt{\frac{F_{new}}{F_{old}}} = f_{old}\sqrt{\frac{4F_{old}}{F_{old}}} = f_{old}\sqrt{4} = (55.0\ \text{Hz})(2) = \boxed{1.10\times10^2\ \text{Hz}}$$

---

27. **REASONING**   According to Equation 17.3, the fundamental ($n = 1$) frequency of a string fixed at both ends is related to the wave speed v by $f_1 = v/2L$, where L is the length of the string.   Thus, the speed of the wave is $v = 2Lf_1$.   Combining this with Equation 16.2, $v = \sqrt{F/(m/L)}$, we have, after some rearranging,

$$\frac{F}{L^2} = 4f_1^2(m/L)$$

Since the strings have the same tension and the same lengths between their fixed ends, we have

$$f_{1E}^2(m/L)_E = f_{1G}^2(m/L)_G$$

where the symbols "E" and "G" represent the E and G strings on the violin. This equation can be solved for the linear density of the G string.

SOLUTION   The linear density of the string is

$$(m/L)_G = \frac{f_{1E}^2}{f_{1G}^2}(m/L)_E = \left(\frac{f_{1E}}{f_{1G}}\right)^2(m/L)_E$$

$$= \left(\frac{659.3\ \text{Hz}}{196.0\ \text{Hz}}\right)^2(3.47\times10^{-4}\ \text{kg/m}) = \boxed{3.93\times10^{-3}\ \text{kg/m}}$$

---

31. **REASONING**   We can find the extra length that the D-tuner adds to the E-string by calculating the length of the D-string and then subtracting from it the length of the E string. For standing waves on a string that is fixed at both ends, Equation 17.3 gives the frequencies as $f_n = n(v/2L)$.   The ratio of the fundamental frequency of the D-string to that of the E-string is

$$\frac{f_{\text{D}}}{f_{\text{E}}} = \frac{v/(2L_{\text{D}})}{v/(2L_{\text{E}})} = \frac{L_{\text{E}}}{L_{\text{D}}}$$

This expression can be solved for the length L_{D} of the D-string in terms of quantities given in the problem statement.

SOLUTION The length of the D-string is

$$L_{\text{D}} = L_{\text{E}}\left(\frac{f_{\text{E}}}{f_{\text{D}}}\right) = (0.628 \text{ m})\left(\frac{41.2 \text{ Hz}}{36.7 \text{ Hz}}\right) = 0.705 \text{ m}$$

The length of the E-string is extended by the D-tuner by an amount

$$L_{\text{D}} - L_{\text{E}} = 0.705 \text{ m} - 0.628 \text{ m} = \boxed{0.077 \text{ m}}$$

---

33. **WWW** *REASONING* The natural frequencies of the cord are, according to Equation 17.3, $f_n = nv/(2L)$, where $n = 1, 2, 3, \ldots$. The speed v of the waves on the cord is, according to Equation 16.2, $v = \sqrt{F/(m/L)}$, where F is the tension in the cord. Combining these two expressions, we have

$$f_n = \frac{nv}{2L} = \frac{n}{2L}\sqrt{\frac{F}{m/L}} \qquad \text{or} \qquad \left(\frac{f_n 2L}{n}\right)^2 = \frac{F}{m/L}$$

Applying Newton's second law of motion, $\Sigma F = ma$, to the forces that act on the block and are parallel to the incline gives

$$F - Mg\sin\theta = Ma = 0 \qquad \text{or} \qquad F = Mg\sin\theta$$

where $Mg\sin\theta$ is the component of the block's weight that is parallel to the incline. Substituting this value for the tension into the equation above gives

$$\left(\frac{f_n 2L}{n}\right)^2 = \frac{Mg\sin\theta}{m/L}$$

This expression can be solved for the angle θ and evaluated at the various harmonics. The answer can be chosen from the resulting choices.

SOLUTION Solving this result for $\sin\theta$ shows that

$$\sin\theta = \frac{(m/L)}{Mg}\left(\frac{f_n 2L}{n}\right)^2 = \frac{1.20\times10^{-2}\ \text{kg}/\text{m}}{(15.0\ \text{kg})(9.80\ \text{m}/\text{s}^2)}\left[\frac{(165\ \text{Hz})2(0.600\ \text{m})}{n}\right]^2 = \frac{3.20}{n^2}$$

Thus, we have

$$\theta = \sin^{-1}\left(\frac{3.20}{n^2}\right)$$

Evaluating this for the harmonics corresponding to the range of n from $n = 2$ to $n = 4$, we have

$$\theta = \sin^{-1}\left(\frac{3.20}{2^2}\right) = 53.1°\ \text{for}\ n = 2$$

$$\theta = \sin^{-1}\left(\frac{3.20}{3^2}\right) = 20.8°\ \text{for}\ n = 3$$

$$\theta = \sin^{-1}\left(\frac{3.20}{4^2}\right) = 11.5°\ \text{for}\ n = 4$$

The angles between 15.0° and 90.0° are $\boxed{\theta = 20.8°}$ and $\boxed{\theta = 53.1°}$.

---

37. ***REASONING AND SOLUTION***   The distance between one node and an adjacent antinode is $\lambda/4$. Thus, we must first determine the wavelength of the standing wave. A tube open at only one end can develop standing waves only at the odd harmonic frequencies. Thus, for a tube of length L producing sound at the third harmonic ($n = 3$), $L = 3(\lambda/4)$. Therefore, the wavelength of the standing wave is

$$\lambda = \tfrac{4}{3}L = \tfrac{4}{3}(1.5\ \text{m}) = 2.0\ \text{m}$$

and the distance between one node and the adjacent antinode is $\lambda/4 = \boxed{0.50\ \text{m}}$.

---

39. ***REASONING AND SOLUTION***   Since both tubes have the same length, the wavelength of the fundamental frequency is the same for both gases. From Equation 16.1 we have $f_H = v_H / \lambda$ and $f_N = v_N / \lambda$, where the subscript "H" refers to helium and "N" refers to neon. Dividing the first equation by the second gives

$$f_H / f_N = v_H / v_N$$

We know from Equation 16.5 that the speed of sound in a monatomic ideal gas is given by $v = \sqrt{\gamma kT / m}$, where $\gamma = 1.40$ and m is the mass of one atom of the gas. Therefore,

$$\frac{v_H}{v_N} = \sqrt{\frac{m_N}{m_H}}$$

Substituting the relation above into $f_H/f_N = v_H/v_N$, and using the fact that the atomic mass of neon is $m_N = 20.179$ u and that of helium is $m_H = 4.0026$ u, yields

$$f_H = f_N \sqrt{\frac{m_N}{m_H}} = (268 \text{ Hz}) \sqrt{\frac{20.179 \text{ u}}{4.0026 \text{ u}}} = \boxed{602 \text{ Hz}}$$

---

43. $\boxed{\text{WWW}}$ *REASONING* According to Equation 11.4, the absolute pressure at the bottom of the mercury is $P = P_{atm} + \rho g h$, where the height h of the mercury column is the original length L_0 of the air column minus the shortened length L. Hence,

$$P = P_{atm} + \rho g (L_0 - L)$$

SOLUTION From Equation 17.5, the fundamental ($n = 1$) frequency f_1 of the shortened tube is $f_1 = 1(v/4L)$, where L is the length of the air column in the tube. Likewise, the frequency f_3 of the third ($n = 3$) harmonic in the original tube is $f_3 = 3(v/4L_0)$, where L_0 is the length of the air column in the original tube. Since $f_1 = f_3$, we have that

$$1\left(\frac{v}{4L}\right) = 3\left(\frac{v}{4L_0}\right) \quad or \quad L = \tfrac{1}{3}L_0$$

The pressure at the bottom of the mercury is
$$P = P_{atm} + \rho g\left(\tfrac{2}{3}L_0\right)$$

$$= 1.01 \times 10^5 \text{ Pa} + (13\ 600 \text{ kg/m}^3)(9.80 \text{ m/s}^2)\left(\tfrac{2}{3} \times 0.75 \text{ m}\right) = \boxed{1.68 \times 10^5 \text{ Pa}}$$

---

47. *REASONING* The tones from the two speakers will produce destructive interference with the smallest frequency when the path length difference at C is one-half of a wavelength. From Figure 17.7, we see that the path length difference is $\Delta s = s_{AC} - s_{BC}$. From Example 1, we know that $s_{AC} = 4.00$ m, and from Figure 17.7, $s_{BC} = 2.40$ m. Therefore, the path length difference is $\Delta s = 4.00 \text{ m} - 2.40 \text{ m} = 1.60 \text{ m}$.

SOLUTION Thus, destructive interference will occur when

$$\frac{\lambda}{2} = 1.60 \text{ m} \qquad or \qquad \lambda = 3.20 \text{ m}$$

This corresponds to a frequency of

$$f = \frac{v}{\lambda} = \frac{343 \text{ m/s}}{3.20 \text{ m}} = \boxed{107 \text{ Hz}}$$

---

51. **REASONING**   The fundamental frequency f_1 is given by Equation 17.3 with $n = 1$: $f_1 = v/(2L)$. Since values for f_1 and L are given in the problem statement, we can use this expression to find the speed of the waves on the cello string. Once the speed is known, the tension F in the cello string can be found by using Equation 16.2, $v = \sqrt{F/(m/L)}$.

SOLUTION   Combining Equations 17.3 and 16.2 yields

$$2Lf_1 = \sqrt{\frac{F}{m/L}}$$

Solving for F, we find that the tension in the cello string is

$$F = 4L^2 f_1^2 (m/L) = 4(0.800 \text{ m})^2 (65.4 \text{ Hz})^2 (1.56 \times 10^{-2} \text{ kg/m}) = \boxed{171 \text{ N}}$$

---

55. **WWW**   **REASONING**   The beat frequency produced when the piano and the other instrument sound the note (three octaves higher than middle C) is $f_{\text{beat}} = f - f_0$, where f is the frequency of the piano and f_0 is the frequency of the other instrument ($f_0 = 2093 \text{ Hz}$). We can find f by considering the temperature effects and the mechanical effects that occur when the temperature drops from 25.0 °C to 20.0 °C.

SOLUTION   The fundamental frequency f_0 of the wire at 25.0 °C is related to the tension F_0 in the wire by

$$f_0 = \frac{v}{2L_0} = \frac{\sqrt{F_0/(m/L)}}{2L_0} \tag{1}$$

where Equations 17.3 and 16.2 have been combined.

The amount ΔL by which the piano wire attempts to contract is (see Equation 12.2) $\Delta L = \alpha L_0 \Delta T$, where α is the coefficient of linear expansion of the wire, L_0 is its length at 25.0 °C, and ΔT is the amount by which the temperature drops. Since the wire is prevented from contracting, there must be a stretching force exerted at each end of the wire. According to Equation 10.17, the magnitude of this force is

$$\Delta F = Y\left(\frac{\Delta L}{L_0}\right)A$$

where Y is the Young's modulus of the wire, and A is its cross-sectional area. Combining this relation with Equation 12.2, we have

$$\Delta F = Y\left(\frac{\alpha L_0 \Delta T}{L_0}\right)A = \alpha(\Delta T)YA$$

Thus, the frequency f at the lower temperature is

$$f = \frac{v}{2L_0} = \frac{\sqrt{(F_0 + \Delta F)/(m/L)}}{2L_0} = \frac{\sqrt{[F_0 + \alpha(\Delta T)YA]/(m/L)}}{2L_0} \qquad (2)$$

Using Equations (1) and (2), we find that the frequency f is

$$f = f_0 \frac{\sqrt{[F_0 + \alpha(\Delta T)YA]/(m/L)}}{\sqrt{F_0/(m/L)}} = f_0\sqrt{\frac{F_0 + \alpha(\Delta T)YA}{F_0}}$$

$$f = (2093 \text{ Hz})\sqrt{\frac{818.0 \text{ N} + (12 \times 10^{-6}/\text{C}°)(5.0 \text{ C}°)(2.0 \times 10^{11} \text{ N}/\text{m}^2)(7.85 \times 10^{-7} \text{ m}^2)}{818.0 \text{ N}}}$$

$$= 2105 \text{ Hz}$$

Therefore, the beat frequency is $2105 \text{ Hz} - 2093 \text{ Hz} = \boxed{12 \text{ Hz}}$.

CHAPTER 18 | *ELECTRIC FORCES AND ELECTRIC FIELDS*

1. ***REASONING AND SOLUTION*** The charge on a single electron is -1.60×10^{-19} C. In order to give a neutral silver dollar a charge of $+2.4\ \mu$C, we must remove an amount of negative charge equal to $-2.4\ \mu$C. This corresponds to

$$\left(-2.4 \times 10^{-6}\ \text{C}\right)\left(\frac{1\ \text{electron}}{-1.60 \times 10^{-19}\ \text{C}}\right) = \boxed{1.5 \times 10^{13}\ \text{electrons}}$$

5. ***REASONING*** Identical conducting spheres equalize their charge upon touching. When spheres A and B touch, an amount of charge $+q$, flows from A and instantaneously neutralizes the $-q$ charge on B leaving B momentarily neutral. Then, the remaining amount of charge, equal to $+4q$, is equally split between A and B, leaving A and B each with equal amounts of charge $+2q$. Sphere C is initially neutral, so when A and C touch, the $+2q$ on A splits equally to give $+q$ on A and $+q$ on C. When B and C touch, the $+2q$ on B and the $+q$ on C combine to give a total charge of $+3q$, which is then equally divided between the spheres B and C; thus, B and C are each left with an amount of charge $+1.5q$.

SOLUTION Taking note of the initial values given in the problem statement, and summarizing the final results determined in the *Reasoning* above, we conclude the following:

a. Sphere C ends up with an amount of charge equal to $\boxed{+1.5q}$.

b. The charges on the three spheres before they were touched, are, according to the problem statement, $+5q$ on sphere A, $-q$ on sphere B, and zero charge on sphere C. Thus, the total charge on the spheres is $+5q - q + 0 = \boxed{+4q}$.

c. The charges on the spheres after they are touched are $+q$ on sphere A, $+1.5q$ on sphere B, and $+1.5q$ on sphere C. Thus, the total charge on the spheres is $+q + 1.5q + 1.5q = \boxed{+4q}$.

7. $\boxed{\text{WWW}}$ ***REASONING*** Initially, the two spheres are neutral. Since negative charge is removed from the sphere which loses electrons, it then carries a net positive charge. Furthermore, the neutral sphere to which the electrons are added is then negatively charged. Once the charge is transferred, there exists an electrostatic force on each of the two spheres, the magnitude of which is given by Coulomb's law (Equation 18.1), $F = kq_1q_2 / r^2$.

SOLUTION

a. Since each electron carries a charge of -1.60×10^{-19} C, the amount of negative charge removed from the first sphere is

$$\left(3.0 \times 10^{13} \text{ electrons}\right) \left(\frac{1.60 \times 10^{-19} \text{ C}}{1 \text{ electron}}\right) = 4.8 \times 10^{-6} \text{ C}$$

Thus, the first sphere carries a charge $+4.8 \times 10^{-6}$ C, while the second sphere carries a charge -4.8×10^{-6} C. The magnitude of the electrostatic force that acts on each sphere is, therefore,

$$F = \frac{kq_1 q_2}{r^2} = \frac{(8.99 \times 10^9 \text{ N} \cdot \text{m}^2/\text{C}^2)(4.8 \times 10^{-6} \text{ C})^2}{(0.50 \text{ m})^2} = \boxed{0.83 \text{ N}}$$

b. Since the spheres carry charges of opposite sign, the force is $\boxed{\text{attractive}}$.

---

11. ***REASONING AND SOLUTION*** The net electrostatic force on charge 3 at $x = +3.0$ m is the vector sum of the forces on charge 3 due to the other two charges, 1 and 2. According to Coulomb's law (Equation 18.1), the magnitude of the force on charge 3 due to charge 1 is

$$F_{13} = \frac{kq_1 q_3}{r_{13}^2}$$

where the distance between charges 1 and 3 is r_{13}.

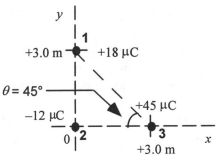

Figure 1

According to the Pythagorean theorem, $r_{13}^2 = x^2 + y^2$. Therefore,

$$F_{13} = \frac{(8.99 \times 10^9 \text{ N} \cdot \text{m}^2/\text{C}^2)(18 \times 10^{-6} \text{ C})(45 \times 10^{-6} \text{ C})}{(3.0 \text{ m})^2 + (3.0 \text{ m})^2} = 0.405 \text{ N}$$

Charges 1 and 3 are equidistant from the origin, so that $\theta = 45°$ (see Figure 1). Since charges 1 and 3 are both positive, the force on charge 3 due to charge 1 is repulsive and along the line that connects them, as shown in Figure 2. The components of F_{13} are:

$$F_{13x} = F_{13} \cos 45° = 0.286 \text{ N} \qquad \text{and} \qquad F_{13y} = -F_{13} \sin 45° = -0.286 \text{ N}$$

The second force on charge 3 is the attractive force (opposite signs) due to its interaction with charge 2 located at the origin. The magnitude of the force on charge 3 due to charge 2 is, according to Coulomb's law,

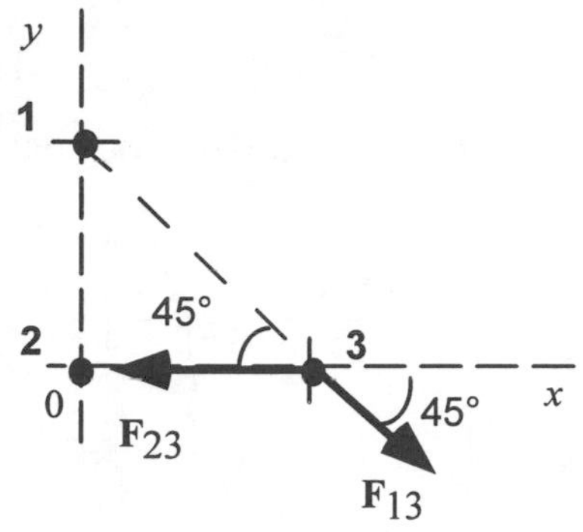

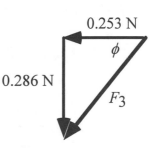

Figure 2

$$F_{23} = \frac{kq_2q_3}{r_{23}^2} = \frac{kq_2q_3}{x^2}$$

$$= \frac{(8.99 \times 10^9 \, \text{N} \cdot \text{m}^2 / \text{C}^2)(12 \times 10^{-6} \, \text{C})(45 \times 10^{-6} \, \text{C})}{(3.0 \, \text{m})^2}$$

$$= 0.539 \, \text{N}$$

Since charges 2 and 3 have opposite signs, they attract each other, and charge 3 experiences a force to the left as shown in Figure 2. Taking up and to the right as the positive directions, we have

$$F_{3x} = F_{13x} + F_{23x} = +0.286 \, \text{N} - 0.539 \, \text{N} = -0.253 \, \text{N}$$

$$F_{3y} = F_{13y} = -0.286 \, \text{N}$$

Using the Pythagorean theorem, we find the magnitude of F_3 to be

$$F_3 = \sqrt{F_{3x}^2 + F_{3y}^2} = \sqrt{(-0.253 \, \text{N})^2 + (-0.286 \, \text{N})^2} = \boxed{0.38 \, \text{N}}$$

The direction of F_3 relative to the $-x$ axis is specified by the angle ϕ, where

Figure 3

$$\phi = \tan^{-1}\left(\frac{0.286 \, \text{N}}{0.253 \, \text{N}}\right) = \boxed{49° \text{ below the } -x \text{ axis}}$$

---

15. $\boxed{\textbf{WWW}}$ ***REASONING***   Each particle will experience an electric force due to the presence of the other charge. According to Coulomb's law (Equation 18.1), the magnitude of the force felt by each particle can be calculated from $F = kq_1q_2 / r^2$, where q_1 and q_2 are the respective charges on particles 1 and 2 and r is the distance between them. According to Newton's second law, the magnitude of the force experienced by each particle is given by $F = ma$, where a is the acceleration of the particle.

SOLUTION
a. Since the two particles have identical positive charges, $q_1 = q_2 = q$, and we have, using the data for particle 1,

$$\frac{kq^2}{r^2} = m_1 a_1$$

Solving for q, we find that

$$q = \sqrt{\frac{m_1 a_1 r^2}{k}} = \sqrt{\frac{(6.00 \times 10^{-6} \text{ kg}) (4.60 \times 10^3 \text{ m/s}^2) (2.60 \times 10^{-2} \text{ m})^2}{8.99 \times 10^9 \text{ N} \cdot \text{m}^2/\text{C}^2}} = \boxed{4.56 \times 10^{-8} \text{ C}}$$

b. Since each particle experiences a force of the same magnitude (From Newton's third law), we can write $F_1 = F_2$, or $m_1 a_1 = m_2 a_2$. Solving this expression for the mass m_2 of particle 2, we have

$$m_2 = \frac{m_1 a_1}{a_2} = \frac{(6.00 \times 10^{-6} \text{ kg})(4.60 \times 10^3 \text{ m/s}^2)}{8.50 \times 10^3 \text{ m/s}^2} = \boxed{3.25 \times 10^{-6} \text{ kg}}$$

---

19. ***REASONING AND SOLUTION*** Before the spheres have been charged, they exert no forces on each other. After the spheres are charged, each sphere experiences a repulsive force F due to the charge on the other sphere, according to Coulomb's law (Equation 18.1). Therefore, since each sphere has the same charge, the magnitude F of this force is

$$F = \frac{kq_1 q_2}{r^2} = \frac{(8.99 \times 10^9 \text{ N} \cdot \text{m}^2/\text{C}^2)(1.60 \times 10^{-6} \text{ C})^2}{(0.100 \text{ m})^2} = 2.30 \text{ N}$$

The repulsive force on each sphere compresses the spring to which it is attached. The magnitude of this repulsive force is related to the amount of compression by Equation 10.1: $F = kx$. Therefore, solving for k, we find

$$k = \frac{F}{x} = \frac{2.30 \text{ N}}{0.0250 \text{ m}} = \boxed{92.0 \text{ N/m}}$$

---

23. ***REASONING*** The charged insulator experiences an electric force due to the presence of the charged sphere shown in the drawing in the text. The forces acting on the insulator are the downward force of gravity (i.e., its weight, $W = mg$), the electrostatic force $F = kq_1 q_2 / r^2$ (see Coulomb's law, Equation 18.1) pulling to the right, and the tension T in the wire pulling up and to the left at an angle θ with respect to the vertical as shown in the drawing in the problem statement. We can analyze the forces to determine the desired quantities θ and T.

SOLUTION

a. We can see from the diagram that

$$T_x = F \qquad \text{which gives} \qquad T\sin\theta = kq_1q_2/r^2$$

and

$$T_y = W \qquad \text{which gives} \qquad T\cos\theta = mg$$

Dividing the first equation by the second yields $(T\sin\theta)/(T\cos\theta) = \tan\theta = kq_1q_2/(mgr^2)$. Solving for θ, we find that

$$\theta = \tan^{-1}\left(\frac{kq_1q_2}{mgr^2}\right)$$

$$= \tan^{-1}\left[\frac{(8.99\times10^9 \text{ N}\cdot\text{m}^2/\text{C}^2)(0.600\times10^{-6} \text{ C})(0.900\times10^{-6} \text{ C})}{(8.00\times10^{-2} \text{ kg})(9.80 \text{ m/s}^2)(0.150 \text{ m})^2}\right] = \boxed{15.4°}$$

b. Since $T\cos\theta = mg$, the tension can be obtained as follows:

$$T = \frac{mg}{\cos\theta} = \frac{(8.00\times10^{-2} \text{ kg})(9.80 \text{ m/s}^2)}{\cos 15.4°} = \boxed{0.813 \text{ N}}$$

---

27. **REASONING AND SOLUTION** The force **F** exerted on a charge q_0 placed in an electric field **E** can be determined from Equation 18.2, the definition of electric field ($\mathbf{E} = \mathbf{F}/q_0$). Writing this in terms of magnitudes, and taking due east as the positive direction, we have, solving for F,

$$F = q_0 E = (+3.0\times10^{-5} \text{ C})(+15\,000 \text{ N/C}) = \boxed{+0.45 \text{ N}}$$

where the plus sign indicates that the force on the charge is directed $\boxed{\text{due east}}$.

---

31. **REASONING** Before the 3.0-μC point charge q is introduced into the region, the region contains a uniform electric field **E** of magnitude 1.6×10^4 N/C. After the 3.0-μC charge is introduced into the region, the net electric field changes. In addition to the uniform electric field **E**, the region will also contain the electric field $\mathbf{E}_q$ due to the point charge q. The field at any point in the region is the vector sum of **E** and $\mathbf{E}_q$. The field $\mathbf{E}_q$ is radial as discussed in the text, and its magnitude at any distance r from the charge q is given by Equation 18.3, $E_q = kq/r^2$. There will be one point P in the region where the net electric field $\mathbf{E}_{net}$ is zero. This point is located where the field **E** has the same magnitude and

points in the direction opposite to the field $\mathbf{E}_q$. We will use this reasoning to find the distance r_0 from the charge q to the point P.

SOLUTION   Let us assume that the field $\mathbf{E}$ points to the right and that the charge q is *negative* (the problem is done the same way if q is positive, although then the relative positions of P and q will be reversed). Since q is negative, its electric field is radially *inward* (i.e., toward q); therefore, in order for the field $\mathbf{E}_q$ to point in the opposite direction to $\mathbf{E}$, the charge q will have to be to the left of the point P where $\mathbf{E}_{net}$ is zero, as shown in the following drawing.

From Equation 18.3, $E_q = kq/r_0^2$, and solving for the distance r_0, we have $r_0 = \sqrt{kq/E_q}$. Since the magnitude E_q must be equal to the magnitude of E at the point P, we have

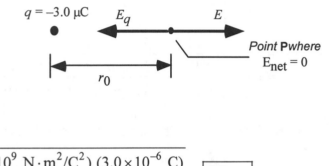

$$r_0 = \sqrt{\frac{kq}{E}} = \sqrt{\frac{(8.99 \times 10^9 \text{ N} \cdot \text{m}^2/\text{C}^2)\ (3.0 \times 10^{-6} \text{ C})}{1.6 \times 10^4 \text{ N/C}}} = \boxed{1.3 \text{ m}}$$

---

35. ***REASONING***   Since the charged droplet (charge = q) is suspended motionless in the electric field $\mathbf{E}$, the net force on the droplet must be zero. There are two forces that act on the droplet, the force of gravity $\mathbf{W} = m\mathbf{g}$, and the electric force $\mathbf{F} = q\mathbf{E}$ due to the electric field. Since the net force on the droplet is zero, we conclude that $mg = qE$. We can use this reasoning to determine the sign and the magnitude of the charge on the droplet.

SOLUTION
a. Since the net force on the droplet is zero, and the weight W points downward, the electric force $F = qE$ must point upward. Since the electric field points upward, the excess charge on the droplet must be $\boxed{\text{positive}}$ in order for the force F to point upward.

b. The excess charge on the droplet is, using the expression $mg = qE$,

$$q = \frac{mg}{E} = \frac{(3.50 \times 10^{-9} \text{ kg})(9.80 \text{ m/s}^2)}{8480 \text{ N/C}} = 4.04 \times 10^{-12} \text{ C}$$

The charge on a proton is 1.60×10^{-19} C, so the excess number of protons is

$$\left(4.04 \times 10^{-12} \text{ C}\right)\left(\frac{1 \text{ proton}}{1.60 \times 10^{-19} \text{ C}}\right) = \boxed{2.53 \times 10^7 \text{ protons}}$$

---

39. **WWW**   *REASONING*   The figure shows the arrangement of the three charges. Let $\mathbf{E}_q$ represent the electric field at the empty corner due to the $-q$ charge. Furthermore, let $\mathbf{E}_1$ and $\mathbf{E}_2$ be the electric fields at the empty corner due to charges $+q_1$ and $+q_2$, respectively.

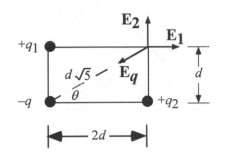

According to the Pythagorean theorem, the distance from the charge $-q$ to the empty corner along the diagonal is given by $\sqrt{(2d)^2 + d^2} = \sqrt{5d^2} = d\sqrt{5}$. The magnitude of each electric field is given by Equation 18.3, $E = kq/r^2$. Thus, the magnitudes of each of the electric fields at the empty corner are given as follows: $E_q = kq/r^2 = kq/\left(d\sqrt{5}\right)^2 = kq/(5d^2)$, since the length of the diagonal is $d\sqrt{5}$; $E_1 = kq_1/(4d^2)$; and $E_2 = kq_2/d^2$. The angle θ that the diagonal makes with the horizontal is $\theta = \tan^{-1}(d/2d) = 26.57°$. Since the net electric field E_{net} at the empty corner is zero, the horizontal component of the net field must be zero, and we have

$$E_1 - E_q \cos 26.57° = 0 \qquad \text{or} \qquad kq_1/\left(4d^2\right) - kq\left(\cos 26.57°\right)/\left(5d^2\right) = 0$$

Similarly, the vertical component of the net field must be zero, and we have

$$E_2 - E_q \sin 26.57° = 0 \qquad \text{or} \qquad kq_2/d^2 - kq\left(\sin 26.57°\right)/\left(5d^2\right) = 0$$

These last two expressions can be solved for the charges q_1 and q_2.

SOLUTION   Solving the last two expressions for the charges q_1 and q_2, we have

$$q_1 = \frac{4}{5}q \cos 26.57° = \boxed{0.716 \, q}$$

$$q_2 = \frac{1}{5}q \sin 26.57° = \boxed{0.0895 \, q}$$

43. ***REASONING*** The net electric field at point P in Figure 1 is the vector sum of the fields $\mathbf{E}_+$ and $\mathbf{E}_-$, which are due, respectively, to the charges $+q$ and $-q$. These fields are shown in Figure 2.

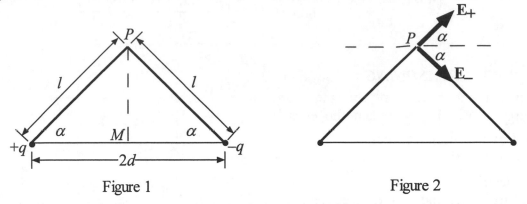

Figure 1   Figure 2

According to Equation 18.3, the magnitudes of the fields E_+ and E_- are the same, since the triangle is an isosceles triangle with equal sides of length ℓ. Therefore, $E_+ = E_- = kq/\ell^2$. The vertical components of these two fields cancel, while the horizontal components reinforce, leading to a total field at point P that is horizontal and has a magnitude of

$$E_P = E_+ \cos \alpha + E_- \cos \alpha = 2\left(\frac{kq}{\ell^2}\right) \cos \alpha$$

At point M in Figure 1, both $\mathbf{E}_+$ and $\mathbf{E}_-$ are horizontal and point to the right. Again using Equation 18.3, we find

$$E_M = E_+ + E_- = \frac{kq}{\ell^2} + \frac{kq}{\ell^2} = \frac{2kq}{\ell^2}$$

Since $E_M/E_P = 9.0$, we have

$$\frac{E_M}{E_P} = \frac{2kq/d^2}{2kq(\cos \alpha)/\ell^2} = \frac{1}{(\cos \alpha)d^2/\ell^2} = 9.0$$

But from Figure 1, we can see that $d/\ell = \cos \alpha$. Thus, it follows that

$$\frac{1}{\cos^3 \alpha} = 9.0 \qquad \text{or} \qquad \cos \alpha = \sqrt[3]{1/9.0} = 0.48$$

The value for α is, then, $\alpha = \cos^{-1}(0.48) = \boxed{61°}$.

47. **REASONING** As discussed in Section 18.9, the electric flux Φ_E through a surface is equal to the component of the electric field that is normal to the surface multiplied by the area of the surface, $\Phi_E = E_\perp A$, where $E_\perp$ is the component of **E** that is normal to the surface of area A. We can use this expression and the figure in the text to determine the flux through the two surfaces.

SOLUTION
a. The flux through surface 1 is

$$\left(\Phi_E\right)_1 = (E\cos 35°)A_1 = (250 \text{ N/C})(\cos 35°)(1.7 \text{ m}^2) = \boxed{350 \text{ N}\cdot\text{m}^2/\text{C}}$$

b. Similarly, the flux through surface 2 is

$$\left(\Phi_E\right)_2 = (E\cos 55°)A_2 = (250 \text{ N/C})(\cos 55°)(3.2 \text{ m}^2) = \boxed{460 \text{ N}\cdot\text{m}^2/\text{C}}$$

---

51. **REASONING** The electric flux through each face of the cube is given by $\Phi_E = (E\cos\phi)A$ (see Section 18.9) where E is the magnitude of the electric field at the face, A is the area of the face, and ϕ is the angle between the electric field and the outward normal of that face. We can use this expression to calculate the electric flux Φ_E through each of the six faces of the cube.

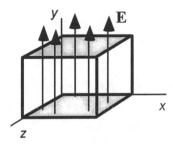

SOLUTION
a. On the bottom face of the cube, the outward normal points parallel to the $-y$ axis, in the opposite direction to the electric field, and $\phi = 180°$. Therefore,

$$\left(\Phi_E\right)_{\text{bottom}} = (1500 \text{ N/C})(\cos 180°)(0.20 \text{ m})^2 = \boxed{-6.0\times 10^1 \text{ N}\cdot\text{m}^2/\text{C}}$$

On the top face of the cube, the outward normal points parallel to the $+y$ axis, and $\phi = 0.0°$. The electric flux is, therefore,

$$(\Phi_E)_{\text{top}} = (1500 \text{ N/C})(\cos 0.0°)(0.20 \text{ m})^2 = \boxed{+6.0\times 10^1 \text{ N}\cdot\text{m}^2/\text{C}}$$

On each of the other four faces, the outward normals are perpendicular to the direction of the electric field, so $\phi = 90°$. So for each of the four side faces,

$$(F_E)_{\text{sides}} = (1500 \text{ N/C})(\cos 90°)(0.20 \text{ m})^2 = \boxed{0 \text{ N}\cdot\text{m}^2/\text{C}}$$

b. The total flux through the cube is

$$(\Phi_E)_{total} = (\Phi_E)_{top} + (\Phi_E)_{bottom} + (\Phi_E)_{side\ 1} + (\Phi_E)_{side\ 2} + (\Phi_E)_{side\ 3} + (\Phi_E)_{side\ 4}$$

Therefore,

$$(\Phi_E)_{total} = (+6.0 \times 10^1\,\text{N m}^2/\text{C}) + (-6.0 \times 10^1\,\text{N m}^2/\text{C}) + 0 + 0 + 0 + 0 = \boxed{0\ \text{N} \cdot \text{m}^2/\text{C}}$$

---

55. ***REASONING AND SOLUTION***   The magnitude of the force of attraction between the charges is given by Coulomb's law (Equation 18.1): $F = kq_1q_2/r^2$, where q_1 and q_2 are the magnitudes of the charges and r is the separation of the charges.   Let F_A and F_B represent the magnitudes of the forces between the charges when the separations are r_A and $r_B = r_A/9$, respectively.   Then

$$\frac{F_B}{F_A} = \frac{kq_1q_2/r_B^2}{kq_1q_2/r_A^2} = \left(\frac{r_A}{r_B}\right)^2 = \left(\frac{r_A}{r_A/9}\right)^2 = (9)^2 = 81$$

Therefore, we can conclude that $F_B = 81F_A = (81)(1.5\ \text{N}) = \boxed{120\ \text{N}}$.

---

59. $\boxed{\text{WWW}}$ ***REASONING***   Two forces act on the charged ball (charge q);   they are the downward force of gravity $m\mathbf{g}$ and the electric force $\mathbf{F}$ due to the presence of the charge q in the electric field $\mathbf{E}$.   In order for the ball to float, these two forces must be equal in magnitude and opposite in direction, so that the net force on the ball is zero (Newton's second law).   Therefore, $\mathbf{F}$ must point upward, which we will take as the positive direction.   According to Equation 18.2, $\mathbf{F} = q\mathbf{E}$.   Since the charge q is negative, the electric field $\mathbf{E}$ must point downward, as the product $q\mathbf{E}$ in the expression $\mathbf{F} = q\mathbf{E}$ must be positive, since the force $\mathbf{F}$ points upward.   The magnitudes of the two forces must be equal, so that $mg = qE$.   This expression can be solved for E.

SOLUTION   The magnitude of the electric field $\mathbf{E}$ is

$$E = \frac{mg}{q} = \frac{(0.012\ \text{kg})(9.80\ \text{m/s}^2)}{18 \times 10^{-6}\ \text{C}} = \boxed{6.5 \times 10^3\ \text{N/C}}$$

As discussed in the reasoning, this electric field points $\boxed{\text{downward}}$.

---

61.   ***REASONING AND SOLUTION***
a. Since the gravitational force between the spheres is one of attraction and the electrostatic force must balance it, the electric force must be one of repulsion.   Therefore, the charges must have $\boxed{\text{the same algebraic signs, both positive or both negative}}$.

b. There are two forces that act on each sphere; they are the gravitational attraction F_G of one sphere for the other, and the repulsive electric force F_E of one sphere on the other. From the problem statement, we know that these two forces balance each other, so that $F_G = F_E$. The magnitude of F_G is given by Newton's law of gravitation (Equation 4.3: $F_G = Gm_1m_2/r^2$), while the magnitude of F_E is given by Coulomb's law (Equation 18.1: $F_E = kq_1q_2/r^2$). Therefore, we have $Gm_1m_2/r^2 = kq_1q_2/r^2$, and since the spheres have the same mass and carry charges of the same magnitude, $Gm^2/r^2 = kq^2/r^2$. Solving for q, we find

$$q = m\sqrt{\frac{G}{k}} = (2.0\times10^{-6} \text{ kg})\sqrt{\frac{6.67\times10^{-11} \text{ N}\cdot\text{m}^2/\text{kg}^2}{8.99\times10^9 \text{ N}\cdot\text{m}^2/\text{C}^2}} = \boxed{1.7\times10^{-16} \text{ C}}$$

---

65. **REASONING**   The charge at A experiences a force $\mathbf{F_A}$ given by $\mathbf{F_A} = q_A\mathbf{E_A}$, where $\mathbf{E_A}$ is the net electric field at A. The net electric field $\mathbf{E_A}$ is the vector sum of the fields due to the presence of the other two charges. In order for the net force on the charge at corner A to point along the vertical, the fourth charge that is placed at the empty corner, must produce an electric field $\mathbf{E_q}$ at A with an x-component that is equal in magnitude and opposite in direction to the x-component of $\mathbf{E_A}$. The x-component of $\mathbf{E_A}$ is due solely to the $+3.0 \text{ }\mu\text{C}$ charge in the lower right corner. According to Equation 18.3, then, $E_{Ax} = kq_{\text{right}}/(16d^2)$, where q_{right} is the charge in the lower right corner. Since q_{right} is positive, E_{Ax} points to the left. The x component of the field $\mathbf{E_q}$ must, therefore, be equal to $E_{qx} = kq_{\text{right}}/(16d^2)$ and point to the right. We will now use this expression to solve for q.

SOLUTION   According to Equation 18.3, the electric field $\mathbf{E_q}$ has a magnitude that is given by

$$E_q = kq/r^2 = kq/(d^2 + 16d^2) = kq/(17d^2)$$

Furthermore, this field must point along the diagonal from the charge q_A toward the empty corner, as shown in the following drawing. Therefore, the charge placed at the empty corner must be negative.

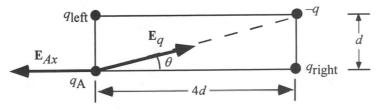

The angle θ is given by $\theta = \tan^{-1}[d/(4d)] = \tan^{-1}(0.250) = 14.0°$. Using the arguments developed in the reasoning, we have $E_{qx} = kq_{\text{right}}/(16d^2)$, or

$$E_{qx} = E_q \cos\theta = \frac{kq}{17d^2} \cos\theta = \frac{kq_{\text{right}}}{16d^2}$$

Solving for the charge magnitude q, we obtain

$$q = \frac{q_{\text{right}}}{\cos\theta}\left(\frac{17}{16}\right) = \left(\frac{3.0 \times 10^{-6}\ \text{C}}{\cos 14.0°}\right)\left(\frac{17}{16}\right) = 3.3 \times 10^{-6}\ \text{C}$$

Thus, the charge at the empty corner is $\boxed{-3.3 \times 10^{-6}\ \text{C}}$.

CHAPTER 19 | *ELECTRIC POTENTIAL ENERGY AND THE ELECTRIC POTENTIAL*

PROBLEMS

---

1. ***REASONING AND SOLUTION*** Combining Equations 19.1 and 19.3, we have

$$W_{AB} = \text{EPE}_A - \text{EPE}_B = q_0(V_A - V_B) = (+1.6 \times 10^{-19} \text{ C})(0.070 \text{ V}) = \boxed{1.1 \times 10^{-20} \text{ J}}$$

---

5. ***REASONING*** The only force acting on the moving electron is the conservative electric force. Therefore, the total energy of the electron (the sum of the kinetic energy KE and the electric potential energy EPE) remains constant throughout the trajectory of the electron. Let the subscripts A and B refer to the initial and final positions, respectively, of the electron. Then,

$$\tfrac{1}{2}mv_A^2 + \text{EPE}_A = \tfrac{1}{2}mv_B^2 + \text{EPE}_B$$

Solving for v_B gives

$$v_B = \sqrt{v_A^2 - \frac{2}{m}(\text{EPE}_B - \text{EPE}_A)}$$

Since the electron starts from rest, $v_A = 0$. The difference in potential energies is related to the difference in potentials by Equation 19.4, $\text{EPE}_B - \text{EPE}_A = q(V_B - V_A)$.

SOLUTION The speed v_B of the electron just before it reaches the screen is

$$v_B = \sqrt{-\frac{2q}{m}(V_B - V_A)} = \sqrt{-\frac{2(-1.6 \times 10^{-19} \text{ C})}{9.11 \times 10^{-31} \text{ kg}}(25\ 000 \text{ V})} = \boxed{9.4 \times 10^7 \text{ m/s}}$$

---

9. $\boxed{\text{WWW}}$ ***REASONING*** The only force acting on the moving charge is the conservative electric force. Therefore, the total energy of the charge remains constant. Applying the principle of conservation of energy between locations A and B, we obtain

$$\tfrac{1}{2}mv_A^2 + \text{EPE}_A = \tfrac{1}{2}mv_B^2 + \text{EPE}_B$$

Since the charged particle starts from rest, $v_A = 0$. The difference in potential energies is related to the difference in potentials by Equation 19.4, $\text{EPE}_B - \text{EPE}_A = q(V_B - V_A)$. Thus, we have

$$q(V_A - V_B) = \tfrac{1}{2}mv_B^2 \tag{1}$$

Similarly, applying the conservation of energy between locations C and B gives

$$q(V_C - V_B) = \tfrac{1}{2}m(2v_B)^2 \tag{2}$$

Dividing Equation (1) by Equation (2) yields

$$\frac{V_A - V_B}{V_C - V_B} = \frac{1}{4}$$

This expression can be solved for V_B.

SOLUTION Solving for V_B, we find that

$$V_B = \frac{4V_A - V_C}{3} = \frac{4(452 \text{ V}) - 791 \text{ V}}{3} = \boxed{339 \text{ V}}$$

---

11. ***REASONING AND SOLUTION*** The electric potential V at a distance r from a point charge q is given by Equation 19.6, $V = kq/r$. Solving this expression for q, we find that

$$q = \frac{rV}{k} = \frac{(0.25 \text{ m})(+130 \text{ V})}{9.0 \times 10^9 \text{ N} \cdot \text{m}^2 / \text{C}^2} = \boxed{+3.6 \times 10^{-9} \text{ C}}$$

---

15. $\boxed{\text{WWW}}$ ***REASONING*** Initially, suppose that one charge is at C and the other charge is held fixed at B. The charge at C is then moved to position A. According to Equation 19.4, the work W_{CA} done by the electric force as the charge moves from C to A is $W_{CA} = q(V_C - V_A)$, where, from Equation 19.6, $V_C = kq/d$ and $V_A = kq/r$. From the figure at the right we see that $d = \sqrt{r^2 + r^2} = \sqrt{2}r$. Therefore, we find that

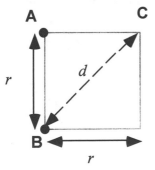

$$W_{CA} = q\left(\frac{kq}{\sqrt{2}r} - \frac{kq}{r}\right) = \frac{kq^2}{r}\left(\frac{1}{\sqrt{2}} - 1\right)$$

SOLUTION Substituting values, we obtain

$$W_{CA} = \frac{(8.99 \times 10^9 \text{ N} \cdot \text{m}^2 / \text{C}^2)(3.0 \times 10^{-6} \text{ C})^2}{0.500 \text{ m}}\left(\frac{1}{\sqrt{2}} - 1\right) = \boxed{-4.7 \times 10^{-2} \text{ J}}$$

19. **REASONING** The only force acting on the moving charge is the conservative electric force. Therefore, the sum of the kinetic energy KE and the electric potential energy EPE is the same at points A and B:

$$\tfrac{1}{2}mv_A^2 + \text{EPE}_A = \tfrac{1}{2}mv_B^2 + \text{EPE}_B$$

Since the particle comes to rest at B, $v_B = 0$. Combining Equations 19.3 and 19.6, we have

$$\text{EPE}_A = qV_A = q\left(\frac{kq_1}{d}\right)$$

and

$$\text{EPE}_B = qV_B = q\left(\frac{kq_1}{r}\right)$$

where d is the initial distance between the fixed charge and the moving charged particle, and r is the distance between the charged particles after the moving charge has stopped. Therefore, the expression for the conservation of energy becomes

$$\tfrac{1}{2}mv_A^2 + \frac{kqq_1}{d} = \frac{kqq_1}{r}$$

This expression can be solved for r. Once r is known, the distance that the charged particle moves can be determined.

SOLUTION Solving the expression above for r gives

$$r = \frac{kqq_1}{\tfrac{1}{2}mv_A^2 + \dfrac{kqq_1}{d}}$$

$$= \frac{(8.99\times10^9\ \text{N}\cdot\text{m}^2/\text{C}^2)(-8.00\times10^{-6}\ \text{C})(-3.00\times10^{-6}\ \text{C})}{\tfrac{1}{2}(7.20\times10^{-3}\ \text{kg})(65.0\ \text{m/s})^2 + \dfrac{(8.99\times10^9\ \text{N}\cdot\text{m}^2/\text{C}^2)(-8.00\times10^{-6}\ \text{C})(-3.00\times10^{-6}\ \text{C})}{0.0450\ \text{m}}}$$

$$= 0.0108\ \text{m}$$

Therefore, the charge moves a distance of $0.0450\ \text{m} - 0.0108\ \text{m} = \boxed{0.0342\ \text{m}}$.

---

23. **REASONING** Initially, the three charges are infinitely far apart. We will proceed as in Example 8 by adding charges to the triangle, one at a time, and determining the electric potential energy at each step. According to Equation 19.3, the electric potential energy EPE is the product of the charge q and the electric potential V at the spot where the charge is placed, EPE = qV. The total electric potential energy of the group is the sum of the energies of each step in assembling the group.

SOLUTION Let the corners of the triangle be numbered clockwise as 1, 2 and 3, starting with the top corner. When the first charge ($q_1 = 8.00$ µC) is placed at a corner 1, the charge has no electric potential energy, $EPE_1 = 0$. This is because the electric potential V_1 produced by the other two charges at corner 1 is zero, since they are infinitely far away.

Once the 8.00-µC charge is in place, the electric potential V_2 that it creates at corner 2 is

$$V_2 = \frac{kq_1}{r_{21}}$$

where $r_{21} = 5.00$ m is the distance between corners 1 and 2, and $q_1 = 8.00$ µC. When the 20.0-µC charge is placed at corner 2, its electric potential energy EPE_2 is

$$EPE_2 = q_2 V_2 = q_2 \left(\frac{kq_1}{r_{21}} \right)$$

$$= (20.0 \times 10^{-6} \text{ C}) \left[\frac{(8.99 \times 10^9 \text{ N} \cdot \text{m}^2 / \text{C}^2)(8.00 \times 10^{-6} \text{ C})}{5.00 \text{ m}} \right] = 0.288 \text{ J}$$

The electric potential V_3 at the remaining empty corner is the sum of the potentials due to the two charges that are already in place on corners 1 and 2:

$$V_3 = \frac{kq_1}{r_{31}} + \frac{kq_2}{r_{32}}$$

where $q_1 = 8.00$ µC, $r_{31} = 3.00$ m, $q_2 = 20.0$ µC, and $r_{32} = 4.00$ m. When the third charge ($q_3 = -15.0$ µC) is placed at corner 3, its electric potential energy EPE_3 is

$$EPE_3 = q_3 V_3 = q_3 \left(\frac{kq_1}{r_{31}} + \frac{kq_2}{r_{32}} \right) = q_3 k \left(\frac{q_1}{r_{31}} + \frac{q_2}{r_{32}} \right)$$

$$= (-15.0 \times 10^{-6} \text{ C})(8.99 \times 10^9 \text{ N} \cdot \text{m}^2 / \text{C}^2) \left(\frac{8.00 \times 10^{-6} \text{ C}}{3.00 \text{ m}} + \frac{20.0 \times 10^{-6} \text{ C}}{4.00 \text{ m}} \right) = -1.034 \text{ J}$$

The electric potential energy of the entire array is given by

$$EPE = EPE_1 + EPE_2 + EPE_3 = 0 + 0.288 \text{ J} + (-1.034 \text{ J}) = \boxed{-0.746 \text{ J}}$$

27. ***REASONING AND SOLUTION*** Since all points on the equipotential surface are the same distance r from the point charge, the potential is given by Equation 19.6,

$$V = \frac{kq}{r} = \frac{(8.99 \times 10^9 \text{ N} \cdot \text{m}^2 / \text{C}^2)(3.0 \times 10^{-7} \text{ C})}{0.15 \text{ m}} = \boxed{18\ 000 \text{ V}}$$

---

31. ***REASONING AND SOLUTION*** Equation 19.7 gives the result directly:

$$E = -\frac{\Delta V}{\Delta s} = -\frac{V_B - V_A}{\Delta s} = -\frac{28 \text{ V} - 95 \text{ V}}{0.016 \text{ m}} = \boxed{4.2 \times 10^3 \text{ V} / \text{m}}$$

The electric field points from high potential to low potential. Thus, it points from $\boxed{A \text{ to } B}$.

---

35. ***REASONING*** The total energy of the particle on the equipotential surface A is $E_A = \frac{1}{2}mv_A^2 + qV_A$. Similarly, the total energy of the particle when it reaches equipotential surface B is $E_B = \frac{1}{2}mv_B^2 + qV_B$. According to Equation 6.8, the work W done by the outside force is equal to the final total energy E_B minus the initial total energy E_A, $W = E_B - E_A$.

SOLUTION The work done by the outside force in moving the particle from A to B is,

$$W = E_B - E_A = \left(\tfrac{1}{2}mv_B^2 + qV_B\right) - \left(\tfrac{1}{2}mv_A^2 + qV_A\right) = \tfrac{1}{2}m\left(v_B^2 - v_A^2\right) + q\left(V_B - V_A\right)$$

$$= \tfrac{1}{2}(5.00 \times 10^{-2} \text{ kg})\left[(3.00 \text{ m/s})^2 - (2.00 \text{ m/s})^2\right]$$

$$+ (4.00 \times 10^{-5} \text{ C})\left[7850 \text{ V} - 5650 \text{ V}\right] = \boxed{0.213 \text{ J}}$$

---

37. ***REASONING*** According to Equation 19.11, the energy stored in a capacitor with capacitance C and potential V across its plates is $\text{Energy} = \frac{1}{2}CV^2$.

SOLUTION Therefore, solving Equation 19.11 for V, we have

$$V = \sqrt{\frac{2(\text{Energy})}{C}} = \sqrt{\frac{2(73 \text{ J})}{120 \times 10^{-6} \text{ F}}} = \boxed{1.1 \times 10^3 \text{ V}}$$

---

39. ***REASONING*** The charge that resides on the outer surface of the cell membrane is $q = CV$, according to Equation 19.8. Before we can use this expression, however, we must first determine the capacitance of the membrane. If we assume that the cell membrane behaves like a parallel plate capacitor filled with a dielectric, Equation 19.10 ($C = \kappa \varepsilon_0 A / d$) applies as well.

SOLUTION The capacitance of the cell membrane is

$$C = \frac{\kappa \varepsilon_0 A}{d} = \frac{(5.0)(8.85 \times 10^{-12} \text{ F/m})(5.0 \times 10^{-9} \text{ m}^2)}{1.0 \times 10^{-8} \text{ m}} = 2.2 \times 10^{-11} \text{ F}$$

a. The charge on the outer surface of the membrane is, therefore,

$$q = CV = (2.2 \times 10^{-11} \text{ F})(60.0 \times 10^{-3} \text{ V}) = \boxed{1.3 \times 10^{-12} \text{ C}}$$

b. If the charge in part (a) is due to K^+ ions with charge $+e$ ($e = 1.6 \times 10^{-19}$ C), the number of ions present on the outer surface of the membrane is

$$\frac{\text{Number of}}{K^+ \text{ ions}} = \frac{1.3 \times 10^{-12} \text{ C}}{1.6 \times 10^{-19} \text{ C}} = \boxed{8.1 \times 10^6}$$

---

43. *REASONING*   According to Equation 19.11, the energy stored in a capacitor with a capacitance C and potential V across its plates is Energy $= \frac{1}{2}CV^2$. Once we determine how much energy is required to operate a 75-W light bulb for one minute, we can then use the expression for the energy to solve for V.

SOLUTION   The energy stored in the capacitor, which is equal to the energy required to operate a 75-W bulb for one minute (= 60 s), is

$$\text{Energy} = Pt = (75 \text{ W})(60 \text{ s}) = 4500 \text{ J}$$

Therefore, solving Equation 19.11 for V, we have

$$V = \sqrt{\frac{2(\text{Energy})}{C}} = \sqrt{\frac{2(4500 \text{ J})}{3.3 \text{ F}}} = \boxed{52 \text{ V}}$$

---

47. $\boxed{\text{WWW}}$ *REASONING*   If we assume that the motion of the proton and the electron is horizontal in the $+x$ direction, the motion of the proton is determined by Equation 2.8, $x = v_0 t + \frac{1}{2} a_p t^2$, where x is the distance traveled by the proton, v_0 is its initial speed, and a_p is its acceleration. If the distance between the capacitor places is d, then this relation becomes $\frac{1}{2}d = v_0 t + \frac{1}{2} a_p t^2$, or

$$d = 2v_0 t + a_p t^2 \qquad\qquad (1)$$

We can solve Equation (1) for the initial speed v_0 of the proton, but, first, we must determine the time t and the acceleration a_p of the proton . Since the proton strikes the negative plate at the same instant the electron strikes the positive plate, we can use the motion of the electron to determine the time t.

For the electron, $\frac{1}{2}d = \frac{1}{2}a_e t^2$, where we have taken into account the fact that the electron is released from rest. Solving this expression for t we have $t = \sqrt{d/a_e}$. Substituting this expression into Equation (1), we have

$$d = 2v_0 \sqrt{\frac{d}{a_e}} + \left(\frac{a_p}{a_e}\right) d \qquad (2)$$

The accelerations can be found by noting that the magnitudes of the forces on the electron and proton are equal, since these particles have the same magnitude of charge. The force on the electron is $F = eE = eV/d$, and the acceleration of the electron is, therefore,

$$a_e = \frac{F}{m_e} = \frac{eV}{m_e d} \qquad (3)$$

Newton's second law requires that $m_e a_e = m_p a_p$, so that

$$\frac{a_p}{a_e} = \frac{m_e}{m_p} \qquad (4)$$

Combining Equations (2), (3) and (4) leads to the following expression for v_0, the initial speed of the proton:

$$v_0 = \frac{1}{2}\left(1 - \frac{m_e}{m_p}\right)\sqrt{\frac{eV}{m_e}}$$

SOLUTION Substituting values into the expression above, we find

$$v_0 = \frac{1}{2}\left(1 - \frac{9.11 \times 10^{-31} \text{ kg}}{1.67 \times 10^{-27} \text{ kg}}\right)\sqrt{\frac{(1.6 \times 10^{-19} \text{ C})(175 \text{ V})}{9.11 \times 10^{-31} \text{ kg}}} = \boxed{2.77 \times 10^6 \text{ m/s}}$$

---

51. *REASONING AND SOLUTION* Equation 19.10 gives the capacitance for a parallel plate capacitor filled with a dielectric of constant κ: $C = \kappa \varepsilon_0 A/d$. Solving for κ, we have

$$\kappa = \frac{Cd}{\varepsilon_0 A} = \frac{(7.0 \times 10^{-6} \text{ F})(1.0 \times 10^{-5} \text{ m})}{(8.85 \times 10^{-12} \text{ F/m})(1.5 \text{ m}^2)} = \boxed{5.3}$$

---

55. WWW   *REASONING AND SOLUTION*   As described in the problem statement, the charges jump between your hand and a doorknob. If we assume that the electric field is uniform, Equation 19.7 applies, and we have

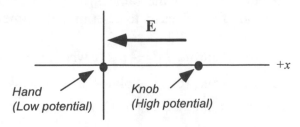

$$E = -\frac{\Delta V}{\Delta s} = -\frac{V_{knob} - V_{hand}}{\Delta s}$$

Therefore, solving for the potential difference between your hand and the doorknob, we have

$$V_{knob} - V_{hand} = -E\Delta s = -(-3.0 \times 10^6 \, \text{N/C})(3.0 \times 10^{-3} \, \text{m}) = \boxed{+9.0 \times 10^3 \, \text{V}}$$

---

59. **REASONING**   According to Equation 19.10, the capacitance of a parallel plate capacitor filled with a dielectric is $C = \kappa \varepsilon_0 A / d$, where κ is the dielectric constant, A is the area of one plate, and d is the distance between the plates.

From the definition of capacitance (Equation 19.8), $q = CV$. Thus, using Equation 19.10, we see that the charge q on a parallel plate capacitor that contains a dielectric is given by $q = (\kappa \varepsilon_0 A / d)V$. Since each dielectric occupies one-half of the volume between the plates, the area of each plate in contact with each material is $A/2$. Thus,

$$q_1 = \frac{\kappa_1 \varepsilon_0 (A/2)}{d} V = \frac{\kappa_1 \varepsilon_0 A}{2d} V \qquad \text{and} \qquad q_2 = \frac{\kappa_2 \varepsilon_0 (A/2)}{d} V = \frac{\kappa_2 \varepsilon_0 A}{2d} V$$

According to the problem statement, the total charge stored by the capacitor is

$$q_1 + q_2 = CV \tag{1}$$

where q_1 and q_2 are the charges on the plates in contact with dielectrics 1 and 2, respectively. Using the expressions for q_1 and q_2 above, Equation (1) becomes

$$CV = \frac{\kappa_1 \varepsilon_0 A}{2d} V + \frac{\kappa_2 \varepsilon_0 A}{2d} V = \frac{\kappa_1 \varepsilon_0 A + \kappa_2 \varepsilon_0 A}{2d} V = \frac{(\kappa_1 + \kappa_2) \varepsilon_0 A}{2d} V$$

This expression can be solved for C.

SOLUTION   Solving for C, we obtain $\boxed{C = \frac{\varepsilon_0 A (\kappa_1 + \kappa_2)}{2d}}$

CHAPTER 20 | *ELECTRIC CIRCUITS*

1. **REASONING** Since current is defined as charge per unit time, the current used by the portable compact disc player is equal to the charge provided by the battery pack (180 C) divided by the time in which the charge is delivered (2.0 h).

SOLUTION The amount of current that the player uses in operation is determined from Equation 20.1:

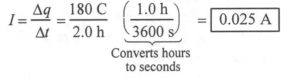

$$I = \frac{\Delta q}{\Delta t} = \frac{180 \text{ C}}{2.0 \text{ h}} \left(\frac{1.0 \text{ h}}{3600 \text{ s}} \right) = \boxed{0.025 \text{ A}}$$

Converts hours to seconds

5. **REASONING AND SOLUTION** Ohm's law (Equation 20.2, $V = IR$) gives the result directly

$$I = \frac{V}{R} = \frac{240 \text{ V}}{11 \text{ }\Omega} = \boxed{22 \text{ A}}$$

9. **REASONING** The number N of protons that strike the target is equal to the amount of electric charge Δq striking the target divided by the charge e of a proton, $N = (\Delta q)/e$. From Equation 20.1, the amount of charge is equal to the product of the current I and the time Δt. We can combine these two relations to find the number of protons that strike the target in 15 seconds.

The heat Q that must be supplied to change the temperature of the aluminum sample of mass m by an amount ΔT is given by Equation 12.4 as $Q = cm\Delta T$, where c is the specific heat capacity of aluminum. The heat is provided by the kinetic energy of the protons and is equal to the number of protons that strike the target times the kinetic energy per proton. Using this reasoning, we can find the change in temperature of the block for the 15 second-time interval.

SOLUTION
a. The number N of protons that strike the target is

$$N = \frac{\Delta q}{e} = \frac{I\Delta t}{e} = \frac{(0.50 \times 10^{-6} \text{ A})(15 \text{ s})}{1.6 \times 10^{-19} \text{ C}} = \boxed{4.7 \times 10^{13}}$$

b. The amount of heat Q provided by the kinetic energy of the protons is

$$Q = (4.7 \times 10^{13} \text{ protons})(4.9 \times 10^{-12} \text{ J/proton}) = 230 \text{ J}$$

Since $Q = cm\Delta T$ and since Table 12.2 gives the specific heat of aluminum as $c = 9.00 \times 10^2$ J/(kg·C°), the change in temperature of the block is

$$\Delta T = \frac{Q}{cm} = \frac{230 \text{ J}}{\left(9.00 \times 10^2 \ \dfrac{\text{J}}{\text{kg} \cdot \text{C}°}\right)(15 \times 10^{-3} \text{ kg})} = \boxed{17 \text{ C}°}$$

---

11. $\boxed{\text{WWW}}$ **REASONING** The resistance of a metal wire of length L, cross-sectional area A and resistivity ρ is given by Equation 20.3: $R = \rho L / A$. Solving for A, we have $A = \rho L / R$. We can use this expression to find the ratio of the cross-sectional area of the aluminum wire to that of the copper wire.

SOLUTION Forming the ratio of the areas and using resistivity values from Table 20.1, we have

$$\frac{A_{\text{aluminum}}}{A_{\text{copper}}} = \frac{\rho_{\text{aluminum}} L / R}{\rho_{\text{copper}} L / R} = \frac{\rho_{\text{aluminum}}}{\rho_{\text{copper}}} = \frac{2.82 \times 10^{-8} \ \Omega \cdot \text{m}}{1.72 \times 10^{-8} \ \Omega \cdot \text{m}} = \boxed{1.64}$$

---

17. $\boxed{\text{WWW}}$ **REASONING** The resistance of a metal wire of length L, cross-sectional area A and resistivity ρ is given by Equation 20.3: $R = \rho L / A$. The volume V_2 of the new wire will be the same as the original volume V_1 of the wire, where volume is the product of length and cross-sectional area. Thus, $V_1 = V_2$ or $A_1 L_1 = A_2 L_2$. Since the new wire is three times longer than the first wire, we can write

$$A_1 L_1 = A_2 L_2 = A_2 (3L_1) \quad \text{or} \quad A_2 = A_1 / 3$$

We can form the ratio of the resistances, use this expression for the area A_2, and find the new resistance.

SOLUTION The resistance of the new wire is determined as follows:

$$\frac{R_2}{R_1} = \frac{\rho L_2 / A_2}{\rho L_1 / A_1} = \frac{L_2 A_1}{L_1 A_2} = \frac{(3L_1) A_1}{L_1 (A_1 / 3)} = 9$$

Solving for R_2, we find that

$$R_2 = 9 R_1 = 9(21.0 \ \Omega) = \boxed{189 \ \Omega}$$

19. WWW *REASONING*   We will ignore any changes in length due to thermal expansion. Although the resistance of each section changes with temperature, the total resistance of the composite does not change with temperature. Therefore,

$$\underbrace{\left(R_{\text{tungsten}}\right)_0 + \left(R_{\text{carbon}}\right)_0}_{\text{At room temperature}} = \underbrace{R_{\text{tungsten}} + R_{\text{carbon}}}_{\text{At temperature } T}$$

From Equation 20.5, we know that the temperature dependence of the resistance for a wire of resistance R_0 at temperature T_0 is given by $R = R_0[1 + \alpha(T - T_0)]$, where α is the temperature coefficient of resistivity. Thus,

$$\left(R_{\text{tungsten}}\right)_0 + \left(R_{\text{carbon}}\right)_0 = \left(R_{\text{tungsten}}\right)_0 (1 + \alpha_{\text{tungsten}} \Delta T) + \left(R_{\text{carbon}}\right)_0 (1 + \alpha_{\text{carbon}} \Delta T)$$

Since ΔT is the same for each wire, this simplifies to

$$\left(R_{\text{tungsten}}\right)_0 \alpha_{\text{tungsten}} = -\left(R_{\text{carbon}}\right)_0 \alpha_{\text{carbon}} \tag{1}$$

This expression can be used to find the ratio of the resistances. Once this ratio is known, we can find the ratio of the lengths of the sections with the aid of Equation 20.3 ($L = RA/\rho$).

SOLUTION   From Equation (1), the ratio of the resistances of the two sections of the wire is

$$\frac{\left(R_{\text{tungsten}}\right)_0}{\left(R_{\text{carbon}}\right)_0} = -\frac{\alpha_{\text{carbon}}}{\alpha_{\text{tungsten}}} = -\frac{-0.0005 \ [(\text{C}°)^{-1}]}{0.0045 \ [(\text{C}°)^{-1}]} = \frac{1}{9}$$

Thus, using Equation 20.3, we find the ratio of the tungsten and carbon lengths to be

$$\frac{L_{\text{tungsten}}}{L_{\text{carbon}}} = \frac{\left(R_0 A / \rho\right)_{\text{tungsten}}}{\left(R_0 A / \rho\right)_{\text{carbon}}} = \frac{\left(R_{\text{tungsten}}\right)_0}{\left(R_{\text{carbon}}\right)_0} \left(\frac{\rho_{\text{carbon}}}{\rho_{\text{tungsten}}}\right) = \left(\frac{1}{9}\right)\left(\frac{3.5 \times 10^{-5} \ \Omega \cdot \text{m}}{5.6 \times 10^{-8} \ \Omega \cdot \text{m}}\right) = \boxed{70}$$

where we have used resistivity values from Table 20.1 and the fact that the two sections have the same cross-sectional areas.

---

23. *REASONING*   According to Equation 6.10b, the energy used is Energy $= Pt$, where P is the power and t is the time. According to Equation 20.6a, the power is $P = IV$, where I is the current and V is the voltage. Thus, Energy $= IVt$, and we apply this result first to the drier and then to the computer.

SOLUTION   The energy used by the drier is

$$\text{Energy} = Pt = IVt = (16 \text{ A})(240 \text{ V})(45 \text{ min}) \underbrace{\left(\frac{60 \text{ s}}{1.00 \text{ min}}\right)}_{\substack{\text{Converts minutes} \\ \text{to seconds}}} = 1.04 \times 10^7 \text{ J}$$

For the computer, we have

$$\text{Energy} = 1.04 \times 10^7 \text{ J} = IVt = (2.7 \text{ A})(120 \text{ V})t$$

Solving for t we find

$$t = \frac{1.04 \times 10^7 \text{ J}}{(2.7 \text{ A})(120 \text{ V})} = 3.21 \times 10^4 \text{ s} = \left(3.21 \times 10^4 \text{ s}\right)\left(\frac{1.00 \text{ h}}{3600 \text{ s}}\right) = \boxed{8.9 \text{ h}}$$

---

27.   **REASONING AND SOLUTION**   As a function of temperature, the resistance of the wire is given by Equation 20.5: $R = R_0\left[1 + \alpha(T - T_0)\right]$, where α is the temperature coefficient of resistivity.   From Equation 20.6c, we have $P = V^2/R$.   Combining these two equations, we have

$$P = \frac{V^2}{R_0\left[1 + \alpha\left(T - T_0\right)\right]} = \frac{P_0}{1 + \alpha\left(T - T_0\right)}$$

where $P_0 = V^2/R_0$, since the voltage is constant.   But $P = \frac{1}{2}P_0$, so we find

$$\frac{P_0}{2} = \frac{P_0}{1 + \alpha\left(T - T_0\right)} \qquad \text{or} \qquad 2 = 1 + \alpha\left(T - T_0\right)$$

Solving for T, we find

$$T = \frac{1}{\alpha} + T_0 = \frac{1}{0.0045 \text{ (C}°)^{-1}} + 28° = \boxed{250 \text{ °C}}$$

---

33.   **REASONING AND SOLUTION**
a. According to Equation 20.15c, the average power consumed by the iron is

$$\overline{P} = \frac{V^2_{\text{rms}}}{R} = \frac{(120 \text{ V})^2}{16 \text{ }\Omega} = \boxed{9.0 \times 10^2 \text{ W}}$$

b. The peak power is $P_{\text{peak}} = 2\overline{P} = \boxed{1.8 \times 10^3 \text{ W}}$.

---

37. **REASONING** According to Equation 6.10b, the energy supplied to the water in the form of heat is $Q = Pt$, where P is the power and t is the time. According to Equation 20.6c, the power is $P = V^2/R$, so that $Q = (V^2/R)t$, where V is the voltage and R is the resistance. But for a mass m of water with a specific heat capacity c, the heat Q is related to the change in temperature ΔT by $Q = mc\Delta T$ according to Equation 12.4. Our solution uses these two expressions for Q.

SOLUTION Combining the two expressions for Q, we have $Q = mc\Delta T = V^2 t/R$. Solving for t, we have

$$t = \frac{Rcm\Delta T}{V^2} \tag{1}$$

The mass of the water can be determined from its volume V_{water}. From Equation 11.1, we have $m = \rho_{water} V_{water}$, where $\rho_{water} = 1.000 \times 10^3$ kg/m^3 is the mass density of water (see Table 11.1). Therefore, Equation (1) becomes

$$t = \frac{Rc\rho_{water} V_{water} \Delta T}{V^2} \tag{2}$$

The value (see Table 12.2) for the specific heat of water is 4186 J/(kg•C°). The volume of water is

$$V_{water} = (52 \text{ gal})\left(\frac{3.79 \times 10^{-3} \text{ m}^3}{1.00 \text{ gal}}\right) = 0.197 \text{ m}^3$$

Substituting values into Equation (2), we obtain

$$t = \frac{(3.0 \ \Omega)[4186 \text{ J/(kg} \cdot \text{C°)}](1.000 \times 10^3 \text{ kg/m}^3)(0.197 \text{ m}^3)(53 \text{ °C} - 11 \text{ °C})}{(120 \text{ V})^2} = 7220 \text{ s}$$

$$= (7220 \text{ s})\left(\frac{1.00 \text{ h}}{3600 \text{ s}}\right) = \boxed{2.0 \text{ h}}$$

---

39. **REASONING** The equivalent series resistance R_s is the sum of the resistances of the three resistors. The potential difference V can be determined from Ohm's law as $V = IR_s$.

SOLUTION
a. The equivalent resistance is

$$R_s = 25 \ \Omega + 45 \ \Omega + 75 \ \Omega = \boxed{145 \ \Omega}$$

b. The potential difference across the three resistors is

$$V = IR_s = (0.51 \text{ A})(145 \ \Omega) = \boxed{74 \text{ V}}$$

---

43. ***REASONING***   Using Ohm's law (Equation 20.2) we can write an expression for the voltage across the original circuit as $V = I_0 R_0$. When the additional resistor R is inserted in series, assuming that the battery remains the same, the voltage across the new combination is given by $V = I(R + R_0)$.   Since V is the same in both cases, we can write $I_0 R_0 = I(R + R_0)$. This expression can be solved for R_0.

SOLUTION   Solving for R_0, we have $I_0 R_0 - IR_0 = IR$ or $R_0(I_0 - I) = IR$; therefore,

$$R_0 = \frac{IR}{I_0 - I} = \frac{(12.0 \text{ A})(8.00 \ \Omega)}{15.0 \text{ A} - 12.0 \text{ A}} = \boxed{32 \ \Omega}$$

---

49. ***REASONING AND SOLUTION***   Since the circuit elements are in parallel, the equivalent resistance can be obtained directly from Equation 20.17:

$$\frac{1}{R_p} = \frac{1}{R_1} + \frac{1}{R_2} = \frac{1}{16 \ \Omega} + \frac{1}{8.0 \ \Omega} \qquad \text{or} \qquad \boxed{R_p = 5.3 \ \Omega}$$

---

51. ***REASONING***   Since the resistors are connected in parallel, the voltage across each one is the same and can be calculated from Ohm's Law (Equation 20.2: $V = IR$). Once the voltage across each resistor is known, Ohm's law can again be used to find the current in the second resistor. The total power consumed by the parallel combination can be found calculating the power consumed by each resistor from Equation 20.6b: $P = I^2 R$. Then, the total power consumed is the sum of the power consumed by each resistor.

SOLUTION   Using data for the second resistor, the voltage across the resistors is equal to

$$V = IR = (3.00 \text{ A})(64.0 \ \Omega) = 192 \ \Omega$$

a. The current through the 42.0-Ω resistor is

$$I = \frac{V}{R} = \frac{192 \text{ V}}{42.0 \ \Omega} = \boxed{4.57 \text{ A}}$$

b. The power consumed by the 42.0-Ω resistor is

$$P = I^2 R = (4.57 \text{ A})^2 (42.0 \ \Omega) = 877 \text{ W}$$

while the power consumed by the 64.0-Ω resistor is

$$P = I^2 R = (3.00 \text{ A})^2 (64.0 \text{ } \Omega) = \boxed{576 \text{ W}}$$

Therefore the total power consumed by the two resistors is 877 W + 576 W = $\boxed{1450 \text{ W}}$.

---

55. **REASONING**   The equivalent resistance of the three devices in parallel is R_p, and we can find the value of R_p by using our knowledge of the total power consumption of the circuit; the value of R_p can be found from Equation 20.6c, $P = V^2 / R_p$. Ohm's law (Equation 20.2, $V = IR$) can then be used to find the current through the circuit.

SOLUTION
a. The total power used by the circuit is P = 1650 W + 1090 W +1250 W = 3990 W. The equivalent resistance of the circuit is

$$R_p = \frac{V^2}{P} = \frac{(120 \text{ V})^2}{3990 \text{ W}} = \boxed{3.6 \text{ } \Omega}$$

b. The total current through the circuit is

$$I = \frac{V}{R_p} = \frac{120 \text{ V}}{3.6 \text{ } \Omega} = \boxed{33 \text{ A}}$$

This current is larger than the rating of the circuit breaker; therefore, the $\boxed{\text{breaker will open}}$.

---

59. **REASONING**   We will analyze the combination in parts. The 65 and 85-Ω resistors are in parallel. This parallel combination is in series with the 63-Ω resistor.

SOLUTION   The equivalent resistance for the parallel combination of the 65 and 85-Ω resistors can be determined as follows, using Equation 20.17:

$$\frac{1}{R_p} = \frac{1}{65 \text{ } \Omega} + \frac{1}{85 \text{ } \Omega} = 0.027 \text{ } \Omega^{-1} \qquad \text{or} \qquad R_p = \frac{1}{0.027 \text{ } \Omega^{-1}} = 37 \text{ } \Omega$$

This equivalent resistance is in series with the 63-Ω resistance, the two giving the following total resistance between points A and B, according to Equation 20.16:

$$R_s = R_{AB} = 63 \text{ } \Omega + 37 \text{ } \Omega = \boxed{1.0 \times 10^2 \text{ } \Omega}$$

61. **_REASONING_** When two or more resistors are in series, the equivalent resistance is given by Equation 20.16: $R_s = R_1 + R_2 + R_3 + \dots$. Likewise, when resistors are in parallel, the expression to be solved to find the equivalent resistance is given by Equation 20.17: $\frac{1}{R_p} = \frac{1}{R_1} + \frac{1}{R_2} + \frac{1}{R_3} + \dots$. We will successively apply these to the individual resistors in the figure in the text beginning with the resistors on the right side of the figure.

SOLUTION Since the 4.0-Ω and the 6.0-Ω resistors are in series, the equivalent resistance of the combination of those two resistors is 10.0 Ω. The 9.0-Ω and 8.0-Ω resistors are in parallel; their equivalent resistance is 4.24 Ω. The equivalent resistances of the parallel combination (9.0 Ω and 8.0 Ω) and the series combination (4.0 Ω and the 6.0 Ω) are in parallel; therefore, their equivalent resistance is 2.98 Ω. The 2.98-Ω combination is in series with the 3.0-Ω resistor, so that equivalent resistance is 5.98 Ω. Finally, the 5.98-Ω combination and the 20.0-Ω resistor are in parallel, so the equivalent resistance between the points A and B is $\boxed{4.6\ \Omega}$.

---

65. $\boxed{\text{WWW}}$ **_REASONING_** Since we know that the current in the 8.00-Ω resistor is 0.500 A, we can use Ohm's law ($V = IR$) to find the voltage across the 8.00-Ω resistor. The 8.00-Ω resistor and the 16.0-Ω resistor are in parallel; therefore, the voltages across them are equal. Thus, we can also use Ohm's law to find the current through the 16.0-Ω resistor. The currents that flow through the 8.00-Ω and the 16.0-Ω resistors combine to give the total current that flows through the 20.0-Ω resistor. Similar reasoning can be used to find the current through the 9.00-Ω resistor.

SOLUTION
a. The voltage across the 8.00-Ω resistor is $V_8 = (0.500\text{ A})(8.00\ \Omega) = 4.00\text{ V}$. Since this is also the voltage that is across the 16.0-Ω resistor, we find that the current through the 16.0-Ω resistor is $I_{16} = (4.00\text{ V})/(16.0\ \Omega) = 0.250\text{ A}$. Therefore, the total current that flows through the 20.0-Ω resistor is

$$I_{20} = 0.500\text{ A} + 0.250\text{ A} = \boxed{0.750\text{ A}}$$

b. The 8.00-Ω and the 16.0-Ω resistors are in parallel, so their equivalent resistance can be obtained from Equation 20.17, $\frac{1}{R_p} = \frac{1}{R_1} + \frac{1}{R_2} + \frac{1}{R_3} + \dots$, and is equal to 5.33 Ω. Therefore, the equivalent resistance of the upper branch of the circuit is $R_{\text{upper}} = 5.33\ \Omega + 20.0\ \Omega = 25.3\ \Omega$, since the 5.33-Ω resistance is in series with the 20.0-Ω resistance. Using Ohm's law, we find that the voltage across the upper branch must be $V = (0.750\text{ A})(25.3\ \Omega) = 19.0\text{ V}$. Since the lower branch is in parallel with the upper branch, the voltage across both branches must be the same. Therefore, the current through the 9.00-Ω resistor is, from Ohm's law,

$$I_9 = \frac{V_{\text{lower}}}{R_9} = \frac{19.0 \text{ V}}{9.00 \ \Omega} = \boxed{2.11 \text{ A}}$$

---

69. **REASONING** The terminal voltage of the battery is given by $V_{\text{terminal}} = \text{Emf} - Ir$, where r is the internal resistance of the battery. Since the terminal voltage is observed to be one-half of the emf of the battery, we have $V_{\text{terminal}} = \text{Emf}/2$ and $I = \text{Emf}/(2r)$. From Ohm's law, the equivalent resistance of the circuit is $R = \text{emf}/I = 2r$. We can also find the equivalent resistance of the circuit by considering that the identical bulbs are in parallel across the battery terminals, so that the equivalent resistance of the N bulbs is found from

$$\frac{1}{R_p} = \frac{N}{R_{\text{bulb}}} \qquad \text{or} \qquad R_p = \frac{R_{\text{bulb}}}{N}$$

This equivalent resistance is in series with the battery, so we find that the equivalent resistance of the circuit is

$$R = 2r = \frac{R_{\text{bulb}}}{N} + r$$

This expression can be solved for N.

SOLUTION Solving the above expression for N, we have

$$N = \frac{R_{\text{bulb}}}{2r - r} = \frac{R_{\text{bulb}}}{r} = \frac{15 \ \Omega}{0.50 \ \Omega} = \boxed{30}$$

---

73. **REASONING** The current I can be found by using Kirchhoff's loop rule. Once the current is known, the voltage between points A and B can be determined.

SOLUTION

a. We assume that the current is directed clockwise around the circuit. Starting at the upper-left corner and going clockwise around the circuit, we set the potential drops equal to the potential rises:

$$\underbrace{(5.0 \ \Omega)I + (27 \ \Omega)I + 10.0 \text{ V} + (12 \ \Omega)I + (8.0 \ \Omega)I}_{\text{Potential drops}} = \underbrace{30.0 \text{ V}}_{\text{Potential rises}}$$

Solving for the current gives $\boxed{I = 0.38 \text{ A}}$.

b. The voltage between points A and B is

$$V_{AB} = 30.0 \text{ V} - (0.38 \text{ A})(27 \text{ }\Omega) = \boxed{2.0 \times 10^1 \text{ V}}$$

c. $\boxed{\text{Point B}}$ is at the higher potential.

---

77. **REASONING** We begin by labeling the currents in the three resistors. The drawing below shows the directions chosen for these currents. The directions are arbitrary, and if any of them is incorrect, then the analysis will show that the corresponding value for the current is negative.

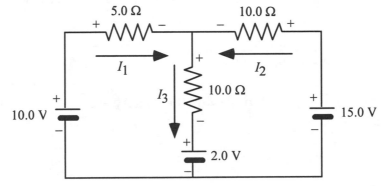

We then mark the resistors with the plus and minus signs that serve as an aid in identifying the potential drops and rises for the loop rule, recalling that conventional current is always directed from a higher potential (+) toward a lower potential (–). Thus, given the directions chosen for I_1, I_2, and I_3, the plus and minus signs *must* be those shown in the drawing. We can now use Kirchhoff's rules to find the voltage across the 5.0-Ω resistor.

SOLUTION Applying the loop rule to the left loop (and suppressing units for onvenience) gives

$$5.0 I_1 + 10.0 I_3 + 2.0 = 10.0 \tag{1}$$

Similarly, for the right loop,

$$10.0 I_2 + 10.0 I_3 + 2.0 = 15.0 \tag{2}$$

If we apply the junction rule to the upper junction, we obtain

$$I_1 + I_2 = I_3 \tag{3}$$

Subtracting Equation (2) from Equation (1) gives

$$5.0 I_1 - 10.0 I_2 = -5.0 \tag{4}$$

We now multiply Equation (3) by 10 and add the result to Equation (2); the result is

$$10.0 I_1 + 20.0 I_2 = 13.0 \tag{5}$$

If we then multiply Equation (4) by 2 and add the result to Equation (5), we obtain $20.0 I_1 = 3.0$, or solving for I_1, we obtain $I_1 = 0.15 \text{ A}$. The fact that I_1 is positive means that the current in the drawing has the correct direction. The voltage across the 5.0-Ω resistor can be found from Ohm's law:

$$V = (0.15 \text{ A})(5.0 \ \Omega) = \boxed{0.75 \text{ V}}$$

Current flows from the higher potential to the lower potential, and the current through the 5.0-Ω flows from left to right, so the $\boxed{\text{left end of the resistor}}$ is at the higher potential.

---

81. ***REASONING AND SOLUTION***   According to Ohm's law, the voltage across the series combination of the galvanometer and the resistor, under full-scale conditions is $V = I(R + R_g)$ where R_g is the resistance of the galvanometer. Therefore, solving for R, we have

$$R = \frac{V}{I} - R_g = \frac{30.0 \text{ V}}{8.30 \times 10^{-3} \text{ A}} - 180 \ \Omega = \boxed{3.43 \times 10^3 \ \Omega}$$

---

85. ***REASONING AND SOLUTION***
    a. According to Ohm's law (Equation 20.2, $V = IR$) the current in the circuit is

$$I = \frac{V}{R + R} = \frac{V}{2R}$$

The voltage across either resistor is IR, so that we find

$$IR = \left(\frac{V}{2R}\right) R = \frac{V}{2} = \frac{60.0 \text{ V}}{2} = \boxed{30.0 \text{ V}}$$

b. The voltmeter's resistance is $R_v = V / I = (60.0 \text{ V})/(5.00 \times 10^{-3} \text{ A}) = 12.0 \times 10^3 \ \Omega$, and this resistance is in parallel with the resistance $R = 1550 \ \Omega$. The equivalent resistance of this parallel combination can be obtained as follows:

$$\frac{1}{R_p} = \frac{1}{R_v} + \frac{1}{R} \quad \text{or} \quad R_p = \frac{R R_v}{R + R_v} = \frac{(1550 \ \Omega)(12.0 \times 10^3 \ \Omega)}{1550 \ \Omega + 12.0 \times 10^3 \ \Omega} = 1370 \ \Omega$$

The voltage registered by the voltmeter is $I R_p$, where I is the current supplied by the battery to the series combination of the other 1550-Ω resistor and R_p. According to Ohm's law

$$I = \frac{60.0 \text{ V}}{1550 \ \Omega + 1370 \ \Omega} = 0.0205 \text{ A}$$

Thus, the voltage registered by the voltmeter is

$$IR_p = (0.0205 \text{ A})(1370 \ \Omega) = \boxed{28.1 \text{ V}}$$

---

89. **REASONING** The magnitude q of the charge on each plate of a capacitor is related to the voltage V across the capacitor plates and the capacitance C by Equation 19.8: $q = CV$. When two or more capacitors are in series, the equivalent capacitance of the combination can be obtained from Equation 20.19: $\frac{1}{C_s} = \frac{1}{C_1} + \frac{1}{C_2} + \frac{1}{C_3} \cdots$. Equation 20.18 gives the equivalent capacitance for two or more capacitors in parallel: $C_p = C_1 + C_2 + C_3 + \cdots$.

In this problem two of the capacitors are in series, while the third capacitor is in parallel with the series combination. We proceed by first finding the equivalent capacitance of the circuit, and then we can use Equation 19.8 to find the total charge that the battery delivers to all the capacitors.

SOLUTION The equivalent capacitance of the series combination can be obtained as follows:

$$\frac{1}{C_s} = \frac{1}{3.0 \ \mu\text{F}} + \frac{1}{4.0 \ \mu\text{F}} = 0.583 \ (\mu\text{F})^{-1} \qquad \text{or} \qquad C_s = \frac{1}{0.583 \ (\mu\text{F})^{-1}} = 1.72 \ \mu\text{F}$$

The equivalent capacitance of the entire circuit is

$$C_{\text{equivalent}} = C_s = 1.72 \ \mu\text{F} + 10.0 \ \mu\text{F} = 11.72 \ \mu\text{F}$$

Therefore the total charge delivered by the battery is

$$q = C_{\text{equivalent}} V = (11.72 \times 10^{-6} \text{ F})(40.0 \text{ V}) = \boxed{4.69 \times 10^{-4} \text{ C}}$$

---

93. **REASONING** When two or more capacitors are in series, the equivalent capacitance of the combination can be obtained from Equation 20.19, $\frac{1}{C_s} = \frac{1}{C_1} + \frac{1}{C_2} + \frac{1}{C_3} \cdots$. Equation 20.18 gives the equivalent capacitance for two or more capacitors in parallel: $C_p = C_1 + C_2 + C_3 + \cdots$. The energy stored in a capacitor is given by $\frac{1}{2} CV^2$, according to Equation 19.11. Thus, the energy stored in the series combination is $\frac{1}{2} C_s V_s^2$, where

$$\frac{1}{C_s} = \frac{1}{7.0\ \mu F} + \frac{1}{3.0\ \mu F} = 0.476\ (\mu F)^{-1} \quad \text{or} \quad C_s = \frac{1}{0.476\ (\mu F)^{-1}} = 2.10\ \mu F$$

Similarly, the energy stored in the parallel combination is $\frac{1}{2} C_p V_p^2$ where

$$C_p = 7.0\ \mu F + 3.0\ \mu F = 10.0\ \mu F$$

The voltage required to charge the parallel combination of the two capacitors to the same total energy as the series combination can be found by equating the two energy expressions and solving for V_p.

SOLUTION   Equating the two expressions for the energy, we have

$$\tfrac{1}{2} C_s V_s^2 = \tfrac{1}{2} C_p V_p^2$$

Solving for V_p, we obtain the result

$$V_p = V_s \sqrt{\frac{C_s}{C_p}} = (24\ \text{V}) \sqrt{\frac{2.10\ \mu F}{10.0\ \mu F}} = \boxed{11\text{V}}$$

---

95. **REASONING**   The charge q on a discharging capacitor in a RC circuit is given by Equation 20.22: $q = q_0 e^{-t/RC}$, where q_0 is the original charge at time $t = 0$ s. Once t (time for one pulse) and the ratio q/q_0 are known, this expression can be solved for C.

SOLUTION   Since the pacemaker delivers 81 pulses per minute, the time for one pulse is $[1\ \text{min}/(81\ \text{pulses})] \times [60.0\ \text{s}/(1.00\ \text{min})] = 0.74\ \text{s}$.   Since one pulse is delivered every time the fully-charged capacitor loses 63.2% of its original charge, the charge remaining is 36.8% of the original charge.   Thus, we have $q = (0.368) q_0$, or $q/q_0 = 0.368$.

From Equation 20.22, we have

$$\frac{q}{q_0} = e^{-t/RC}$$

Taking the natural logarithm of both sides, we have,

$$\ln\left(\frac{q}{q_0}\right) = -\frac{t}{RC}$$

Solving for C, we find

$$C = \frac{-t}{R \ln(q/q_0)} = \frac{-(0.74 \text{ s})}{(1.8 \times 10^6 \ \Omega) \ln(0.368)} = \boxed{4.1 \times 10^{-7} \text{ F}}$$

---

101. ***REASONING AND SOLUTION***   The equivalent capacitance of the combination can be obtained immediately from Equation 20.19, $\dfrac{1}{C_s} = \dfrac{1}{C_1} + \dfrac{1}{C_2} + \dfrac{1}{C_3} \ldots$

$$\frac{1}{C_s} = \frac{1}{3.0 \ \mu\text{F}} + \frac{1}{7.0 \ \mu\text{F}} + \frac{1}{9.0 \ \mu\text{F}} = 0.587 \left(\mu\text{F}\right)^{-1} \qquad \text{or} \qquad C_s = \frac{1}{0.587 \left(\mu\text{F}\right)^{-1}} = \boxed{1.7 \ \mu\text{F}}$$

---

103. ***REASONING AND SOLUTION***   Ohm's law (Equation 20.2), $V = IR$, gives the result directly:

$$R = \frac{V}{I} = \frac{9.0 \text{ V}}{0.11 \text{ A}} = \boxed{82 \ \Omega}$$

---

109. ***REASONING*** When two or more resistors are in series, the equivalent resistance is given by Equation 20.16: $R_s = R_1 + R_2 + R_3 + \ldots$.   When resistors are in parallel, the expression to be solved to find the equivalent resistance is Equation 20.17: $\dfrac{1}{R_p} = \dfrac{1}{R_1} + \dfrac{1}{R_2} + \dfrac{1}{R_3} + \ldots$.

We will use these relations to determine the eight different values of resistance that can be obtained by connecting together the three resistors: 1.00, 2.00, and 3.00 Ω.

SOLUTION   When all the resistors are connected in series, the equivalent resistance is the sum of all three resistors and the equivalent resistance is $\boxed{6.00 \ \Omega}$.   When all three are in parallel, we have from Equation 20.17, the equivalent resistance is $\boxed{0.545 \ \Omega}$.

We can also connect two of the resistors in parallel and connect the parallel combination in series with the third resistor. When the 1.00 and 2.00-Ω resistors are connected in parallel and the 3.00-Ω resistor is connected in series with the parallel combination, the equivalent resistance is $\boxed{3.67 \ \Omega}$.   When the 1.00 and 3.00-Ω resistors are connected in parallel and the 2.00-Ω resistor is connected in series with the parallel combination, the equivalent resistance is $\boxed{2.75 \ \Omega}$.   When the 2.00 and 3.00-Ω resistors are connected in parallel and the 1.00-Ω resistor is connected in series with the parallel combination, the equivalent resistance is $\boxed{2.20 \ \Omega}$.

We can also connect two of the resistors in series and put the third resistor in parallel with the series combination. When the 1.00 and 2.00-Ω resistors are connected in series and the

3.00-Ω resistor is connected in parallel with the series combination, the equivalent resistance is $\boxed{1.50 \ \Omega}$. When the 1.00 and 3.00-Ω resistors are connected in series and the 2.00-Ω resistor is connected in parallel with the series combination, the equivalent resistance is $\boxed{1.33 \ \Omega}$. Finally, when the 2.00 and 3.00-Ω resistors are connected in series and the 1.00-Ω resistor is connected in parallel with the series combination, the equivalent resistance is $\boxed{0.833 \ \Omega}$.

---

111. **REASONING**   The resistance of one of the wires in the extension cord is given by Equation 20.3: $R = \rho L / A$, where the resistivity of copper is $\rho = 1.72 \times 10^{-8} \ \Omega \cdot m$, according to Table 20.1. Since the two wires in the cord are in series with each other, their total resistance is $R_{cord} = R_{wire \ 1} + R_{wire \ 2} = 2\rho L / A$.   Once we find the equivalent resistance of the entire circuit (extension cord + trimmer), Ohm's law can be used to find the voltage applied to the trimmer.

SOLUTION
a. The resistance of the extension cord is

$$R_{cord} = \frac{2\rho L}{A} = \frac{2(1.72 \times 10^{-8} \ \Omega \cdot m)(46 \ m)}{1.3 \times 10^{-6} \ m^2} = \boxed{1.2 \ \Omega}$$

b. The total resistance of the circuit (cord + trimmer) is, since the two are in series,

$$R_s = 1.2 \ \Omega + 15.0 \ \Omega = 16.2 \ \Omega$$

Therefore from Ohm's law (Equation 20.2: $V = IR$), the current in the circuit is

$$I = \frac{V}{R_s} = \frac{120 \ V}{16.2 \ \Omega} = 7.4 \ A$$

Finally, the voltage applied to the trimmer alone is (again using Ohm's law),

$$V_{trimmer} = (7.4 \ A)(15.0 \ \Omega) = \boxed{110 \ V}$$

---

115. **REASONING**   The foil effectively converts the capacitor into two capacitors in series. Equation 19.10 gives the expression for the capacitance of a capacitor of plate area A and plate separation d (no dielectric): $C_0 = \varepsilon_0 A / d$. We can use this expression to determine the capacitance of the individual capacitors created by the presence of the foil. Then using the fact that the "two capacitors" are in series, we can use Equation 20.19 to find the equivalent capacitance of the system.

SOLUTION   Since the foil is placed one-third of the way from one plate of the original capacitor to the other, we have $d_1 = (2/3)d$, and $d_2 = (1/3)d$. Then

$$C_1 = \frac{\varepsilon_0 A}{(2/3)d} = \frac{3\varepsilon_0 A}{2d}$$

and

$$C_2 = \frac{\varepsilon_0 A}{(1/3)d} = \frac{3\varepsilon_0 A}{d}$$

Since these two capacitors are effectively in series, it follows that

$$\frac{1}{C_s} = \frac{1}{C_1} + \frac{1}{C_2} = \frac{1}{3\varepsilon_0 A/(2d)} + \frac{1}{3\varepsilon_0 A/d} = \frac{3d}{3\varepsilon_0 A} = \frac{d}{\varepsilon_0 A}$$

But $C_0 = \varepsilon_0 A / d$, so that $d/(\varepsilon_0 A) = 1/C_0$, and we have

$$\frac{1}{C_s} = \frac{1}{C_0} \qquad \text{or} \qquad \boxed{C_s = C_0}$$

CHAPTER 21 | *MAGNETIC FORCES AND MAGNETIC FIELDS*

PROBLEMS

1. ***REASONING AND SOLUTION*** According to Equation 21.1, the magnitude B of the magnetic field is

$$B = \frac{F}{q_0 v \sin\theta} = \frac{8.7 \times 10^{-3}\,\text{N}}{(12 \times 10^{-6}\,\text{C})(9.0 \times 10^6\,\text{m/s})\sin 90.0°} = \boxed{8.1 \times 10^{-5}\,\text{T}}$$

---

5. ***REASONING*** According to Equation 21.1, the magnitude of the magnetic force on a moving charge is $F = q_0 vB \sin\theta$. Since the magnetic field points due north and the proton moves eastward, $\theta = 90.0°$. Furthermore, since the magnetic force on the moving proton balances its weight, we have $mg = q_0 vB \sin\theta$, where m is the mass of the proton. This expression can be solved for the speed v.

SOLUTION Solving for the speed v, we have

$$v = \frac{mg}{q_0 B \sin\theta} = \frac{(1.67 \times 10^{-27}\,\text{kg})(9.80\,\text{m/s}^2)}{(1.6 \times 10^{-19}\,\text{C})(2.5 \times 10^{-5}\,\text{T})\sin 90.0°} = \boxed{4.1 \times 10^{-3}\,\text{m/s}}$$

---

9. **WWW** ***REASONING*** The direction in which the electrons are deflected can be determined using Right-Hand Rule No. 1 and reversing the direction of the force (RHR-1 applies to positive charges, and electrons are negatively charged).

Each electron experiences an acceleration a given by Newton's second law of motion, $a = F/m$, where F is the net force and m is the mass of the electron. The only force acting on the electron is the magnetic force, $F = q_0 vB \sin\theta$, so it is the net force. The speed v of the electron is related to its kinetic energy KE by the relation $KE = \frac{1}{2}mv^2$. Thus, we have enough information to find the acceleration.

SOLUTION
a. According to RHR-1, if you extend your right hand so that your fingers point along the direction of the magnetic field **B** and your thumb points in the direction of the velocity **v** of a *positive* charge, your palm will face in the direction of the force **F** on the positive charge.

For the electron in question, the fingers of the right hand should be oriented downward (direction of **B**) with the thumb pointing to the east (direction of **v**). The palm of the right hand points due north (the direction of **F** on a positive charge). Since the electron is negatively charged, it will be deflected ⬚due south⬚.

b. The acceleration of an electron is given by Newton's second law, where the net force is the magnetic force. Thus,

$$a = \frac{F}{m} = \frac{q_0 v B \sin\theta}{m}$$

Since the kinetic energy is $KE = \frac{1}{2}mv^2$, the speed of the electron is $v = \sqrt{2(KE)/m}$. Thus, the acceleration of the electron is

$$a = \frac{q_0 v B \sin\theta}{m} = \frac{q_0 \sqrt{\dfrac{2(KE)}{m}} B \sin\theta}{m}$$

$$= \frac{\left(1.60 \times 10^{-19}\ C\right)\sqrt{\dfrac{2\left(2.40 \times 10^{-15}\ J\right)}{9.11 \times 10^{-31}\ kg}}\left(2.00 \times 10^{-5}\ T\right)\sin 90.0°}{9.11 \times 10^{-31}\ kg} = \boxed{2.55 \times 10^{14}\ m/s^2}$$

---

13. ***REASONING AND SOLUTION***
   a. The speed of a proton can be found from Equation 21.2 ($r = mv/qB$),

$$v = \frac{qBr}{m} = \frac{(1.6 \times 10^{-19}\ C)(0.30\ T)(0.25\ m)}{1.67 \times 10^{-27}\ kg} = \boxed{7.2 \times 10^6\ m/s}$$

   b. The magnitude F_c of the centripetal force is given by Equation 5.3,

$$F_c = \frac{mv^2}{r} = \frac{(1.67 \times 10^{-27}\ kg)(7.2 \times 10^6\ m/s)^2}{0.25\ m} = \boxed{3.5 \times 10^{-13}\ N}$$

---

15. ⬚WWW⬚ ***REASONING AND SOLUTION*** In one revolution, the particle moves once around the circumference of the circle. Therefore, it travels a distance of $d = 2\pi r$, where r is the radius of the circle. Since the particle moves at constant speed, $v = d/t$, and the time required for one revolution is $t = d/v$. According to Equation 21.2, $r = mv/qB$, so $v = qBr/m$. Thus, the time required for the particle to complete one revolution is

$$t = \frac{d}{v} = \frac{2\pi r}{qBr/m} = \frac{2\pi}{B(q/m)} = \frac{2\pi}{(0.72\ T)(5.7 \times 10^8\ C/kg)} = \boxed{1.5 \times 10^{-8}\ s}$$

19. ***REASONING*** From the discussion in Section 21.3, we know that when a charged particle moves perpendicular to a magnetic field, the trajectory of the particle is a circle. The drawing shows a particle moving in the plane of the paper (the magnetic field is perpendicular to the paper). If the particle is moving initially through the coordinate origin and to the right (along the +x axis), the subsequent circular path of the particle will intersect the y axis at the greatest possible value, which is equal to twice the radius r of the circle.

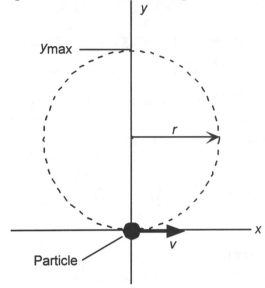

SOLUTION
a. From the drawing above, it can be seen that the largest value of y is equal to the diameter (2r) of the circle. When the particle passes through the coordinate origin its velocity must be parallel to the +x axis. Thus, the angle is $\boxed{\theta = 0°}$.

b. The maximum value of y is twice the radius r of the circle. According to Equation 21.2, the radius of the circular path is $r = mv/qB$. The maximum value y_{max} is, therefore,

$$y_{max} = 2r = 2\left(\frac{mv}{qB}\right) = 2\left[\frac{(3.8 \times 10^{-8} \text{ kg})(44 \text{ m/s})}{(7.3 \times 10^{-6} \text{ C})(1.6 \text{ T})}\right] = \boxed{0.29 \text{ m}}$$

23. ***REASONING AND SOLUTION*** According to Right-Hand Rule No. 1, the magnetic force on the positively charged particle is toward the bottom of the page in the drawing in the text. If the presence of the electric field is to double the magnitude of the net force on the charge, the electric field must also be $\boxed{\text{directed toward the bottom of the page}}$. Note that this results in the electric field being perpendicular to the magnetic field, even though the electric force and the magnetic force are in the same direction.

Furthermore, if the magnitude of the net force on the particle is twice the magnetic force, the electric force must be equal in magnitude to the magnetic force. In other words, combining

Equations 18.2 and 21.1, we find $qE = qvB\sin\theta$, with $\sin\theta = \sin 90.0° = 1.0$. Then, solving for E

$$E = vB\sin\theta = (270 \text{ m/s})(0.52 \text{ T})(1.0) = \boxed{140 \text{ V/m}}$$

---

27. ***REASONING AND SOLUTION*** The magnetic force exerted on the power line is given by Equation 21.3,

$$F = ILB\sin\theta = (1400 \text{ A})(120 \text{ m})(5.0\times10^{-5} \text{ T})(\sin 75°) = \boxed{8.1 \text{ N}}$$

---

31. ***REASONING AND SOLUTION***   The force on a current-carrying wire is given by Equation 21.3: $F = ILB\sin\theta$. Solving for the angle θ, we find that the angle between the wire and the magnetic field is

$$\theta = \sin^{-1}\left(\frac{F}{ILB}\right) = \sin^{-1}\left[\frac{5.46 \text{ N}}{(21.0 \text{ A})(0.655 \text{ m})(0.470 \text{ T})}\right] = \boxed{57.6°}$$

---

33. ***REASONING AND SOLUTION***
a.   From Right-Hand Rule No. 1, if we extend the right hand so that the fingers point in the direction of the magnetic field, and the thumb points in the direction of the current, the palm of the hand faces the direction of the magnetic force on the current.

The springs will stretch when the magnetic force exerted on the copper rod is downward, toward the bottom of the page. Therefore, if you extend your right hand with your fingers pointing out of the page and the palm of your hand facing the bottom of the page, your thumb points left-to-right along the copper rod. Thus, the current flows $\boxed{\text{left-to-right}}$ in the copper rod.

b.   The downward magnetic force exerted on the copper rod is, according to Equation 21.3

$$F = ILB\sin\theta = (12 \text{ A})(0.85 \text{ m})(0.16 \text{ T})\sin 90.0° = 1.6 \text{ N}$$

From Hooke's law, we know that the net force exerted on the rod by the two springs is $F = kx + kx = 2kx$, where k is the spring constant. Solving for x, we find that

$$x = \frac{F}{2k} = \frac{1.6 \text{ N}}{2(75 \text{ N/m})} = \boxed{1.1\times10^{-2} \text{ m}}$$

37. **REASONING AND SOLUTION**

a. The magnetic moment of the coil is

$$Magnetic\ moment = NIA = (50)(15\ A)\left[\pi(0.10\ m)^2\right] = \boxed{24\ A \cdot m^2}$$

where we have used the fact that the area of the circular loop is $A = \pi r^2 = \pi(0.10\ m)^2$.

b. According to Equation 21.4, the torque is the product of the magnetic moment NIA and $B\sin\phi$. However, the maximum torque occurs when $\phi = 90.0°$, so we have

$$\tau = (Magnetic\ moment)(B\sin 90.0°) = (24\ A \cdot m^2)(0.20\ T) = \boxed{4.8\ N \cdot m}$$

---

41. $\boxed{\text{WWW}}$ **REASONING** The torque on the loop is given by Equation 21.4, $\tau = NIAB\sin\phi$. From the drawing in the text, we see that the angle ϕ between the normal to the plane of the loop and the magnetic field is $90° - 35° = 55°$. The area of the loop is $0.70\ m \times 0.50\ m = 0.35\ m^2$.

SOLUTION

a. The magnitude of the net torque exerted on the loop is

$$\tau = NIAB\sin\phi = (75)(4.4\ A)(0.35\ m^2)(1.8\ T)\sin 55° = \boxed{170\ N \cdot m}$$

b. As discussed in the text, when a current-carrying loop is placed in a magnetic field, the loop tends to rotate such that its normal becomes aligned with the magnetic field. The normal to the loop makes an angle of 55° with respect to the magnetic field. Since this angle decreases as the loop rotates, the $\boxed{35°\ \text{angle increases}}$.

---

45. **REASONING** The magnetic moment of the orbiting electron can be found from the expression $Magnetic\ moment = NIA$. For this situation, $N = 1$. Thus, we need to find the current and the area for the orbiting charge.

SOLUTION The current for the orbiting charge is, by definition (see Equation 20.1), $I = \Delta q / \Delta t$, where Δq is the amount of charge that passes a given point during a time interval Δt. Since the charge ($\Delta q = e$) passes by a given point once per revolution, we can find the current by dividing the total orbiting charge by the period T of revolution.

$$I = \frac{\Delta q}{T} = \frac{\Delta q}{2\pi r / v} = \frac{(1.6 \times 10^{-19}\ C)(2.2 \times 10^6\ m/s)}{2\pi(5.3 \times 10^{-11}\ m)} = 1.06 \times 10^{-3}\ A$$

The area of the orbiting charge is

$$A = \pi r^2 = \pi(5.3 \times 10^{-11} \text{ m})^2 = 8.82 \times 10^{-21} \text{ m}^2$$

Therefore, the magnetic moment is

$$\textit{Magnetic moment} = NIA = (1)(1.06 \times 10^{-3} \text{ A})(8.82 \times 10^{-21} \text{ m}^2) = \boxed{9.3 \times 10^{-24} \text{ A} \cdot \text{m}^2}$$

---

49. ***REASONING AND SOLUTION***   The magnetic field inside a long solenoid is given by Equation 21.7:

$$B = \mu_0 nI = (4\pi \times 10^{-7} \text{ T} \cdot \text{m/A})\left(\frac{1400 \text{ turns}}{0.65 \text{ m}}\right)(4.7 \text{ A}) = \boxed{1.3 \times 10^{-2} \text{ T}}$$

---

51. ***REASONING***   The two rods attract each other because they each carry a current I in the same direction. The bottom rod floats because it is in equilibrium. The two forces that act on the bottom rod are the downward force of gravity $m\mathbf{g}$ and the upward magnetic force of attraction to the upper rod. If the two rods are a distance s apart, the magnetic field generated by the top rod at the location of the bottom rod is (see Equation 21.5) $B = \mu_0 I/(2\pi s)$. According to Equation 21.3, the magnetic force exerted on the bottom rod is $F = ILB\sin\theta = \mu_0 I^2 L \sin\theta/(2\pi s)$, where θ is the angle between the magnetic field at the location of the bottom rod and the direction of the current in the bottom rod. Since the rods are parallel, the magnetic field is perpendicular to the direction of the current (RHR-2), and $\theta = 90.0°$, so that $\sin\theta = 1.0$.

SOLUTION   Taking upward as the positive direction, the net force on the bottom rod is

$$\frac{\mu_0 I^2 L \sin\theta}{2\pi s} - mg = 0$$

Solving for I, we find

$$I = \sqrt{\frac{2\pi mgs}{\mu_0 L}} = \sqrt{\frac{2\pi(0.073 \text{ kg})(9.80 \text{ m/s}^2)(8.2 \times 10^{-3} \text{ m})}{(4\pi \times 10^{-7} \text{ T} \cdot \text{m/A})(0.85 \text{ m})}} = \boxed{190 \text{ A}}$$

---

57. ***REASONING***   According to Equation 21.6 the magnetic field at the center of a circular, current-carrying loop of N turns and radius r is $B = N\mu_0 I/(2r)$. The number of turns N in the coil can be found by dividing the total length L of the wire by the circumference after it has been wound into a circle. The current in the wire can be found by using Ohm's law, $I = V/R$.

SOLUTION   The number of turns in the wire is

$$N = \frac{L}{2\pi r}$$

The current in the wire is

$$I = \frac{V}{R} = \frac{12.0 \text{ V}}{\left(5.90 \times 10^{-3} \ \Omega/\text{m}\right)L} = \frac{2.03 \times 10^{3}}{L} \text{ A}$$

Therefore, the magnetic field at the center of the coil is

$$B = N\left(\frac{\mu_0 I}{2r}\right) = \left(\frac{L}{2\pi r}\right)\left(\frac{\mu_0 I}{2r}\right) = \frac{\mu_0 L I}{4\pi r^2}$$

$$= \frac{\left(4\pi \times 10^{-7} \text{ T·m/A}\right)L\left(\dfrac{2.03 \times 10^{3}}{L} \text{ A}\right)}{4\pi(0.140 \text{ m})^2} = \boxed{1.04 \times 10^{-2} \text{ T}}$$

---

61. **REASONING** Since the two wires are next to each other, the net magnetic field is everywhere parallel to $\Delta\ell$ in Figure 21.40. Moreover, the net magnetic field **B** has the same magnitude B at each point along the circular path, because each point is at the same distance from the wires. Thus, in Ampere's law (Equation 21.8), $B_\parallel = B$, $I = I_1 + I_2$, and we have

$$\Sigma B_\parallel \Delta\ell = B\left(\Sigma\Delta\ell\right) = \mu_0\left(I_1 + I_2\right)$$

But $\Sigma\Delta\ell$ is just the circumference $(2\pi r)$ of the circle, so Ampere's law becomes

$$B(2\pi r) = \mu_0\left(I_1 + I_2\right)$$

This expression can be solved for B.

SOLUTION
a. When the currents are in the same direction, we find that

$$B = \frac{\mu_0\left(I_1 + I_2\right)}{2\pi r} = \frac{\left(4\pi \times 10^{-7} \text{ T·m/A}\right)(28 \text{ A} + 12 \text{ A})}{2\pi(0.72 \text{ m})} = \boxed{1.1 \times 10^{-5} \text{ T}}$$

b. When the currents have opposite directions, a similar calculation shows that

$$B = \frac{\mu_0\left(I_1 - I_2\right)}{2\pi r} = \frac{\left(4\pi \times 10^{-7} \text{ T·m/A}\right)(28 \text{ A} - 12 \text{ A})}{2\pi(0.72 \text{ m})} = \boxed{4.4 \times 10^{-6} \text{ T}}$$

65. ***REASONING*** The coil carries a current and experiences a torque when it is placed in an external magnetic field. Thus, when the coil is placed in the magnetic field due to the solenoid, it will experience a torque given by Equation 21.4: $\tau = NIAB\sin\phi$, where N is the number of turns in the coil, A is the area of the coil, B is the magnetic field inside the solenoid, and ϕ is the angle between the normal to the plane of the coil and the magnetic field. The magnetic field in the solenoid can be found from Equation 21.7: $B = \mu_0 nI$, where n is the number of turns per unit length of the solenoid and I is the current.

SOLUTION The magnetic field inside the solenoid is

$$B = \mu_0 nI = (4\pi \times 10^{-7}\ \text{T}\cdot\text{m}/\text{A})(1400\ \text{turns}/\text{m})(3.5\ \text{A}) = 6.2 \times 10^{-3}\ \text{T}$$

The torque exerted on the coil is

$$\tau = NIAB\sin\phi = (50)(0.50\ \text{A})(1.2 \times 10^{-3}\ \text{m}^2)(6.2 \times 10^{-3}\ \text{T})(\sin 90.0°) = \boxed{1.9 \times 10^{-4}\ \text{N}\cdot\text{m}}$$

---

69. ***REASONING AND SOLUTION*** According to Equation 21.1, $B = \dfrac{F}{q_0(v\sin\theta)}$. Solving this for the angle θ, we find

$$\theta = \sin^{-1}\left(\frac{F}{q_0 vB}\right) = \sin^{-1}\left[\frac{2.30 \times 10^{-7}\ \text{N}}{(1.70 \times 10^{-5}\ \text{C})(2.80 \times 10^2\ \text{m}/\text{s})(5.00 \times 10^{-5}\ \text{T})}\right] = \boxed{75.1°\ \text{and}\ 105°}$$

---

73. ⬚WWW ***REASONING*** The particle travels in a semicircular path of radius r, where r is given by Equation 21.2 ($r = mv/qB$). The time spent by the particle in the magnetic field is given by $t = s/v$, where s is the distance traveled by the particle and v is its speed. The distance s is equal to one-half the circumference of a circle ($s = \pi r$).

SOLUTION We find that

$$t = \frac{s}{v} = \frac{\pi r}{v} = \frac{\pi}{v}\left(\frac{mv}{qB}\right) = \frac{\pi m}{qB} = \frac{\pi(6.0 \times 10^{-8}\ \text{kg})}{(7.2 \times 10^{-6}\ \text{C})(3.0\ \text{T})} = \boxed{8.7 \times 10^{-3}\ \text{s}}$$

77. **REASONING**   The magnetic moment of the rotating charge can be found from the expression *Magnetic moment = NIA*. For this situation, $N = 1$. Thus, we need to find the current and the area for the rotating charge. This can be done by resorting to first principles.

SOLUTION   The current for the rotating charge is, by definition (see Equation 20.1), $I = \Delta q / \Delta t$, where Δq is the amount of charge that passes by a given point during a time interval Δt. Since the charge passes by once per revolution, we can find the current by dividing the total rotating charge by the period T of revolution.

$$I = \frac{\Delta q}{T} = \frac{\Delta q}{2\pi/\omega} = \frac{\omega \, \Delta q}{2\pi} = \frac{(150 \text{ rad / s})(4.0 \times 10^{-6} \text{ C})}{2\pi} = 9.5 \times 10^{-5} \text{ A}$$

The area of the rotating charge is $A = \pi r^2 = \pi (0.20 \text{ m})^2 = 0.13 \text{ m}^2$

Therefore, the magnetic moment is

$$\textit{Magnetic moment} = NIA = (1)(9.5 \times 10^{-5} \text{ A})(0.13 \text{ m}^2) = \boxed{1.2 \times 10^{-5} \text{ A} \cdot \text{m}^2}$$

CHAPTER 22 | ELECTROMAGNETIC INDUCTION

1. **REASONING AND SOLUTION** According to Equation 22.1, the motional emf when the velocity **v**, the magnetic field **B**, and the length **L** are mutually perpendicular is given by Emf = vBL. Therefore in order for a spark to jump across the gap, the rod would have to be moving in the magnetic field with speed

$$v = \frac{\text{Emf}}{BL} = \frac{940 \text{ V}}{(4.8 \text{ T})(1.3 \text{ m})} = \boxed{150 \text{ m/s}}$$

5. **WWW** **REASONING AND SOLUTION** For each of the three rods in the drawing, we have the following:

Rod A: The motional emf is $\boxed{\text{zero}}$, because the velocity of the rod is parallel to the direction of the magnetic field, and the charges do not experience a magnetic force.

Rod B: The motional emf is, according to Equation 22.1,

$$\text{Emf} = vBL = (2.7 \text{ m/s})(0.45 \text{ T})(1.3 \text{ m}) = \boxed{1.6 \text{ V}}$$

The positive end of Rod 2 is $\boxed{\text{end 2}}$.

Rod C: The motional emf is $\boxed{\text{zero}}$, because the magnetic force F on each charge is directed perpendicular to the length of the rod. For the ends of the rod to become charged, the magnetic force must be directed parallel to the length of the rod.

7. **REASONING** The moving rod acts as a battery does in supplying an emf or voltage in the circuit. The speed of the rod is related to the emf by Equation 22.1, Emf = vBL. From Equation 20.6c, we have $P = \left(\text{Emf}\right)^2 / R$, where P is the power consumed by the circuit and R is the resistance in the circuit.

SOLUTION
a. Solving Equation 20.6c for the emf, we have

$$\text{Emf} = \sqrt{PR} = \sqrt{(15 \text{ W})(6.0 \text{ }\Omega)} = 9.5 \text{ V}$$

The speed of the rod is

$$v = \frac{\text{Emf}}{LB} = \frac{9.5 \text{ V}}{(2.4 \text{ T})(1.2 \text{ m})} = \boxed{3.3 \text{ m/s}}$$

b.   The force exerted on the rod by the magnetic field is given by Equation 21.3 with $\theta = 90°$, and $I = \text{Emf}/R$. Therefore,

$$F = ILB (\sin 90°) = \left(\frac{\text{Emf}}{R}\right)LB = \left(\frac{9.5 \text{ V}}{6.0 \ \Omega}\right)(1.2 \text{ m})(2.4 \text{ T}) = \boxed{4.6 \text{ N}}$$

---

11. **REASONING**   The general expression for the magnetic flux through an area A is given by Equation 22.2: $\Phi = BA\cos\phi$ where B is the magnitude of the magnetic field and ϕ is the angle of inclination of the magnetic field **B** with respect to the normal to the area.

The magnetic flux through the door is a maximum when the magnetic field lines are perpendicular to the door and $\phi_1 = 0.0°$ so that $\Phi_1 = \Phi_{max} = BA(\cos 0.0°) = BA$.

SOLUTION   When the door rotates through an angle ϕ_2, the magnetic flux that passes through the door decreases from its maximum value to one-third of its maximum value. Therefore $\Phi_2 = \frac{1}{3}\Phi_{max}$, and we have

$$\Phi_2 = BA\cos\phi_2 = \frac{1}{3}BA \quad \text{or} \quad \cos\phi_2 = \frac{1}{3} \quad \text{or} \quad \phi_2 = \cos^{-1}\left(\frac{1}{3}\right) = \boxed{70.5°}$$

---

15. **WWW   REASONING**   The general expression for the magnetic flux through an area A is given by Equation 22.2: $\Phi = BA\cos\phi$, where B is the magnitude of the magnetic field and ϕ is the angle of inclination of the magnetic field **B** with respect to the normal to the surface.

SOLUTION   Since the magnetic field **B** is parallel to the surface for the triangular ends and the bottom surface, the flux through each of these three surfaces is $\boxed{0 \text{ Wb}}$.

The flux through the 1.2 m by 0.30 m face is

$$\Phi = (0.25 \text{ T})(1.2 \text{ m})(0.30 \text{ m}) \cos 0.0° = \boxed{0.090 \text{ Wb}}$$

For the 1.2 m by 0.50 m side, the area makes an angle ϕ with the magnetic field **B**, where

$$\phi = 90° - \tan^{-1}\left(\frac{0.30 \text{ m}}{0.40 \text{ m}}\right) = 53°$$

Therefore,

$$\Phi = (0.25 \text{ T})(1.2 \text{ m})(0.50 \text{ m}) \cos 53° = \boxed{0.090 \text{ Wb}}$$

19. **REASONING** According to Equation 22.3, the average emf induced in a coil of N loops is $\text{Emf} = -N\Delta\Phi/\Delta t$.

SOLUTION For the circular coil in question, the flux through a single turn changes by

$$\Delta\Phi = BA\cos 45° - BA\cos 90° = BA\cos 45°$$

during the interval of $\Delta t = 0.010$ s. Therefore, for N turns, Faraday's law gives the magnitude of the emf (without the minus sign) as

$$\text{Emf} = \frac{NBA\cos 45°}{\Delta t}$$

Since the loops are circular, the area A of each loop is equal to πr^2. Solving for B, we have

$$B = \frac{(\text{Emf})\Delta t}{N\pi r^2 \cos 45°} = \frac{(0.065 \text{ V})(0.010 \text{ s})}{(950)\pi(0.060 \text{ m})^2 \cos 45°} = \boxed{8.6\times10^{-5} \text{ T}}$$

25. 　$\boxed{\text{WWW}}$　**REASONING AND SOLUTION** According to Ohm's law, the emf developed by the coil is $\text{Emf} = IR$, which, when combined with the definition of electric current becomes $\text{Emf} = (\Delta q)R/(\Delta t)$. The magnitude of the emf is also given by Equation 22.3 (Faraday's law) without the minus sign:

$$\text{Emf} = N\frac{\Delta\Phi}{\Delta t} = N\left(\frac{BA\cos 0° - 0}{\Delta t}\right) = \frac{NBA}{\Delta t}$$

Combining the two expressions for the emf, we have

$$\frac{(\Delta q)R}{\Delta t} = \frac{NBA}{\Delta t} \qquad \text{or} \qquad B = \frac{(\Delta q)R}{NA} = \frac{(8.87\times10^{-3} \text{ C})(45.0 \text{ }\Omega)}{(1850)(4.70\times10^{-4} \text{ m}^2)} = \boxed{0.459 \text{ T}}$$

29. **REASONING AND SOLUTION** If the north and south poles of the magnet are interchanged, the currents in the ammeter would simply be reversed in direction.

Therefore, in Figure 22.1b, the current will flow $\boxed{\text{right to left}}$ through the ammeter.

Similarly, in Figure 22.1c, the current will flow $\boxed{\text{left to right}}$ through the ammeter.

31. **REASONING** In solving this problem, we apply Lenz's law, which essentially says that the change in magnetic flux must be opposed by the induced magnetic field.

SOLUTION
a. The magnetic field due to the wire in the vicinity of position 1 is directed out of the paper. The coil is moving closer to the wire into a region of higher magnetic field, so the flux through the coil is increasing. Lenz's law demands that the induced field counteract this increase. The direction of the induced field, therefore, must be into the paper. The current in the coil must be
$\boxed{\text{clockwise}}$.

b. At position 2 the magnetic field is directed into the paper and is decreasing as the coil moves away from the wire. The induced magnetic field, therefore, must be directed into the paper, so the current in the coil must be $\boxed{\text{clockwise}}$.

---

35. $\boxed{\text{WWW}}$ **REASONING AND SOLUTION**
a. From the drawing, we see that the period of the generator (the time for one full cycle) is 0.42 s; therefore, the frequency of the generator is

$$f = \frac{1}{T} = \frac{1}{0.42 \text{ s}} = \boxed{2.4 \text{ Hz}}$$

b. The angular speed of the generator is related to its frequency by $\omega = 2\pi f$, so the angular speed is

$$\omega = 2\pi(2.4 \text{ Hz}) = \boxed{15 \text{ rad/s}}$$

c. The magnitude of the maximum emf induced in a rotating planar coil is given by $(\text{Emf})_0 = NAB\omega$ (see Equation 22.4). The magnitude of the magnetic field can be found by solving this expression for B and noting from the drawing that $(\text{Emf})_0 = 28$ V:

$$B = \frac{(\text{Emf})_0}{NA\omega} = \frac{28 \text{ V}}{(150)(0.020 \text{ m}^2)(15 \text{ rad/s})} = \boxed{0.62 \text{ T}}$$

---

39. **REASONING** We can use the information given in the problem statement to determine the area of the coil A. Since it is square, the length of one side is $\ell = \sqrt{A}$.

SOLUTION According to Equation 22.4, the maximum emf induced in the coil is $(\text{Emf})_0 = NAB\omega$. Therefore, the length of one side of the coil is

$$\ell = \sqrt{A} = \sqrt{\frac{(\text{Emf})_0}{NB\omega}} = \sqrt{\frac{75.0 \text{ V}}{(248)(0.170 \text{ T})(79.1 \text{ rad/s})}} = \boxed{0.150 \text{ m}}$$

---

41. ***REASONING AND SOLUTION*** When the motor just begins to turn the fan blade, there is no back emf. Therefore, Ohm's law indicates that $120.0 \text{ V} = I_0 R$, where I_0 is the current in the motor and R is the resistance of the motor coils. Solving for I_0 gives

$$I_0 = \frac{120.0 \text{ V}}{R} \tag{1}$$

At normal operating speed, there is a back emf, and Equation 22.5 gives the current in the motor as

$$I = \frac{120.0 \text{ V} - \text{Emf}}{R} \tag{2}$$

But we know that $I = 0.150 I_0$. Therefore, using Equations (1) and (2), we find that

$$\frac{120.0 \text{ V} - \text{Emf}}{R} = 0.150 \left(\frac{120.0 \text{ V}}{R} \right)$$

Eliminating R algebraically, we have

$$120.0 \text{ V} - \text{Emf} = 0.150(120.0 \text{ V}) \qquad \text{or} \qquad \text{Emf} = 0.850(120.0 \text{ V}) = \boxed{102 \text{ V}}$$

---

45. ***REASONING AND SOLUTION*** From the results of Example 13, the self-inductance L of a long solenoid is given by $L = \mu_0 n^2 A\ell$. Solving for the number of turns n per unit length gives

$$n = \sqrt{\frac{L}{\mu_0 A\ell}} = \sqrt{\frac{1.4 \times 10^{-3} \text{ H}}{(4\pi \times 10^{-7} \text{T} \cdot \text{m/A})(1.2 \times 10^{-3} \text{ m}^2)(0.052 \text{ m})}} = 4.2 \times 10^3 \text{ turns/m}$$

Therefore, the total number of turns N is the product of n and the length ℓ of the solenoid:

$$N = n\ell = (4.2 \times 10^3 \text{ turns/m})(0.052 \text{ m}) = \boxed{220 \text{ turns}}$$

---

49. ***REASONING*** The energy density is given by Equation 22.11 as

$$\text{Energy density} = \frac{\text{Energy}}{\text{Volume}} = \frac{1}{2\mu_0} B^2$$

The energy stored is the energy density times the volume.

SOLUTION The volume is the area A times the height h. Therefore, the energy stored is

$$\text{Energy} = \frac{B^2 Ah}{2\mu_0} = \frac{(7.0\times10^{-5}\text{ T})^2(5.0\times10^8\text{ m}^2)(1500\text{ m})}{2(4\pi\times10^{-7}\text{ T}\cdot\text{m/A})} = \boxed{1.5\times10^9\text{ J}}$$

---

53. **REASONING** According to Ohm's law, the average current I induced in the coil is given by $I = (\text{Emf})/R$, where R is the resistance of the coil. To find the induced emf, we use Faraday's law of electromagnetic induction.

SOLUTION The magnitude of the induced emf can be found from Faraday's law of electromagnetic induction and is given by Equation 22.3 without the minus sign:

$$\text{Emf} = N\frac{\Delta\Phi}{\Delta t} = N\frac{\Delta(BA\cos 0°)}{\Delta t} = NA\frac{\Delta B}{\Delta t}$$

We have used the fact that the field within a long solenoid is perpendicular to the cross-sectional area A of the solenoid and makes an angle of $0°$ with respect to the normal to the area. The field is given by Equation 21.7 as $B = \mu_0 nI$, so the change ΔB in the field is $\Delta B = \mu_0 n\Delta I$, where ΔI is the change in the current. The induced current is, then,

$$I = \frac{NA\dfrac{\Delta B}{\Delta t}}{R} = \frac{NA\dfrac{\mu_0 n\Delta I}{\Delta t}}{R}$$

$$= \frac{(10)(6.0\times10^{-4}\text{ m}^2)\dfrac{(4\pi\times10^{-7}\text{ T}\cdot\text{m/A})(400\text{ turns/m})(0.40\text{ A})}{(0.050\text{ s})}}{1.5\ \Omega} = \boxed{1.6\times10^{-5}\text{ A}}$$

---

55. **REASONING AND SOLUTION** Since the secondary voltage is less than the primary voltage, we can conclude that the transformer used in the doorbell described in the problem is a $\boxed{\text{step-down transformer}}$. The turns ratio is given by the transformer equation, Equation 22.12:

$$\frac{N_s}{N_p} = \frac{V_s}{V_p} = \frac{10.0\text{ V}}{120\text{ V}} = \frac{1}{12}$$

so the turns ratio is $\boxed{1:12}$.

59. **REASONING** The power lost in heating the wires is given by Equation 20.6b: $P = I^2 R$.
Before we can use this equation, however, we must determine the total resistance R of the wire
and the current that flows through the wire.

SOLUTION
a. The total resistance of one of the wires is equal to the resistance per unit length multiplied
by the length of the wire. Thus, we have

$$(5.0 \times 10^{-2} \ \Omega/\text{km})(7.0 \ \text{km}) = 0.35 \ \Omega$$

and the total resistance of the transmission line is twice this value or 0.70 Ω. According to
Equation 20.6a ($P = IV$), the current flowing into the town is

$$I = \frac{P}{V} = \frac{1.2 \times 10^6 \ \text{W}}{1200 \ \text{V}} = 1.0 \times 10^3 \ \text{A}$$

Thus, the power lost in heating the wire is

$$P = I^2 R = \left(1.0 \times 10^3 \ \text{A}\right)^2 (0.70 \ \Omega) = \boxed{7.0 \times 10^5 \ \text{W}}$$

b. According to the transformer equation (Equation 22.12), the stepped-up voltage is

$$V_s = V_p \left(\frac{N_s}{N_p}\right) = (1200 \ \text{V})\left(\frac{100}{1}\right) = 1.2 \times 10^5 \ \text{V}$$

According to Equation 20.6a ($P = IV$), the current in the wires is

$$I = \frac{P}{V} = \frac{1.2 \times 10^6 \ \text{W}}{1.2 \times 10^5 \ \text{V}} = 1.0 \times 10^1 \ \text{A}$$

The power lost in heating the wires is now

$$P = I^2 R = \left(1.0 \times 10^1 \ \text{A}\right)^2 (0.70 \ \Omega) = \boxed{7.0 \times 10^1 \ \text{W}}$$

---

65. **REASONING AND SOLUTION**   According to Faraday's law, only the relative motion
between the coil and magnet is significant.. It makes no difference if the coil is held fixed and
the magnet is moved or the magnet is held fixed and the coil is moved. The results are the
same. Therefore,

a. The current will flow from $\boxed{\text{left to right}}$ as shown in Figure 22.1b.

b. The current will flow from | right to left | as shown in Figure 22.1c.

---

67. ***REASONING AND SOLUTION***   The emf induced in the coil of an ac generator is given by Equation 22.4 as

$$\text{Emf} = NAB\omega \ \sin\omega t = (500)(1.2\times10^{-2} \ \text{m}^2)(0.13 \ \text{T})(34 \ \text{rad/s}) \sin 27° = \boxed{12 \ \text{V}}$$

---

71. ***REASONING***   According to Equation 22.3, the average emf induced in a single loop is emf $= -\Delta\Phi / \Delta t$. Since the magnitude of the magnetic field is changing, the area of the loop remains constant, and the direction of the field is parallel to the normal to the loop, the change in flux through the loop is given by $\Delta\Phi = (\Delta B)A$. Thus the magnitude of the induced emf in the loop is given by $\text{Emf} = (\Delta B)A / \Delta t$.

Similarly, when the area of the loop is changed and the field B has a given value, we find the magnitude of the induced emf to be $\text{Emf} = B\Delta A / \Delta t$.

SOLUTION
a. The magnitude of the induced emf when the field changes in magnitude is

$$\text{Emf} = \frac{(\Delta B)A}{\Delta t} = (0.20 \ \text{T/s})(0.018 \ \text{m}^2) = \boxed{3.6\times10^{-3} \ \text{V}}$$

b. At a particular value of B (when B is changing), the rate at which the area must change can be obtained from

$$\text{Emf} = \frac{B\Delta A}{\Delta t} \quad \text{or} \quad \frac{\Delta A}{\Delta t} = \frac{\text{Emf}}{B} = \frac{3.6\times10^{-3} \ \text{V}}{1.8 \ \text{T}} = \boxed{2.0\times10^{-3} \ \text{m}^2 / \text{s}}$$

In order for the induced emf to be zero, the magnitude of the magnetic field and the area of the loop must change in such a way that the flux remains constant.   Since the magnitude of the magnetic field is increasing, the area of the loop must decrease, if the flux is to remain constant. Therefore, | the area of the loop must be shrunk | .

CHAPTER 23 | ALTERNATING CURRENT CIRCUITS

PROBLEMS

---

1. **REASONING AND SOLUTION** The capacitive reactance is given by Equation 23.2: $X_C = \dfrac{1}{2\pi f C}$. Solving for the frequency f, we find that

$$f = \frac{1}{2\pi X_C C} = \frac{1}{2\pi (168 \ \Omega)(7.50 \times 10^{-6} \ \text{F})} = \boxed{126 \ \text{Hz}}$$

---

5. **REASONING AND SOLUTION**

a. The equivalent capacitance C_s of two capacitors in series is given by Equation 20.19 as

$$\frac{1}{C_s} = \frac{1}{C_1} + \frac{1}{C_2} = \frac{1}{3.00 \times 10^{-6} \ \text{F}} + \frac{1}{6.00 \times 10^{-6} \ \text{F}} \qquad \text{or} \qquad \boxed{C_s = 2.00 \ \mu\text{F}}$$

b. The current in the circuit can be found by solving Equation 23.1, $V_{rms} = I_{rms} X_C$, for I_{rms}. However, we must first find the capacitive reactance X_C. From Equation 23.2, we have

$$X_C = \frac{1}{2\pi f C_s} = \frac{1}{2\pi (510 \ \text{Hz})(2.00 \times 10^{-6} \ \text{F})} = 156 \ \Omega$$

The current in the circuit is given by

$$I_{rms} = \frac{V_{rms}}{X_C} = \frac{120 \ \text{V}}{156 \ \Omega} = \boxed{0.77 \ \text{A}}$$

---

9. **REASONING** The individual reactances are given by Equations 23.2 and 23.4, respectively,

[Capacitive reactance] $\qquad X_C = \dfrac{1}{2\pi f C}$

[Inductive reactance] $\qquad X_L = 2\pi f L$

When the reactances are equal, we have $X_C = X_L$, from which we find

$$\frac{1}{2\pi f C} = 2\pi f L \qquad \text{or} \qquad 4\pi^2 f^2 L C = 1$$

The last expression may be solved for the frequency f.

SOLUTION   Solving for f with $L = 52 \times 10^{-3}$ H and $C = 76 \times 10^{-6}$ F, we obtain

$$f = \frac{1}{2\pi\sqrt{LC}} = \frac{1}{2\pi\sqrt{(52 \times 10^{-3} \text{ H})(76 \times 10^{-6} \text{ F})}} = \boxed{8.0 \times 10^{1} \text{ Hz}}$$

---

11. $\boxed{\textbf{WWW}}$ **REASONING**   The rms current can be calculated from Equation 23.3, $I_{rms} = V_{rms} / X_L$, provided that the inductive reactance is obtained first. Then the peak value of the current I_0 supplied by the generator can be calculated from the rms current I_{rms} by using Equation 20.12, $I_0 = \sqrt{2}\, I_{rms}$.

SOLUTION   At the frequency of $f = 620$ Hz, we find, using Equations 23.4 and 23.3, that

$$X_L = 2\pi f L = 2\pi(620 \text{ Hz})(8.2 \times 10^{-3} \text{ H}) = 32 \text{ }\Omega$$

$$I_{rms} = \frac{V_{rms}}{X_L} = \frac{10.0 \text{ V}}{32 \text{ }\Omega} = 0.31 \text{ A}$$

Therefore, from Equation 20.12, we find that the *peak value I_0* of the current supplied by the generator must be

$$I_0 = \sqrt{2}\, I_{rms} = \sqrt{2}\,(0.31 \text{ A}) = \boxed{0.44 \text{ A}}$$

---

15.   **REASONING**   The voltage supplied by the generator can be found from Equation 23.6, $V_{rms} = I_{rms} Z$. The value of I_{rms} is given in the problem statement, so we must obtain the impedance of the circuit.

SOLUTION   The impedance of the circuit is, according to Equation 23.7,

$$Z = \sqrt{R^2 + (X_L - X_C)^2} = \sqrt{(275 \text{ }\Omega)^2 + (648 \text{ }\Omega - 415 \text{ }\Omega)^2} = 3.60 \times 10^2 \text{ }\Omega$$

The rms voltage of the generator is

$$V_{rms} = I_{rms} Z = (0.233 \text{ A})(3.60 \times 10^2 \text{ }\Omega) = \boxed{83.9 \text{ V}}$$

19. ***REASONING***   We can use the equations for a series RCL circuit to solve this problem provided that we set $X_C = 0$ since there is no capacitor in the circuit. The current in the circuit can be found from Equation 23.6, $V_{rms} = I_{rms}Z$, once the impedance of the circuit has been obtained. Equation 23.8, $\tan\phi = (X_L - X_C)/R$, can then be used (with $X_C = 0$) to find the phase angle between the current and the voltage.

SOLUTION   The inductive reactance is (Equation 23.4)

$$X_L = 2\pi f L = 2\pi(106\text{ Hz})(0.200\text{ H}) = 133\ \Omega$$

The impedance of the circuit is

$$Z = \sqrt{R^2 + (X_L - X_C)^2} = \sqrt{R^2 + X_L^2} = \sqrt{(215\ \Omega)^2 + (133\ \Omega)^2} = 253\ \Omega$$

a.   The current through each circuit element is, using Equation 23.6,

$$I_{rms} = \frac{V_{rms}}{Z} = \frac{234\text{ V}}{253\ \Omega} = \boxed{0.925\text{ A}}$$

b.   The phase angle between the current and the voltage is, according to Equation 23.8 (with $X_C = 0$),

$$\tan\phi = \frac{X_L - X_C}{R} = \frac{X_L}{R} = \frac{133\ \Omega}{215\ \Omega} = 0.619$$

$$\phi = \tan^{-1}(0.619) = \boxed{31.8°}$$

---

21. $\boxed{\text{WWW}}$ ***REASONING***   We can use the equations for a series RCL circuit to solve this problem, provided that we set $X_L = 0$ since there is no inductance in the circuit. Thus, according to Equations 23.6 and 23.7, the current in the circuit is $I_{rms} = V_{rms}/\sqrt{R^2 + X_C^2}$. When the frequency f is very large, the capacitive reactance is zero, or $X_C = 0$, in which case the current becomes $I_{rms}(\text{large }f) = V_{rms}/R$. When the current I_{rms} in the circuit is one-half the value of $I_{rms}(\text{large }f)$ that exists when the frequency is very large, we have

$$\frac{I_{rms}}{I_{rms}(\text{large }f)} = \frac{1}{2}$$

We can use these expressions to write the ratio above in terms of the resistance and the capacitive reactance. Once the capacitive reactance is known, the frequency can be determined.

SOLUTION The ratio of the currents is

$$\frac{I_{rms}}{I_{rms}\,(\text{large }f)} = \frac{V_{rms}/\sqrt{R^2+X_C^2}}{V_{rms}/R} = \frac{R}{\sqrt{R^2+X_C^2}} = \frac{1}{2} \qquad \text{or} \qquad \frac{R^2}{R^2+X_C^2} = \frac{1}{4}$$

Taking the reciprocal of this result gives

$$\frac{R^2+X_C^2}{R^2} = 4 \qquad \text{or} \qquad 1+\frac{X_C^2}{R^2} = 4$$

Therefore,

$$\frac{X_C}{R} = \sqrt{3}$$

According to Equation 23.2, $X_C = 1/(2\pi f C)$, so it follows that

$$\frac{X_C}{R} = \frac{1/(2\pi f C)}{R} = \sqrt{3}$$

Thus,

$$f = \frac{1}{2\pi RC\sqrt{3}} = \frac{1}{2\pi(85\,\Omega)(4.0\times10^{-6}\text{ F})\sqrt{3}} = \boxed{270\text{ Hz}}$$

---

27. **REASONING** The current in an RCL circuit is given by Equation 23.6, $I_{rms} = V_{rms}/Z$, where the impedance Z of the circuit is given by Equation 23.7 as $Z = \sqrt{R^2+(X_L-X_C)^2}$. The current is a maximum when the impedance is a minimum for a given generator voltage. The minimum impedance occurs when the frequency is f_0, corresponding to the condition that $X_L = X_C$, or $2\pi f_0 L = 1/(2\pi f_0 C)$. Solving for the frequency f_0, called the resonant frequency, we find that

$$f_0 = \frac{1}{2\pi\sqrt{LC}}$$

Note that the resonant frequency depends on the inductance and the capacitance, but does not depend on the resistance.

SOLUTION
a. The frequency at which the current is a maximum is

$$f_0 = \frac{1}{2\pi\sqrt{LC}} = \frac{1}{2\pi\sqrt{(17.0\times10^{-3}\text{ H})(12.0\times10^{-6}\text{ F})}} = \boxed{352\text{ Hz}}$$

b.   The maximum value of the current occurs when $f = f_0$. This occurs when $X_L = X_C$, so that $Z = R$. Therefore, according to Equation 23.6, we have

$$I_{rms} = \frac{V_{rms}}{Z} = \frac{V_{rms}}{R} = \frac{155 \text{ V}}{10.0 \text{ } \Omega} = \boxed{15.5 \text{ A}}$$

---

29.   ***REASONING AND SOLUTION***   The resonant frequency is given by Equation 23.10 as

$$f_0 = \frac{1}{2\pi\sqrt{LC}} = 9.3 \text{ kHz}$$

If the inductance and capacitance of the circuit are each tripled, the new resonant frequency is

$$f_0' = \frac{1}{2\pi\sqrt{(3L)(3C)}} = \frac{1}{2\pi\sqrt{9LC}} = \frac{1}{3}\left(\frac{1}{2\pi\sqrt{LC}}\right) = \tfrac{1}{3}f_0 = \tfrac{1}{3}(9.3 \text{ kHz}) = \boxed{3.1 \text{ kHz}}$$

---

33.   ***REASONING***   At the resonant frequency f_0, we have $C = 1/(4\pi^2 f_0^2 L)$. We want to determine some series combination of capacitors whose equivalent capacitance C_s' is such that $f_0' = 3f_0$. Thus,

$$C_s' = \frac{1}{4\pi^2 f_0'^2 L} = \frac{1}{4\pi^2 (3f_0)^2 L} = \frac{1}{9}\left(\frac{1}{4\pi^2 f_0^2 L}\right) = \frac{1}{9}C$$

The equivalent capacitance of a series combination of capacitors is $1/C_s' = 1/C_1 + 1/C_2 + ...$ If we require that all the capacitors have the same capacitance C, the equivalent capacitance is

$$\frac{1}{C_s'} = \frac{1}{C} + \frac{1}{C} + \cdots = \frac{n}{C}$$

where n is the total number of identical capacitors. Using the result above, we find that

$$\frac{1}{C_s'} = \frac{1}{\tfrac{1}{9}C} = \frac{n}{C} \qquad \text{or} \qquad n = 9$$

Therefore, the number of *additional* capacitors that must be inserted in series in the circuit so that the frequency triples is $n' = n - 1 = \boxed{8}$.

---

37. **REASONING** The voltage across the capacitor reaches its maximum instantaneous value when the generator voltage reaches its maximum instantaneous value. The maximum value of the capacitor voltage first occurs one-fourth of the way, or one-quarter of a period, through a complete cycle (see the voltage curve in Figure 23.4).

SOLUTION The period of the generator is $T = 1/f = 1/(5.00 \text{ Hz}) = 0.200 \text{ s}$. Therefore, the least amount of time that passes before the instantaneous voltage across the capacitor reaches its maximum value is $\frac{1}{4}T = \frac{1}{4}(0.200 \text{ s}) = \boxed{5.00 \times 10^{-2} \text{ s}}$.

---

41. **WWW** **REASONING**   Since the capacitor and the inductor are connected in parallel, the voltage across each of these elements is the same or $V_L = V_C$. Using Equations 23.3 and 23.1, respectively, this becomes $I_{rms}X_L = I_{rms}X_C$. Since the currents in the inductor and capacitor are equal, this relation simplifies to $X_L = X_C$. Therefore, we can find the value of the inductance by equating the expressions (Equations 23.4 and 23.2) for the inductive reactance and the capacitive reactance, and solving for L.

SOLUTION Since $X_L = X_C$, we have

$$2\pi f L = \frac{1}{2\pi f C}$$

Therefore, the value of the inductance is

$$L = \frac{1}{4\pi^2 f^2 C} = \frac{1}{4\pi^2 (60.0 \text{ Hz})^2 (40.0 \times 10^{-6} \text{ F})} = 0.176 \text{ H} = \boxed{176 \text{ mH}}$$

---

43. **WWW** **REASONING**   Since we know the values of the resonant frequency of the circuit, the capacitance, and the generator voltage, we can find the value of the inductance from Equation 23.10, the expression for the resonant frequency. The resistance can be found from energy considerations at resonance; the power factor is given by $\cos\phi$, where the phase angle ϕ is given by Equation 23.8, $\tan\phi = (X_L - X_C)/R$.

SOLUTION
a. Solving Equation 23.10 for the inductance L, we find that

$$L = \frac{1}{4\pi^2 f_0^2 C} = \frac{1}{4\pi^2 (1.30 \times 10^3 \text{ Hz})^2 (5.10 \times 10^{-6} \text{ F})} = \boxed{2.94 \times 10^{-3} \text{ H}}$$

b.   At resonance, $f = f_0$, and the current is a maximum. This occurs when $X_L = X_C$, so that $Z = R$.   Thus, the average power $\overline{P}$ provided by the generator is $\overline{P} = V^2_{rms} / R$, and solving for R we find

$$R = \frac{V^2_{rms}}{\overline{P}} = \frac{(11.0 \text{ V})^2}{25.0 \text{ W}} = \boxed{4.84 \ \Omega}$$

c.   When the generator frequency is 2.31 kHz, the individual reactances are

$$X_C = \frac{1}{2\pi f C} = \frac{1}{2\pi (2.31 \times 10^3 \text{ Hz})(5.10 \times 10^{-6} \text{ F})} = 13.5 \ \Omega$$

$$X_L = 2\pi f L = 2\pi (2.31 \times 10^3 \text{ Hz})(2.94 \times 10^{-3} \text{ H}) = 42.7 \ \Omega$$

The phase angle ϕ is, from Equation 23.8,

$$\phi = \tan^{-1}\left(\frac{X_L - X_C}{R}\right) = \tan^{-1}\left(\frac{42.7 \ \Omega - 13.5 \ \Omega}{4.84 \ \Omega}\right) = 80.6°$$

The power factor is then given by

$$\cos\phi = \cos 80.6° = \boxed{0.163}$$

CHAPTER 24 | *ELECTROMAGNETIC WAVES*

1. ***REASONING AND SOLUTION*** This is a standard exercise in units conversion. We first determine the number of meters in one light-year. The distance that light travels in one year is

$$d = ct = (3.00 \times 10^8 \text{ m/s})(1.00 \text{ year}) \left(\frac{365.25 \text{ days}}{1 \text{ year}} \right) \left(\frac{24 \text{ hours}}{1.0 \text{ day}} \right) \left(\frac{3600 \text{ s}}{1 \text{ hour}} \right) = 9.47 \times 10^{15} \text{ m}$$

Thus, 1 light year $= 9.47 \times 10^{15}$ m. Then, the distance to Alpha Centauri is

$$(4.3 \text{ light - years}) \left(\frac{9.47 \times 10^{15} \text{ m}}{1 \text{ light - year}} \right) = \boxed{4.1 \times 10^{16} \text{ m}}$$

---

5. ***REASONING*** The equation that represents the wave mathematically is $y = A \sin(2\pi f t - 2\pi x / \lambda)$. In this expression the amplitude is $A = 156$ N/C. The wavelength λ can be calculated using Equation 16.1, and we obtain

$$\lambda = \frac{c}{f} = \frac{3.00 \times 10^8 \text{ m/s}}{1.50 \times 10^8 \text{ Hz}} = 2.00 \text{ m}$$

SOLUTION
a. For $t = 0$ s, the wave expression becomes

$$y = A \sin(2\pi f t - 2\pi x / \lambda) = 156 \sin \left[2\pi f(0) - \frac{2\pi x}{2.00} \right] = -156 \sin \left(\frac{2\pi x}{2.00} \right) = -156 \sin(\pi x)$$

In this result, the units are suppressed for convenience. The following table gives the values of the electric field obtained using this version of the wave expression with the given values of the position x. The term πx is in radians when x is in meters, and conversion from radians to degrees is accomplished using the fact that $2\pi \text{ rad} = 360°$.

x	$y = -156 \sin(\pi x)$
0 m	$-156 \sin(0) = -156 \sin(0°) = 0$
0.50 m	$-156 \sin(0.50 \pi) = -156 \sin(90°) = -156$
1.00 m	$-156 \sin(1.00 \pi) = -156 \sin(180°) = 0$
1.50 m	$-156 \sin(1.50 \pi) = -156 \sin(270°) = +156$
2.00 m	$-156 \sin(2.00 \pi) = -156 \sin(360°) = 0$

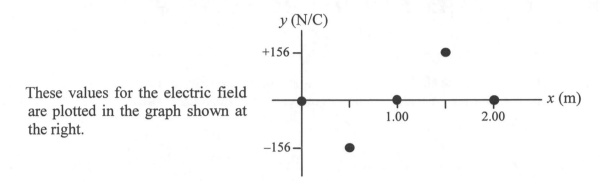

These values for the electric field are plotted in the graph shown at the right.

b. For $t = T/4$, we use the fact that $f = 1/T$, and the wave expression becomes

$$y = A\sin(2\pi ft - 2\pi x/\lambda) = 156\sin\left[2\pi\left(\frac{1}{T}\right)\left(\frac{T}{4}\right) - \frac{2\pi x}{2.00}\right] = 156\sin\left(\frac{\pi}{2} - \pi x\right) = 156\cos(\pi x)$$

In this result, the units are suppressed for convenience. The following table gives the values of the electric field obtained using this version of the wave expression with the given values of the position x. The term πx is in radians when x is in meters, and conversion from radians to degrees is accomplished using the fact that 2π rad $= 360°$.

x	$y = 156\cos(\pi x)$
0 m	$156\cos(0) = 156\cos(0°) = +156$
0.50 m	$156\cos(0.50\,\pi) = 156\cos(90°) = 0$
1.00 m	$156\cos(1.00\,\pi) = 156\cos(180°) = -156$
1.50 m	$156\cos(1.50\,\pi) = 156\cos(270°) = 0$
2.00 m	$156\cos(2.00\,\pi) = 156\cos(360°) = +156$

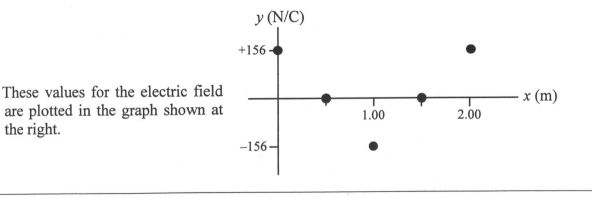

These values for the electric field are plotted in the graph shown at the right.

9. $\boxed{\text{WWW}}$ *REASONING AND SOLUTION* According to Equation 16.1, the wavelength is $\lambda = c/f$. Since the rods are adjusted so that each one has a length of $\lambda/4$, the length of each rod is

$$L = \frac{\lambda}{4} = \frac{(c/f)}{4} = \frac{(3.00 \times 10^8 \text{ m/s})/(60.0 \times 10^6 \text{ Hz})}{4} = \boxed{1.25 \text{ m}}$$

11. ***REASONING AND SOLUTION***   The number of wavelengths that can fit across the width W of your thumb is W/λ. From Equation 16.1, we know that $\lambda = c/f$, so

$$\text{No. of wavelengths} = \frac{W}{\lambda} = \frac{Wf}{c} = \frac{(2.0 \times 10^{-2} \text{ m})(5.5 \times 10^{14} \text{ Hz})}{3.0 \times 10^8 \text{ m/s}} = \boxed{3.7 \times 10^4}$$

15. ***REASONING***   We proceed by first finding the time t for sound waves to travel between the astronauts. Since this is the same time it takes for the electromagnetic waves to travel to earth, the distance between earth and the spaceship is $d_{\text{earth-ship}} = ct$.

SOLUTION   The time it takes for sound waves to travel at 343 m/s through the air between the astronauts is

$$t = \frac{d_{\text{astronaut}}}{v_{\text{sound}}} = \frac{1.5 \text{ m}}{343 \text{ m/s}} = 4.4 \times 10^{-3} \text{ s}$$

Therefore, the distance between the earth and the spaceship is

$$d_{\text{earth-ship}} = ct = (3.0 \times 10^8 \text{ m/s})(4.4 \times 10^{-3} \text{ s}) = \boxed{1.3 \times 10^6 \text{ m}}$$

19. ***REASONING AND SOLUTION***   If d is the distance between the mirrors, the time required for the echo to be heard is $t = 2d/v_{\text{sound}}$. In this time, the light will travel a total distance D where $D = ct = c(2d/v_{\text{sound}})$. Therefore, the number of times that the flash travels the round-trip distance between the mirrors in the time t is

$$N = \frac{D}{2d} = \frac{2cd/v_{\text{sound}}}{2d} = \frac{c}{v_{\text{sound}}} = \frac{3.00 \times 10^8 \text{ m/s}}{343 \text{ m/s}} = \boxed{8.75 \times 10^5}$$

23. ***REASONING***   The rms value E_{rms} of the electric field is related to the average energy density $\bar{u}$ of the microwave radiation according to Equation 24.2b: $\bar{u} = \varepsilon_0 E_{\text{rms}}^2$.

SOLUTION   Solving for E_{rms} gives

$$E_{\text{rms}} = \sqrt{\frac{\bar{u}}{\varepsilon_0}} = \sqrt{\frac{4 \times 10^{-14} \text{ J/m}^3}{8.85 \times 10^{-12} \text{ C}^2/(\text{N} \cdot \text{m}^2)}} = \boxed{0.07 \text{ N/C}}$$

25. ***REASONING AND SOLUTION***

a. According to Equation 24.5b, the average intensity is $\bar{S} = c\varepsilon_0 E_{rms}^2$. In addition, the average intensity is the average power $\bar{P}$ divided by the area A. Therefore,

$$E_{rms} = \sqrt{\frac{\bar{S}}{c\varepsilon_0}} = \sqrt{\frac{\bar{P}}{c\varepsilon_0 A}}$$

$$= \sqrt{\frac{1.20 \times 10^4 \text{ W}}{(3.00 \times 10^8 \text{ m/s})[8.85 \times 10^{-12} \text{C}^2/(\text{N} \cdot \text{m}^2)](135 \text{ m}^2)}} = \boxed{183 \text{ N/C}}$$

b. Then, from Equation 24.3 ($E_{rms} = cB_{rms}$), we have

$$B_{rms} = \frac{E_{rms}}{c} = \frac{183 \text{ N/C}}{3.00 \times 10^8 \text{ m/s}} = \boxed{6.10 \times 10^{-7} \text{ T}}$$

---

29. **WWW** ***REASONING AND SOLUTION*** The sun radiates sunlight (electromagnetic waves) uniformly in all directions, so the intensity at a distance r from the sun is given by Equation 16.9 as $S = P/(4\pi r^2)$, where P is the power radiated by the sun. The power that strikes an area $A_\perp$ oriented perpendicular to the direction in which the sunlight is radiated is $P' = SA_\perp$, according to Equation 16.8. The 0.75-m^2 patch of flat land on the equator at point Q is not perpendicular to the direction of the sunlight, however.

The figure at the right shows that

$$A_\perp = (0.75 \text{ m}^2) \cos 27°$$

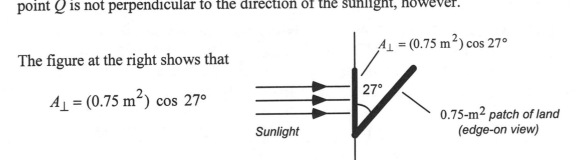

Therefore, the power striking the patch of land is

$$P' = SA_\perp = \left(\frac{P}{4\pi r^2}\right)(0.75 \text{ m}^2) \cos 27°$$

$$= \left[\frac{3.9 \times 10^{26} \text{ W}}{4\pi(1.5 \times 10^{11} \text{ m})^2}\right](0.75 \text{ m}^2) \cos 27° = \boxed{920 \text{ W}}$$

33. **REASONING** The Doppler effect for electromagnetic radiation is given by Equation 24.6, $f_o = f_s(1 \pm v_{rel}/c)$, where f_o is the observed frequency, f_s is the frequency emitted by the source, and v_{rel} is the speed of the source relative to the observer. As discussed in the text, the plus sign applies when the source and the observer are moving toward one another, while the minus sign applies when they are moving apart.

SOLUTION
a. At location A, the galaxy is moving away from the earth with a relative speed of

$$v_{rel} = (1.6 \times 10^6 \text{ m/s}) - (0.4 \times 10^6 \text{ m/s}) = 1.2 \times 10^6 \text{ m/s}$$

Therefore, the minus sign in Equation 24.6 applies and the observed frequency for the light from region A is

$$f_o = f_s\left(1 - \frac{v_{rel}}{c}\right) = (6.200 \times 10^{14} \text{ Hz})\left(1 - \frac{1.2 \times 10^6 \text{ m/s}}{3.0 \times 10^8 \text{ m/s}}\right) = \boxed{6.175 \times 10^{14} \text{ Hz}}$$

b. Similarly, at location B, the galaxy is moving away from the earth with a relative speed of

$$v_{rel} = (1.6 \times 10^6 \text{ m/s}) + (0.4 \times 10^6 \text{ m/s}) = 2.0 \times 10^6 \text{ m/s}$$

The observed frequency for the light from region B is

$$f_o = f_s\left(1 - \frac{v_{rel}}{c}\right) = (6.200 \times 10^{14} \text{ Hz})\left(1 - \frac{2.0 \times 10^6 \text{ m/s}}{3.0 \times 10^8 \text{ m/s}}\right) = \boxed{6.159 \times 10^{14} \text{ Hz}}$$

---

35. **REASONING AND SOLUTION** The average intensity of light leaving the polarizing material is given by Malus' Law (Equation 24.7). Therefore, using Malus' law, we obtain the following results.

a.
$$\frac{\overline{S}}{\overline{S}_0} = \cos^2 25° = \boxed{0.82}$$

b.
$$\frac{\overline{S}}{\overline{S}_0} = \cos^2 65° = \boxed{0.18}$$

---

39. **WWW** **REASONING** The average intensity of light leaving each analyzer is given by Malus' Law (Equation 24.7). Thus, intensity of the light transmitted through the first analyzer is

$$\overline{S}_1 = \overline{S}_0 \cos^2 27°$$

Similarly, the intensity of the light transmitted through the second analyzer is

$$\overline{S}_2 = \overline{S}_1 \cos^2 27° = \overline{S}_0 \cos^4 27°$$

And the intensity of the light transmitted through the third analyzer is

$$\overline{S}_3 = \overline{S}_2 \cos^2 27° = \overline{S}_0 \cos^6 27°$$

If we generalize for the Nth analyzer, we deduce that

$$\overline{S}_N = \overline{S}_{N-1} \cos^2 27° = \overline{S}_0 \cos^{2N} 27°$$

Since we want the light reaching the photocell to have an intensity that is reduced by a least a factor of one hundred relative to the first analyzer, we want $\overline{S}_N / \overline{S}_0 = 0.010$. Therefore, we need to find N such that $\cos^{2N} 27° = 0.010$. This expression can be solved for N.

SOLUTION Taking the common logarithm of both sides of the last expression gives

$$2N \log(\cos 27°) = \log 0.010 \quad \text{or} \quad N = \frac{\log 0.010}{2 \log (\cos 27°)} = \boxed{20}$$

---

43. $\boxed{\text{WWW}}$ *REASONING AND SOLUTION* The actual time t and round-trip distance s are related by $s = ct$. An uncertainty of Δt in the time introduces an uncertainty in the round trip distance Δs, so that $s + \Delta s = c(t + \Delta t)$. Thus, subtraction of these two expression leads to

$$\Delta s = c\Delta t = (3.0 \times 10^8 \text{ m/s})(0.10 \times 10^{-9} \text{ s}) = 0.030 \text{ m}$$

This is the error in the round-trip distance. Therefore, the error in the earth-moon distance is half this distance or $\boxed{0.015 \text{ m}}$.

---

47. *REASONING AND SOLUTION* The average intensity of light leaving each polarizer is given by Malus' Law (Equation 24.7): $\overline{S} = \overline{S}_0 \cos^2 \theta$. Solving for the angle θ and noting that $\overline{S} = 0.10 \, \overline{S}_0$, we have

$$\theta = \cos^{-1} \sqrt{\frac{0.100 \, \overline{S}_0}{\overline{S}_0}} = \boxed{71.6°}$$

---

51. ***REASONING AND SOLUTION*** Since the sun emits radiation uniformly in all directions, at a distance r from the sun's center, the energy spreads out over a sphere of surface area $4\pi r^2$. Therefore, according to Equation 16.9, $S = P/(4\pi r^2)$, the total power radiated by the sun is

$$P = S(4\pi r^2) = (1390 \text{ W/m}^2)(4\pi)(1.50 \times 10^{11} \text{ m})^2 = \boxed{3.93 \times 10^{26} \text{ W}}$$

CHAPTER 25 | *THE REFLECTION OF LIGHT: MIRRORS*

PROBLEMS

1. ***REASONING*** The geometry is shown below. According to the law of reflection, the incident ray, the reflected ray, and the normal to the surface all lie in the same plane, and the angle of reflection θ_r equals the angle of incidence θ_i . We can use the law of reflection and the properties of triangles to determine the angle θ at which the ray leaves M_2.

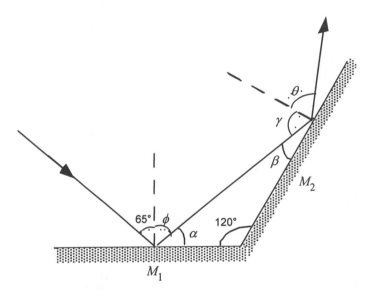

SOLUTION From the law of reflection, we know that $\phi = 65°$. We see from the figure that $\phi + \alpha = 90°$, or $\alpha = 90° - \phi = 90° - 65° = 25°$. From the figure and the fact that the sum of the interior angles in any triangle is 180°, we have $\alpha + \beta + 120° = 180°$. Solving for β, we find that $\beta = 180° - (120° + 25°) = 35°$. Therefore, since $\beta + \gamma = 90°$, we find that the angle γ is given by $\gamma = 90° - \beta = 90° - 35° = 55°$. Since γ is the angle of incidence of the ray on mirror M_2, we know from the law of reflection that $\boxed{\theta = 55°}$.

5. $\boxed{\text{WWW}}$ ***REASONING AND SOLUTION*** The drawing at the right shows a ray diagram in which the reflected rays have been projected behind the mirror. We can see by inspection of this drawing that, after the rays reflect from the plane mirror, the angle α between them is still $\boxed{10°}$.

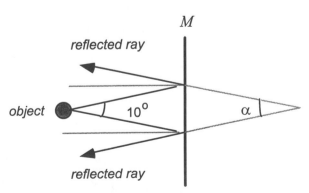

9. **REASONING**   The time of travel is proportional to the total distance for each path. Therefore, using the distances identified in the drawing, we have

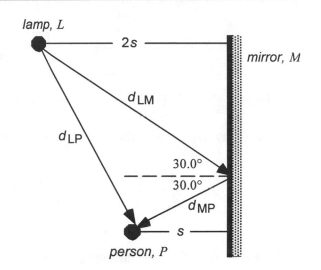

$$\frac{t_{reflected}}{t_{direct}} = \frac{d_{LM} + d_{MP}}{d_{LP}} \qquad (1)$$

We know that the angle of incidence is equal to the angle of reflection. We also know that the lamp L is a distance $2s$ from the mirror, while the person P is a distance s from the mirror.

Therefore, it follows that if $d_{MP} = d$, then $d_{LM} = 2d$. We may use the law of cosines (see Appendix E) to express the distance d_{LP} as

$$d_{LP} = \sqrt{(2d)^2 + d^2 - 2(2d)(d)\cos 2(30.0°)}$$

SOLUTION   Substituting the expressions for d_{MP}, d_{LM}, and d_{LP} into Equation (1), we find that the ratio of the travel times is

$$\frac{t_{reflected}}{t_{direct}} = \frac{2d + d}{\sqrt{(2d)^2 + d^2 - 2(2d)(d)\cos 2(30.0°)}} = \frac{3}{\sqrt{5 - 4\cos 60.0°}} = \boxed{1.73}$$

13. **REASONING AND SOLUTION**   The ray diagram is shown below. (Note: $f = -50.0$ cm and $d_0 = 25.0$ cm)

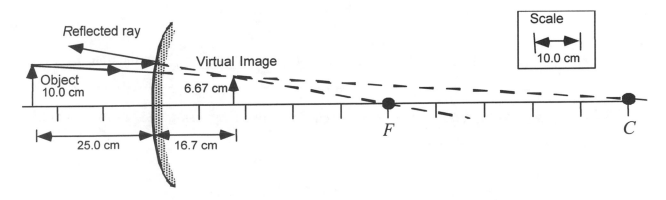

a. The ray diagram indicates that the image is 16.7 cm behind the mirror .

b. The ray diagram indicates that the image height is 6.67 cm .

---

17. **REASONING** This problem can be solved using the mirror equation, Equation 25.3.

SOLUTION Using the mirror equation with $d_i = +97$ cm and $f = 42$ cm, we find

$$\frac{1}{d_o} = \frac{1}{f} - \frac{1}{d_i} = \frac{1}{42 \text{ cm}} - \frac{1}{97 \text{ cm}} \qquad \text{or} \qquad d_o = +74 \text{ cm}$$

---

21. **REASONING** We have seen that a convex mirror always forms a *virtual image* as shown in Figure 25.22a of the text, where the image is *upright* and *smaller* than the object. These characteristics should bear out in the results of our calculations.

SOLUTION The radius of curvature of the convex mirror is 68 cm. Therefore, the focal length is, from Equation 25.2, $f = -(1/2)R = -34$ cm. Since the image is virtual, we know that $d_i = -22$ cm.

a. With $d_i = -22$ cm and $f = -34$ cm, the mirror equation gives

$$\frac{1}{d_o} = \frac{1}{f} - \frac{1}{d_i} = \frac{1}{-34 \text{ cm}} - \frac{1}{-22 \text{ cm}} \qquad \text{or} \qquad d_o = +62 \text{ cm}$$

b. According to the magnification equation, the magnification is

$$m = -\frac{d_i}{d_o} = -\frac{-22 \text{ cm}}{62 \text{ cm}} = \boxed{+0.35}$$

c. Since the magnification m is positive, the image is upright .

d. Since the magnification m is less than one, the image is smaller than the object.

---

23. WWW **REASONING** This problem can be solved by using the mirror equation, Equation 25.3, and the magnification equation, Equation 25.4.

SOLUTION
a. Using the mirror equation with $d_i = d_o$ and $f = R/2$, we have

$$\frac{1}{d_o} = \frac{1}{f} - \frac{1}{d_i} = \frac{1}{R/2} - \frac{1}{d_o} \qquad \text{or} \qquad \frac{2}{d_o} = \frac{2}{R}$$

Therefore, we find that $\boxed{d_o = R}$.

b. According to the magnification equation, the magnification is

$$m = -\frac{d_i}{d_o} = -\frac{d_o}{d_o} = \boxed{-1}$$

c. Since the magnification m is negative, the image is $\boxed{\text{inverted}}$.

---

27. $\boxed{\text{WWW}}$ *REASONING AND SOLUTION*   The magnification equation, Equation 25.4, indicates that $d_i = -md_o$. Substituting this result into the mirror equation, we find that

$$\frac{1}{d_o} + \frac{1}{d_i} = \frac{1}{d_o} + \frac{1}{-md_o} = \frac{1}{f}$$

Solving for the object distance d_o, we find

$$d_o = \frac{(m-1)f}{m}$$

After the object is moved to its new position, its object distance becomes d_o' and its magnification becomes $m' = 2m$. Therefore, the amount by which the object is moved is

$$d_o - d_o' = \frac{(m-1)f}{m} - \frac{(m'-1)f}{m'} = \frac{(m-1)f}{m} - \frac{(2m-1)f}{2m}$$

$$= -\frac{f}{2m} = -\frac{-24.0 \text{ cm}}{2(0.150)} = \boxed{+80.0 \text{ cm}}$$

The positive answer means that the initial object distance is larger than the final object distance, so that $\boxed{\text{the object is moved toward the mirror}}$.

---

31. $\boxed{\text{WWW}}$ *REASONING*   When paraxial light rays that are parallel to the principal axis strike a convex mirror, the reflected rays diverge after being reflected, and appear to originate from the focal point F behind the mirror (see Figure 25.16). We can treat the sun as being infinitely far from the mirror, so it is reasonable to treat the incident rays as paraxial rays that are parallel to the principal axis.

SOLUTION

a.   Since the sun is infinitely far from the mirror and its image is a virtual image that lies *behind* the mirror, we can conclude that the mirror is a ⎡ convex mirror ⎤.

b.   With $d_i = -12.0$ cm and $d_o = \infty$, the mirror equation (Equation 25.3) gives

$$\frac{1}{f} = \frac{1}{d_o} + \frac{1}{d_i} = \frac{1}{\infty} + \frac{1}{d_i} = \frac{1}{d_i}$$

Therefore, the focal length f lies 12.0 cm behind the mirror (this is consistent with the reasoning above that states that, after being reflected, the rays appear to originate from the focal point behind the mirror).   In other words, $f = -12.0$ cm.   Then, according to Equation 25.2, $f = -\frac{1}{2}R$, and the radius of curvature is

$$R = -2f = -2(-12.0 \text{ cm}) = \boxed{24.0 \text{ cm}}$$

---

35.   **REASONING** This problem can be solved using the mirror equation, Equation 25.3.

SOLUTION Using the mirror equation with $d_i = +26$ cm and $f = 12$ cm, we find

$$\frac{1}{d_o} = \frac{1}{f} - \frac{1}{d_i} = \frac{1}{12 \text{ cm}} - \frac{1}{26 \text{ cm}} \qquad \text{or} \qquad \boxed{d_o = +22 \text{ cm}}$$

---

39.   **REASONING** We need to know the focal length of the mirror and can obtain it from the mirror equation, Equation 25.3, as applied to the first object:

$$\frac{1}{d_{o1}} + \frac{1}{d_{i1}} = \frac{1}{14.0 \text{ cm}} + \frac{1}{-7.00 \text{ cm}} = \frac{1}{f} \qquad \text{or} \qquad f = -14.0 \text{ cm}$$

According to the magnification equation, Equation 25.4, the image height h_i is related to the object height h_o as follows: $h_i = m h_o = \left(-d_i / d_o\right)h_o$.

SOLUTION Applying this result to each object, we find that $h_{i2} = h_{i1}$, or

$$\left(\frac{-d_{i2}}{d_{o2}}\right)h_{o2} = \left(\frac{-d_{i1}}{d_{o1}}\right)h_{o1}$$

Therefore,

$$d_{i2} = d_{o2} \left(\frac{d_{i1}}{d_{o1}} \right) \left(\frac{h_{o1}}{h_{o2}} \right)$$

Using the fact that $h_{o2} = 2h_{o1}$, we have

$$d_{i2} = d_{o2} \left(\frac{d_{i1}}{d_{o1}} \right) \left(\frac{h_{o1}}{h_{o2}} \right) = d_{o2} \left(\frac{-7.00 \text{ cm}}{14.0 \text{ cm}} \right) \left(\frac{h_{o1}}{2h_{o1}} \right) = -0.250 \, d_{o2}$$

Using this result in the mirror equation, as applied to the second object, we find that

$$\frac{1}{d_{o2}} + \frac{1}{d_{i2}} = \frac{1}{f}$$

or

$$\frac{1}{d_{o2}} + \frac{1}{-0.250 \, d_{o2}} = \frac{1}{-14.0 \text{ cm}}$$

Therefore,

$$\boxed{d_{o2} = +42.0 \text{ cm}}$$

CHAPTER 26 | *THE REFRACTION OF LIGHT: LENSES AND OPTICAL INSTRUMENTS*

1. ***REASONING AND SOLUTION*** The speed of light in benzene v is related to the speed of light in vacuum c by the index of refraction n. The index of refraction is defined by Equation 26.1 ($n = c/v$). According to Table 26.1, the index of refraction of benzene is 1.501. Therefore, solving for v, we have

$$v = \frac{c}{n} = \frac{3.00 \times 10^8 \text{ m/s}}{1.501} = \boxed{2.00 \times 10^8 \text{ m/s}}$$

5. WWW ***REASONING*** Since the light will travel in glass at a constant speed v, the time it takes to pass perpendicularly through the glass is given by $t = d/v$, where d is the thickness of the glass. The speed v is related to the vacuum value c by Equation 26.1: $n = c/v$.

SOLUTION Substituting for v from Equation 26.1 and substituting values, we obtain

$$t = \frac{d}{v} = \frac{nd}{c} = \frac{(1.5)(4.0 \times 10^{-3} \text{ m})}{3.00 \times 10^8 \text{ m/s}} = \boxed{2.0 \times 10^{-11} \text{ s}}$$

9. ***REASONING*** We begin by using Snell's law (Equation 26.2: $n_1 \sin \theta_1 = n_2 \sin \theta_2$) to find the index of refraction of the material. Then we will use Equation 26.1, the definition of the index of refraction ($n = c/v$) to find the speed of light in the material.

SOLUTION From Snell's law, the index of refraction of the material is

$$n_2 = \frac{n_1 \sin \theta_1}{\sin \theta_2} = \frac{(1.000) \sin 63.0°}{\sin 47.0°} = 1.22$$

Then, from Equation 26.1, we find that the speed of light v in the material is

$$v = \frac{c}{n_2} = \frac{3.00 \times 10^8 \text{ m/s}}{1.22} = \boxed{2.46 \times 10^8 \text{ m/s}}$$

11. **REASONING** We will use the geometry of the situation to determine the angle of incidence. Once the angle of incidence is known, we can use Snell's law to find the index of refraction of the unknown liquid. The speed of light v in the liquid can then be determined.

SOLUTION From the drawing in the text, we see that the angle of incidence at the liquid-air interface is

$$\theta_1 = \tan^{-1}\left(\frac{5.00 \text{ cm}}{6.00 \text{ cm}}\right) = 39.8°$$

The drawing also shows that the angle of refraction is 90.0°. Thus, according to Snell's law (Equation 26.2: $n_1 \sin \theta_1 = n_2 \sin \theta_2$), the index of refraction of the unknown liquid is

$$n_1 = \frac{n_2 \sin\theta_2}{\sin\theta_1} = \frac{(1.000)(\sin 90.0°)}{\sin 39.8°} = 1.56$$

From Equation 26.1 ($n = c/v$), we find that the speed of light in the unknown liquid is

$$v = \frac{c}{n_1} = \frac{3.00 \times 10^8 \text{ m/s}}{1.56} = \boxed{1.92 \times 10^8 \text{ m/s}}$$

---

15. $\boxed{\text{WWW}}$ The drawing at the right shows the geometry of the situation using the same notation as that in Figure 26.7. In addition to the text's notation, let t represent the thickness of the pane, let L represent the length of the ray in the pane, let x (shown twice in the figure) equal the displacement of the ray, and let the difference in angles $\theta_1 - \theta_2$ be given by ϕ.

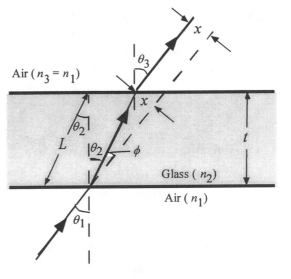

We wish to find the amount x by which the emergent ray is displaced relative to the incident ray. This can be done by applying Snell's law at each interface, and then making use of the geometric and trigonometric relations in the drawing.

SOLUTION If we apply Snell's law (see Equation 26.2) to the bottom interface we obtain $n_1 \sin \theta_1 = n_2 \sin \theta_2$. Similarly, if we apply Snell's law at the top interface where the ray emerges, we have $n_2 \sin \theta_2 = n_3 \sin \theta_3 = n_1 \sin \theta_3$. Comparing this with Snell's law at the bottom face, we see that $n_1 \sin \theta_1 = n_1 \sin \theta_3$, from which we can conclude that $\theta_3 = \theta_1$. Therefore, the emerging ray is parallel to the incident ray.

From the geometry of the ray and thickness of the pane, we see that $L \cos\theta_2 = t$, from which it follows that $L = t/\cos\theta_2$. Furthermore, we see that $x = L \sin\phi = L \sin(\theta_1 - \theta_2)$. Substituting for L, we find

$$x = L \sin(\theta_1 - \theta_2) = \frac{t \sin(\theta_1 - \theta_2)}{\cos\theta_2}$$

Before we can use this expression to determine a numerical value for x, we must find the value of θ_2. Solving the expression for Snell's law at the bottom interface for θ_2, we have

$$\sin\theta_2 = \frac{n_1 \sin\theta_1}{n_2} = \frac{(1.000)\,(\sin 30.0°)}{1.52} = 0.329 \qquad \text{or} \qquad \theta_2 = \sin^{-1} 0.329 = 19.2°$$

Therefore, the amount by which the emergent ray is displaced relative to the incident ray is

$$x = \frac{t \sin(\theta_1 - \theta_2)}{\cos\theta_2} = \frac{(6.00 \text{ mm}) \sin(30.0° - 19.2°)}{\cos 19.2°} = \boxed{1.19 \text{ mm}}$$

---

21. **REASONING**   The drawing at the right shows the situation. As discussed in the text, when the observer is directly above, the apparent depth d' of the object is related to the actual depth by Equation 26.3:

$$d' = d\left(\frac{n_2}{n_1}\right)$$

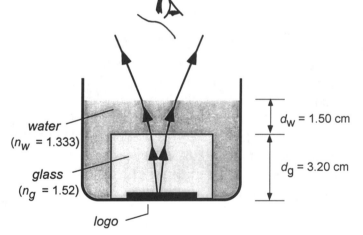

water
(n_w = 1.333)

glass
(n_g = 1.52)

logo

d_W = 1.50 cm

d_g = 3.20 cm

In this case, we must apply Equation 26.3 twice; once for the rays in the glass, and once again for the rays in the water.

SOLUTION   We refer to the drawing for our notation and begin at the logo. To an observer *in the water* directly above the logo, the apparent depth of the logo is $d'_g = d_g(n_w/n_g)$. When viewed directly from above in air, the logo's apparent depth is, $d'_w = (d_w + d'_g)(n_{air}/n_w)$, where we have used the fact that when viewed from air, the logo's actual depth appears to be $d_w + d'_g$. Substituting the expression for d'_g into the expression for d'_w, we obtain

$$d'_w = \left(d_w + d'_g\right)\left(\frac{n_{air}}{n_w}\right) = d_w\left(\frac{n_{air}}{n_w}\right) + d_g\left(\frac{n_w}{n_g}\right)\left(\frac{n_{air}}{n_w}\right) = d_w\left(\frac{n_{air}}{n_w}\right) + d_g\left(\frac{n_{air}}{n_g}\right)$$

$$= (1.50\ \text{cm})\left(\frac{1.000}{1.333}\right) + (3.20\ \text{cm})\left(\frac{1.000}{1.52}\right) = \boxed{3.23\ \text{cm}}$$

---

23. ***REASONING AND SOLUTION*** According to Equation 26.4, the critical angle is related to the refractive indices n_1 and n_2 by $\sin\theta_c = n_2/n_1$, where $n_1 > n_2$. Solving for n_1, we find

$$n_1 = \frac{n_2}{\sin\theta_c} = \frac{1.000}{\sin 40.5°} = \boxed{1.54}$$

---

27. ***REASONING AND SOLUTION*** If a person's eyes are very close to the surface of the water, a light ray coming from the shark will be seen even when it is refracted through an angle of 90.0° as it enters the air. In this situation, the ray strikes the water-air interface at the critical angle. The critical angle θ_c is given by Equation 26.4 as

$$\theta_c = \sin^{-1}\left(\frac{1.00}{1.333}\right) = 48.6°$$

where we have used $n = 1.333$ for the refractive index of water (see Table 26.1). The horizontal distance x of the shark from the boat is related to the depth (4.5 m) of the shark and the critical angle by trigonometry:

$$x = (4.5\ \text{m})\tan 48.6° = \boxed{5.1\ \text{m}}$$

If the shark is farther than 5.1 m from the boat, a light ray from the shark will strike the water-air interface at an angle that is greater than the critical angle. The ray will be totally reflected back into the water, and the person will not see the shark.

---

31. ***REASONING*** Since the light reflected from the coffee table is completely polarized parallel to the surface of the glass, the angle of incidence must be the Brewster angle ($\theta_B = 56.7°$) for the air-glass interface. We can use Brewster's law (Equation 26.5: $\tan\theta_B = n_2/n_1$) to find the index of refraction n_2 of the glass.

SOLUTION Solving Brewster's law for n_2, we find that the refractive index of the glass is

$$n_2 = n_1\tan\theta_B = (1.000)(\tan 56.7°) = \boxed{1.52}$$

35. WWW *REASONING* Brewster's law (Equation 26.5: $\tan \theta_B = n_2/n_1$) relates the angle of incidence θ_B at which the reflected ray is completely polarized parallel to the surface to the indices of refraction n_1 and n_2 of the two media forming the interface. We can use Brewster's law for light incident from above to find the ratio of the refractive indices n_2/n_1. This ratio can then be used to find the Brewster angle for light incident from below on the same interface.

SOLUTION The index of refraction for the medium in which the incident ray occurs is designated by n_1. For the light striking from above $n_2/n_1 = \tan \theta_B = \tan 65.0° = 2.14$. The same equation can be used when the light strikes from below if the indices of refraction are interchanged

$$\theta_B = \tan^{-1}\left(\frac{n_1}{n_2}\right) = \tan^{-1}\left(\frac{1}{n_2/n_1}\right) = \tan^{-1}\left(\frac{1}{2.14}\right) = \boxed{25.0°}$$

---

39. *REASONING* Because the refractive index of the glass depends on the wavelength (i.e., the color) of the light, the rays corresponding to different colors are bent by different amounts in the glass. We can use Snell's law (Equation 26.2: $n_1 \sin \theta_1 = n_2 \sin \theta_2$) to find the angle of refraction for the violet ray and the red ray. The angle between these rays can be found by the subtraction of the two angles of refraction.

SOLUTION In Table 26.2 the index of refraction for violet light in crown glass is 1.538, while that for red light is 1.520. According to Snell's law, then, the sine of the angle of refraction for the violet ray in the glass is $\sin \theta_2 = (1.000/1.538) \sin 45.00° = 0.4598$, so that

$$\theta_2 = \sin^{-1}(0.4598) = 27.37°$$

Similarly, for the red ray, $\sin \theta_2 = (1.000/1.520) \sin 45.00° = 0.4652$, from which it follows that

$$\theta_2 = \sin^{-1}(0.4652) = 27.72°$$

Therefore, the angle between the violet ray and the red ray in the glass is

$$27.72° - 27.37° = \boxed{0.35°}$$

---

43. *REASONING* We can use Snell's law (Equation 26.2: $n_1 \sin \theta_1 = n_2 \sin \theta_2$) at each face of the prism. At the first interface where the ray enters the prism, $n_1 = 1.000$ for air and $n_2 = n_g$ for glass. Thus, Snell's law gives

$$(1) \sin 60.0° = n_g \sin \theta_2 \qquad \text{or} \qquad \sin \theta_2 = \frac{\sin 60.0°}{n_g} \tag{1}$$

We will represent the angles of incidence and refraction at the second interface as θ_1' and θ_2', respectively. Since the triangle is an equilateral triangle, the angle of incidence at the second interface, where the ray emerges back into air, is $\theta_1' = 60.0° - \theta_2$. Therefore, at the second interface, where $n_1 = n_g$ and $n_2 = 1.000$, Snell's law becomes

$$n_g \sin (60.0° - \theta_2) = (1) \sin \theta_2' \tag{2}$$

We can now use Equations (1) and (2) to determine the angles of refraction θ_2' at which the red and violet rays emerge into the air from the prism.

SOLUTION
Red Ray: The index of refraction of flint glass at the wavelength of red light is $n_g = 1.662$. Therefore, using Equation (1), we can find the angle of refraction for the red ray as it enters the prism:

$$\sin \theta_2 = \frac{\sin 60.0°}{1.662} = 0.521 \qquad \text{or} \qquad \theta_2 = \sin^{-1} 0.521 = 31.4°$$

Substituting this value for θ_2 into Equation (2), we can find the angle of refraction at which the red ray emerges from the prism:

$$\sin \theta_2' = 1.662 \sin (60.0° - 31.4°) = 0.796 \qquad \text{or} \qquad \theta_2' = \sin^{-1} 0.796 = \boxed{52.7°}$$

Violet ray: For violet light, the index of refraction for glass is $n_g = 1.698$. Again using Equation (1), we find

$$\sin \theta_2 = \frac{\sin 60.0°}{1.698} = 0.510 \qquad \text{or} \qquad \theta_2 = \sin^{-1} 0.510 = 30.7°$$

Using Equation (2), we find

$$\sin \theta_2' = 1.698 \sin (60.0° - 30.7°) = 0.831 \qquad \text{or} \qquad \theta_2' = \sin^{-1} 0.831 = \boxed{56.2°}$$

---

45. **REASONING AND SOLUTION** Equation 26.6 gives the thin-lens equation which relates the object and image distances d_o and d_i, respectively, to the focal length f of the lens: $(1/d_o) + (1/d_i) = (1/f)$.

The optical arrangement is similar to that in Figure 26.26. The problem statement gives values for the focal length ($f = 50.0 \text{ mm}$) and the maximum lens-to-film distance ($d_i = 275 \text{ mm}$). Therefore, the maximum distance that the object can be located in front of the lens is

$$\frac{1}{d_o} = \frac{1}{f} - \frac{1}{d_i} = \frac{1}{50.0 \text{ mm}} - \frac{1}{275 \text{ mm}} \quad \text{or} \quad \boxed{d_o = 61.1 \text{ mm}}$$

---

49. **REASONING** The ray diagram is constructed by drawing the paths of two rays from a point on the object. For convenience, we will choose the top of the object. The ray that is parallel to the principal axis will be refracted by the lens so that it passes through the focal point on the right of the lens. The ray that passes through the center of the lens passes through undeflected. The image is formed at the intersection of these two rays. In this case, the rays do not intersect on the right of the lens. However, if they are extended backwards they intersect on the left of the lens, locating a virtual, upright, and enlarged image.

SOLUTION
a. The ray-diagram, drawn to scale, is shown below.

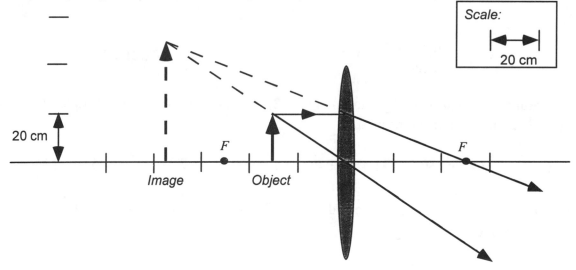

From the diagram, we see that the image distance is $\boxed{d_i = -75 \text{ cm}}$ and the magnification is $\boxed{+2.5}$. The negative image distance indicates that the image is virtual. The positive magnification indicates that the image is larger than the object.

b. From the thin-lens equation [Equation 26.6: $(1/d_o) + (1/d_i) = (1/f)$], we obtain

$$\frac{1}{d_i} = \frac{1}{f} - \frac{1}{d_o} = \frac{1}{50.0 \text{ m}} - \frac{1}{30.0 \text{ cm}} \quad \text{or} \quad \boxed{d_i = -75.0 \text{ cm}}$$

The magnification equation (Equation 26.7) gives the magnification to be

$$m = -\frac{d_i}{d_o} = -\frac{-75.0 \text{ cm}}{30.0 \text{ cm}} = \boxed{+2.50}$$

---

53. **REASONING**   The magnification equation (Equation 26.7) relates the object and image distances d_o and d_i, respectively, to the relative size of the of the image and object: $m = -(d_i / d_o)$. We consider two cases: in case 1, the object is placed 18 cm in front of a diverging lens. The magnification for this case is given by m_1. In case 2, the object is moved so that the magnification m_2 is reduced by a factor of 2 compared to that in case 1. In other words, we have $m_2 = \frac{1}{2} m_1$. Using Equation 26.7, we can write this as

$$-\frac{d_{i2}}{d_{o2}} = -\frac{1}{2}\left(\frac{d_{i1}}{d_{o1}}\right) \tag{1}$$

This expression can be solved for d_{o2}. First, however, we must find a numerical value for d_{i1}, and we must eliminate the variable d_{i2}.

SOLUTION
The image distance for case 1 can be found from the thin-lens equation [Equation 26.6: $(1/d_o) + (1/d_i) = (1/f)$]. The problem statement gives the focal length as $f = -12 \text{ cm}$. Since the object is 18 cm in front of the diverging lens, $d_{o1} = 18 \text{ cm}$. Solving for d_{i1}, we find

$$\frac{1}{d_{i1}} = \frac{1}{f} - \frac{1}{d_{o1}} = \frac{1}{-12 \text{ cm}} - \frac{1}{18 \text{ cm}} \quad \text{or} \quad d_{i1} = -7.2 \text{ cm}$$

where the minus sign indicates that the image is virtual. Solving Equation (1) for d_{o2}, we have

$$d_{o2} = 2d_{i2}\left(\frac{d_{o1}}{d_{i1}}\right) \tag{2}$$

To eliminate d_{i2} from this result, we note that the thin-lens equation applied to case 2 gives

$$\frac{1}{d_{i2}} = \frac{1}{f} - \frac{1}{d_{o2}} = \frac{d_{o2} - f}{f d_{o2}} \quad \text{or} \quad d_{i2} = \frac{f d_{o2}}{d_{o2} - f}$$

Substituting this expression for d_{i2} into Equation (2), we have

$$d_{o2} = \left(\frac{2fd_{o2}}{d_{o2}-f}\right)\left(\frac{d_{o1}}{d_{i1}}\right) \qquad \text{or} \qquad d_{o2} - f = 2f\left(\frac{d_{o1}}{d_{i1}}\right)$$

Solving for d_{o2}, we find

$$d_{o2} = f\left[2\left(\frac{d_{o1}}{d_{i1}}\right)+1\right] = (-12\text{cm})\left[2\left(\frac{18\text{ cm}}{-7.2\text{ cm}}\right)+1\right] = \boxed{48\text{ cm}}$$

---

55. $\boxed{\text{WWW}}$ *REASONING* The optical arrangement is similar to that in Figure 26.26. We begin with the thin-lens equation, [Equation 26.6: $(1/d_o)+(1/d_i)=(1/f)$]. Since the distance between the moon and the camera is so large, the object distance d_o is essentially infinite, and $1/d_o = 1/\infty = 0$. Therefore the thin-lens equation becomes $1/d_i = 1/f$ or $d_i = f$. The diameter of the moon's imagine on the slide film is equal to the image height h_i, as given by the magnification equation (Equation 26.7: $h_i/h_o = -d_i/d_o$).

When the slide is projected onto a screen, the situation is similar to that in Figure 26.27. In this case, the thin-lens and magnification equations can be used in their usual forms.

SOLUTION

a. Solving the magnification equation for h_i gives

$$h_i = -h_o\frac{d_i}{d_o} = (-3.48\times10^6\text{ m})\left(\frac{50.0\times10^{-3}\text{ m}}{3.85\times10^8\text{ m}}\right) = -4.52\times10^{-4}\text{ m}$$

The diameter of the moon's image on the slide film is, therefore, $\boxed{4.52\times10^{-4}\text{ m}}$.

b. From the magnification equation, $h_i = -h_o\left(d_i/d_o\right)$. We need to find the ratio d_i/d_o. Beginning with the thin-lens equation, we have

$$\frac{1}{d_o}+\frac{1}{d_i}=\frac{1}{f} \qquad \text{or} \qquad \frac{1}{d_o}=\frac{1}{f}-\frac{1}{d_i} \qquad \text{which leads to} \qquad \frac{d_i}{d_o}=\frac{d_i}{f}-\frac{d_i}{d_i}=\frac{d_i}{f}-1$$

Therefore,

$$h_i = -h_o\left(\frac{d_i}{f}-1\right) = -(4.52\times10^{-4}\text{ m})\left(\frac{15.0\text{ m}}{110.0\times10^{-3}\text{ m}}-1\right) = -6.12\times10^{-2}\text{ m}$$

The diameter of the image on the screen is $\boxed{6.12\times10^{-2}\text{ m}}$.

59. **REASONING** The problem can be solved using the thin-lens equation [Equation 26.6: $(1/d_o)+(1/d_i)=(1/f)$] twice in succession. We begin by using the thin lens-equation to find the location of the image produced by the converging lens; this image becomes the object for the diverging lens.

SOLUTION
a. The image distance for the converging lens is determined as follows:

$$\frac{1}{d_{i1}}=\frac{1}{f}-\frac{1}{d_{o1}}=\frac{1}{12.0\text{ cm}}-\frac{1}{36.0\text{ cm}} \quad \text{or} \quad d_{i1}=18.0\text{ cm}$$

This image acts as the object for the diverging lens. Therefore,

$$\frac{1}{d_{i2}}=\frac{1}{f}-\frac{1}{d_{o2}}=\frac{1}{-6.00\text{ cm}}-\frac{1}{(30.0\text{ cm}-18.0\text{ cm})} \quad \text{or} \quad d_{i2}=-4.00\text{ cm}$$

Thus, the final image is located $\boxed{4.00\text{ cm to the left of the diverging lens}}$.

b. The magnification equation (Equation 26.7: $h_i/h_o=-d_i/d_o$) gives

$$\underbrace{m_c=-\frac{d_{i1}}{d_{o1}}=-\frac{18.0\text{ cm}}{36.0\text{ cm}}=-0.500}_{\text{Converging lens}} \qquad \underbrace{m_d=-\frac{d_{i2}}{d_{o2}}=-\frac{-4.00\text{ cm}}{12.0\text{ cm}}=0.333}_{\text{Diverging lens}}$$

Therefore, the overall magnification is given by the product $m_c m_d = \boxed{-0.167}$.

c. Since the final image distance is negative, we can conclude that the image is $\boxed{\text{virtual}}$.

d. Since the overall magnification of the image is negative, the image is $\boxed{\text{inverted}}$.

e. The magnitude of the overall magnification is less than one; therefore, the final image is $\boxed{\text{smaller}}$.

---

65. **REASONING** We begin by using the thin-lens equation [Equation 26.6: $(1/d_o)+(1/d_i)=(1/f)$] to locate the image produced by the lens. This image is then treated as the object for the mirror.

SOLUTION
a. The image distance from the diverging lens can be determined as follows:

$$\frac{1}{d_i} = \frac{1}{f} - \frac{1}{d_o} = \frac{1}{-8.00 \text{ cm}} - \frac{1}{20.0 \text{ cm}} \qquad \text{or} \qquad d_i = -5.71 \text{ cm}$$

The image produced by the lens is 5.71 cm to the left of the lens. The distance between this image and the concave mirror is 5.71 cm + 30.0 cm = 35.7 cm. The mirror equation [Equation 25.3: $(1/d_o) + (1/d_i) = (1/f)$] gives the image distance from the mirror:

$$\frac{1}{d_i} = \frac{1}{f} - \frac{1}{d_o} = \frac{1}{12.0 \text{ cm}} - \frac{1}{35.7 \text{ cm}} \qquad \text{or} \qquad \boxed{d_i = 18.1 \text{ cm}}$$

b. The image is $\boxed{\text{real}}$, because d_i is a positive number, indicating that the final image lies to the left of the concave mirror.

c. The image is $\boxed{\text{inverted}}$, because a diverging lens always produces an upright image, and the concave mirror produces an inverted image when the object distance is greater than the focal length of the mirror.

---

69. **REASONING** The far point is 5.0 m from the right eye, and 6.5 m from the left eye. For an object infinitely far away ($d_o = \infty$), the image distances for the corrective lenses are then −5.0 m for the right eye and −6.5 m for the left eye, the negative sign indicating that the images are virtual images formed to the left of the lenses. The thin-lens equation [Equation 26.6: $(1/d_o) + (1/d_i) = (1/f)$] can be used to find the focal length. Then, Equation 26.8 can be used to find the refractive power for the lens for each eye.

SOLUTION Since the object distance d_o is essentially infinite, $1/d_o = 1/\infty = 0$, and the thin-lens equation becomes $1/d_i = 1/f$, or $d_i = f$. Therefore, for the right eye, $f = -5.0 \text{ m}$, and the refractive power is (see Equation 26.8)

Right eye $\qquad$ Refractive power (in diopters) $= \dfrac{1}{f} = \dfrac{1}{(-5.0 \text{ m})} = \boxed{-0.20 \text{ diopters}}$

Similarly, for the left eye, $f = -6.5 \text{ m}$, and the refractive power is

Left eye $\qquad$ Refractive power (in diopters) $= \dfrac{1}{f} = \dfrac{1}{(-6.5 \text{ m})} = \boxed{-0.15 \text{ diopters}}$

---

77. **REASONING** The angular size of a distant object in radians is approximately equal to the diameter of the object divided by the distance from the eye. We will use this definition to calculate the angular size of the quarter, and then, calculate the angular size of the sun; we can then form the ratio $\theta_{quarter} / \theta_{sun}$.

SOLUTION The angular sizes are

$$\theta_{quarter} \approx \frac{2.4 \text{ cm}}{70.0 \text{ cm}} = 0.034 \text{ rad} \qquad \text{and} \qquad \theta_{sun} \approx \frac{1.39 \times 10^9 \text{ m}}{1.50 \times 10^{11} \text{ m}} = 0.0093 \text{ rad}$$

Therefore, the ratio of the angular sizes is

$$\frac{\theta_{quarter}}{\theta_{sun}} = \frac{0.034 \text{ rad}}{0.0093 \text{ rad}} = \boxed{3.7}$$

---

81. **REASONING** The angular magnification of a magnifying glass is given by Equation 26.10: $M \approx \left[(1/f) - (1/d_i)\right]N$, where N is the distance from the eye to the near-point. For maximum magnification, the closest to the eye that the image can be is at the near point, with $d_i = -N$ (where the minus sign indicates that the image lies to the left of the lens and is virtual). In this case, Equation 26.10 becomes $M_{max} \approx N/f + 1$. At minimum magnification, the image is as far from the eye as it can be ($d_i = -\infty$); this occurs when the object is placed at the focal point of the lens. In this case, Equation 26.10 simplifies to $M_{min} \approx N/f$.

   Since the woman observes that for clear vision, the maximum angular magnification is 1.25 times larger than the minimum angular magnification, we have $M_{max} = 1.25 M_{min}$. This equation can be written in terms of N and f using the above expressions, and then solved for f.

   SOLUTION We have
   $$\frac{N}{f} + 1 = 1.25 \frac{N}{f}$$
   Solving for f, we find that

   $$f = 0.25 N = (0.25)(25 \text{ cm}) = \boxed{6.3 \text{ cm}}$$

---

85. **REASONING** The angular magnification of a compound microscope is given by Equation 26.11:

   $$M \approx -\frac{(L - f_e)N}{f_o f_e}$$

   where f_o is the focal length of the objective, f_e is the focal length of the eyepiece, and L is the separation between the two lenses. This expression can be solved for f_o, the focal length of the objective.

SOLUTION Solving for f_0, we find that the focal length of the objective is

$$f_0 = -\frac{(L - f_e)N}{f_e M} = -\frac{(16.0 \text{ cm} - 1.4 \text{ cm})(25 \text{ cm})}{(1.4 \text{ cm})(-320)} = \boxed{0.81 \text{ cm}}$$

---

89. *REASONING AND SOLUTION* The angular magnification of an astronomical telescope, is given by Equation 26.12 as $M \approx -f_0 / f_e$. Solving for the focal length of the eyepiece, we find

$$f_e \approx -\frac{f_0}{M} = -\frac{48.0 \text{ cm}}{(-184)} = \boxed{0.261 \text{ cm}}$$

---

93. *REASONING AND SOLUTION*

a. The lens with the largest focal length should be used for the objective of the telescope. Since the refractive power is the reciprocal of the focal length (in meters), the lens with the smallest refractive power is chosen as the objective, namely, the $\boxed{1.3 \text{ - diopter lens}}$.

b. According to Equation 26.8, the refractive power is related to the focal length f by Refractive power (in diopters) = $1/[f(\text{in meters})]$. Since we know the refractive powers of the two lenses, we can solve Equation 26.8 for the focal lengths of the objective and the eyepiece. We find that $f_0 = 1/(1.3 \text{ diopters}) = 0.77 \text{ m}$. Similarly, for the eyepiece, $f_e = 1/(11 \text{ diopters}) = 0.091 \text{ m}$. Therefore, the distance between the lenses should be

$$L \approx f_0 + f_e = 0.77 \text{ m} + 0.091 \text{ m} = \boxed{0.86 \text{ m}}$$

c. The angular magnification of the telescope is given by Equation 26.12 as

$$M \approx -\frac{f_0}{f_e} = -\frac{0.77 \text{ m}}{0.091 \text{ m}} = \boxed{-8.5}$$

---

97. *REASONING* The ray diagram is constructed by drawing the paths of two rays from a point on the object. For convenience, we choose the top of the object. The ray that is parallel to the principal axis will be refracted by the lens and pass through the focal point on the right side. The ray that passes through the center of the lens passes through undeflected. The image is formed at the intersection of these two rays on the right side of the lens.

SOLUTION The following ray diagram (to scale) shows that $\boxed{d_i = 18 \text{ cm}}$ and reveals a real, inverted, and enlarged image.

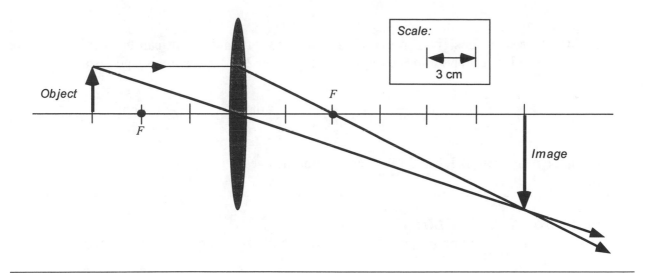

---

101. **REASONING AND SOLUTION**   Since the far point is 220 cm, we know that the image distance is $d_i = -220$ cm, when the object is infinitely far from the lens ($d_o = \infty$). Thus, the thin-lens equation (Equation 26.6) becomes

$$\frac{1}{d_o} + \frac{1}{d_i} = \frac{1}{\infty} + \frac{1}{-220 \text{ cm}} = \frac{1}{-220 \text{ cm}} = \frac{1}{f} \qquad \text{or} \qquad \boxed{f = -220 \text{cm}}$$

---

105. **REASONING AND SOLUTION**

a. We know from the law of reflection (Section 25.2), that the angle of reflection is equal to the angle of incidence, so the reflected ray is reflected at $\boxed{43°}$.

b. Snell's law of refraction (Equation 26.2: $n_1 \sin \theta_1 = n_2 \sin \theta_2$ can be used to find the angle of refraction. Table 26.1 indicates that the index of refraction of water is 1.333. Solving for θ_2 and substituting values, we find that

$$\sin \theta_2 = \frac{n_1 \sin \theta_1}{n_2} = \frac{(1.000)(\sin 43°)}{1.333} = 0.51 \qquad \text{or} \qquad \theta_2 = \sin^{-1} 0.51 = \boxed{31°}$$

---

107. **REASONING AND SOLUTION**   An optometrist prescribes contact lenses that have a focal length of 55.0 cm.

a. The focal length is positive (+55.0 cm); therefore, we can conclude that the lenses are $\boxed{\text{converging}}$.

b. As discussed in the text (see Section 26.10), farsightedness is corrected by converging lenses. Therefore, the person who wears these lens is $\boxed{\text{farsighted}}$.

c. If the lenses are designed so that objects no closer than 35.0 cm can be seen clearly, we have $d_o = 35.0$ cm. The thin-lens equation (Equation 26.6) gives the image distance:

$$\frac{1}{d_i} = \frac{1}{f} - \frac{1}{d_o} = \frac{1}{55.0\text{ cm}} - \frac{1}{35.0\text{ cm}} \qquad \text{or} \qquad d_i = -96.3\text{ cm}$$

Thus, the near point is located $\boxed{96.3\text{ cm}}$ from the eyes.

---

111. **REASONING AND SOLUTION**

a. A real image must be projected on the drum; therefore, the lens in the copier must be a $\boxed{\text{converging lens}}$.

b. If the document and its copy have the same size, but are inverted with respect to one another, the magnification equation (Equation 26.7) indicates that $m = -d_i/d_o = -1$. Therefore, $d_i/d_o = 1$ or $d_i = d_o$. Then, the thin-lens equation (Equation 26.6) gives

$$\frac{1}{d_i} + \frac{1}{d_o} = \frac{1}{f} = \frac{2}{d_o} \qquad \text{or} \qquad d_o = d_i = 2f$$

Therefore the document is located at a distance $\boxed{2f}$ from the lens.

c. Furthermore, the image is located at a distance of $\boxed{2f}$ from the lens.

---

113. **REASONING** A contact lens is placed directly on the eye. Therefore, the object distance, which is the distance from the book to the lens, is 25.0 cm. The near point can be determined from the thin-lens equation [Equation 26.6: $(1/d_o) + (1/d_i) = (1/f)$].

SOLUTION

a. Using the thin-lens equation, we have

$$\frac{1}{d_i} = \frac{1}{f} - \frac{1}{d_o} = \frac{1}{65.0\text{ cm}} - \frac{1}{25.0\text{ cm}} \qquad \text{or} \qquad d_i = -40.6\text{ cm}$$

In other words, at age 40, the man's near point is 40.6 cm. Similarly, when the man is 45, we have

$$\frac{1}{d_i} = \frac{1}{f} - \frac{1}{d_o} = \frac{1}{65.0\text{ cm}} - \frac{1}{29.0\text{ cm}} \qquad \text{or} \qquad d_i = -52.4\text{ cm}$$

and his near point is 52.4 cm.   Thus, the man's near point has changed by 52.4 cm − 40.6 cm = $\boxed{11.8 \text{ cm}}$.

b. With $d_0 = 25.0$ cm and $d_i = -52.4$ cm, the focal length of the lens is found as follows:

$$\frac{1}{f} = \frac{1}{d_0} + \frac{1}{d_i} = \frac{1}{25.0 \text{ cm}} + \frac{1}{(-52.4 \text{ cm})} \qquad \text{or} \qquad \boxed{f = 47.8 \text{ cm}}$$

---

115. ***REASONING***   The angular magnification of a refracting telescope is 32 8000 times larger when you look through the correct end of the telescope than when you look through the wrong end. We wish to find the angular magnification, $M = -f_0 / f_e$ (see Equation 26.12) of the telescope. Thus, we proceed by finding the ratio of the focal lengths of the objective and the eyepiece and using Equation 26.12 to find M.

SOLUTION   When you look through the correct end of the telescope, the angular magnification of the telescope is $M_c = -f_0 / f_e$. If you look through the wrong end, the roles of the objective and eyepiece lenses are interchanged, so that the angular magnification would be $M_w = -f_e / f_0$. Therefore,

$$\frac{M_c}{M_w} = \frac{-f_0 / f_e}{-f_e / f_0} = \left(\frac{f_0}{f_e}\right)^2 = 32\,800 \qquad \text{or} \qquad \frac{f_0}{f_e} = \pm\sqrt{32\,800} = \pm 181$$

The angular magnification of the telescope is negative, so we choose the positive root and obtain $M = -f_0 / f_e = -(+181) = \boxed{-181}$.

CHAPTER 27 | INTERFERENCE AND THE WAVE NATURE OF LIGHT

PROBLEMS

1. **REASONING AND SOLUTION** To decide whether constructive or destructive interference occurs, we need to determine the wavelength of the wave. For electromagnetic waves, Equation 16.1 can be written $f\lambda = c$, so that

$$\lambda = \frac{c}{f} = \frac{3.00 \times 10^8 \text{ m}}{536 \times 10^3 \text{ Hz}} = 5.60 \times 10^2 \text{ m}$$

Since the two wave sources are in phase, constructive interference occurs when the path difference is an integer number of wavelengths, and destructive interference occurs when the path difference is an odd number of half wavelengths. We find that the path difference is 8.12 km − 7.00 km = 1.12 km.   The number of wavelengths in this path difference is $(1.12 \times 10^3 \text{ m})/(5.60 \times 10^2 \text{ m}) = 2.00$. Therefore, $\boxed{\text{constructive interference occurs}}$.

---

5. **REASONING AND SOLUTION** The angular position θ of the bright fringes of a double slit are given by Equation 27.1 as $\sin\theta = m\lambda / d$, with the order of the fringe specified by $m = 0, 1, 2, 3, \ldots$. Solving for λ, we have

$$\lambda = \frac{d \sin\theta}{m} = \frac{(3.8 \times 10^{-5} \text{ m})\sin 2.0°}{2} = 6.6 \times 10^{-7} \text{ m} = \boxed{660 \text{ nm}}$$

---

9. $\boxed{\text{WWW}}$ **REASONING** The light that travels through the plastic has a different path length than the light that passes through the unobstructed slit. Since the center of the screen now appears dark, rather than bright, destructive interference, rather than constructive interference occurs there.   This means that the difference between the number of wavelengths in the plastic sheet and that in a comparable thickness of air is $\frac{1}{2}$.

SOLUTION The wavelength of the light in the plastic sheet is given by Equation 27.3 as

$$\lambda_{\text{plastic}} = \frac{\lambda_{\text{vacuum}}}{n} = \frac{586 \times 10^{-9} \text{ m}}{1.60} = \boxed{366 \times 10^{-9} \text{ m}}$$

The number of wavelengths contained in a plastic sheet of thickness t is

$$N_{\text{plastic}} = \frac{t}{\lambda_{\text{plastic}}} = \frac{t}{366 \times 10^{-9}\ \text{m}}$$

The number of wavelengths contained in an equal thickness of air is

$$N_{\text{air}} = \frac{t}{\lambda_{\text{air}}} = \frac{t}{586 \times 10^{-9}\ \text{m}}$$

where we have used the fact that $\lambda_{\text{air}} \approx \lambda_{\text{vacuum}}$. Destructive interference occurs when the difference, $N_{\text{plastic}} - N_{\text{air}}$, in the number of wavelengths is $\frac{1}{2}$:

$$N_{\text{plastic}} - N_{\text{air}} = \frac{1}{2}$$

$$\frac{t}{366 \times 10^{-9}\ \text{m}} - \frac{t}{586 \times 10^{-9}\ \text{m}} = \frac{1}{2}$$

Solving this equation for t yields $t = 487 \times 10^{-9}\ \text{m} = \boxed{487\ \text{nm}}$.

---

11. ***REASONING***   To solve this problem, we must express the condition for destructive interference in terms of the film thickness t and the wavelength λ_{film} of the light as it passes through the magnesium fluoride coating. We must also take into account any phase changes that occur upon reflection.

SOLUTION   Since the coating is intended to be nonreflective, its thickness must be chosen so that destructive interference occurs between waves 1 and 2 in the drawing. For destructive interference, the combined phase difference between the two waves must be an odd integer number of half wavelengths. The phase change for wave 1 is equivalent to one-half of a wavelength, since this light travels from a smaller refractive index ($n_{\text{air}} = 1.00$) toward a larger refractive index ($n_{\text{film}} = 1.38$).

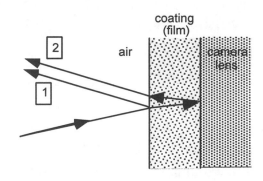

Similarly, there is a phase change when wave 2 reflects from the right surface of the film, since this light also travels from a smaller refractive index ($n_{\text{film}} = 1.38$) toward a larger one ($n_{\text{lens}} = 1.52$). Therefore, a phase change of one-half wavelength occurs at both boundaries, so the net phase change between waves 1 and 2 due to reflection is zero. Since wave 2 travels back and forth through the film and, and since the light is assumed to be at nearly normal incidence, the extra distance traveled by wave 2 compared to wave 1 is twice the film thickness, or $2t$. Thus, in this case, the minimum condition for destructive interference is

$$2t = \tfrac{1}{2}\lambda_{\text{film}}$$

The wavelength of light in the coating is

$$\lambda_{\text{film}} = \frac{\lambda_{\text{vacuum}}}{n} = \frac{565 \text{ nm}}{1.38} = 409 \text{ nm} \tag{27.3}$$

Solving the above expression for t, we find that the minimum thickness that the coating can have is

$$t = \tfrac{1}{4}\lambda_{\text{film}} = \tfrac{1}{4}(409 \text{ nm}) = \boxed{102 \text{ nm}}$$

---

15. $\boxed{\text{WWW}}$ *REASONING*   To solve this problem, we must express the condition for constructive interference in terms of the film thickness t and the wavelength λ_{film} of the light in the soap film. We must also take into account any phase changes that occur upon reflection.

SOLUTION   For the reflection at the top film surface, the light travels from air, where the refractive index is smaller ($n = 1.00$), toward the film, where the refractive index is larger ($n = 1.33$). Associated with this reflection there is a phase change that is equivalent to one-half of a wavelength. For the reflection at the bottom film surface, the light travels from the film, where the refractive index is larger ($n = 1.33$), toward air, where the refractive index is smaller ($n = 1.00$). Associated with this reflection, there is no phase change. As a result of these two reflections, there is a net phase change that is equivalent to one-half of a wavelength. To obtain the condition for constructive interference, this net phase change must be added to the phase change that arises because of the film thickness t, which is traversed twice by the light that penetrates it. For constructive interference we find that

$$2t + \tfrac{1}{2}\lambda_{\text{film}} = \lambda_{\text{film}}, 2\lambda_{\text{film}}, 3\lambda_{\text{film}}, \dots$$

or

$$2t = \left(m + \tfrac{1}{2}\right)\lambda_{\text{film}}, \qquad \text{where } m = 0, 1, 2, \dots$$

Equation 27.3 indicates that $\lambda_{\text{film}} = \lambda_{\text{vacuum}}/n$. Using this expression and the fact that $m = 0$ for the minimum thickness t, we find that the condition for constructive interference becomes

$$2t = \left(m + \tfrac{1}{2}\right)\lambda_{\text{film}} = \left(0 + \tfrac{1}{2}\right)\left(\frac{\lambda_{\text{vacuum}}}{n}\right)$$

or

$$t = \frac{\lambda_{\text{vacuum}}}{4n} = \frac{611 \text{ nm}}{4(1.33)} = \boxed{115 \text{ nm}}$$

19. **REASONING**   This problem can be solved by using Equation 27.4 for the value of the angle θ when $m = 1$ (first dark fringe).

SOLUTION

a.   When the slit width is $W = 1.8 \times 10^{-4}$ m and $\lambda = 675$ nm $= 675 \times 10^{-9}$ m, we find, according to Equation 27.4,

$$\theta = \sin^{-1}\left(m\frac{\lambda}{W}\right) = \sin^{-1}\left[(1)\frac{675 \times 10^{-9} \text{ m}}{1.8 \times 10^{-4} \text{ m}}\right] = \boxed{0.21°}$$

b.   Similarly, when the slit width is $W = 1.8 \times 10^{-6}$ m and $\lambda = 675 \times 10^{-9}$ m, we find

$$\theta = \sin^{-1}\left[(1)\frac{675 \times 10^{-9} \text{ m}}{1.8 \times 10^{-6} \text{ m}}\right] = \boxed{22°}$$

---

23.   **WWW** **REASONING**   According to Equation 27.4, the angles at which the dark fringes occur are given by $\sin\theta = m\lambda / W$, where W is the slit width. In Figure 27.24, we see from the trigonometry of the situation that $\tan\theta = y/L$. Therefore, the latter expression can be used to determine the angle θ, and Equation 27.4 can be used to find the wavelength λ.

SOLUTION   The angle θ is given by

$$\theta = \tan^{-1}\left(\frac{y}{L}\right) = \tan^{-1}\left(\frac{3.5 \times 10^{-3} \text{ m}}{4.0 \text{ m}}\right) = 0.050°$$

The wavelength of the light is

$$\lambda = W \sin\theta = \left(5.6 \times 10^{-4} \text{ m}\right)\sin 0.050° = 4.9 \times 10^{-7} \text{ m} = \boxed{490 \text{ nm}}$$

---

25. **REASONING AND SOLUTION**   Using Equation 27.4 for the first-order dark fringes ($m = 1$) and referring to Figure 27.24 in the text, we see that

$$\sin\theta = (1)\frac{\lambda}{W} = \frac{y}{\sqrt{L^2 + y^2}}$$

Since the distance L between the slit and the screen equals the width $2y$ of the central bright fringe, this equation becomes

$$\frac{\lambda}{W} = \frac{y}{\sqrt{(2y)^2 + y^2}} = \frac{1}{\sqrt{5}} = \boxed{0.447}$$

31. **REASONING AND SOLUTION**   According to Rayleigh's criterion (Equation 27.6), the minimum angular separation of the two objects is

$$\theta_{min} \text{ (in rad)} \approx 1.22\frac{\lambda}{D} = 1.22\left(\frac{565 \times 10^{-9} \text{ m}}{1.02 \text{ m}}\right) = 6.76 \times 10^{-7} \text{ rad}$$

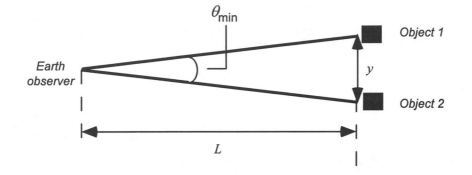

Therefore, from the figure above, the separation y of the two objects is

$$y = L\theta_{min} = (3.75 \times 10^4 \text{ m})(6.76 \times 10^{-7} \text{ rad}) = \boxed{0.0254 \text{ m}}$$

35. WWW **REASONING**   When light of wavelength λ passes through a circular opening of diameter D, a circular diffraction pattern results, and Equation 27.5 [$\sin\theta = 1.22(\lambda/D)$] locates the first circular dark fringe relative to the central bright spot. Therefore, Equation 27.5 can be used to find the angular separation θ of the central bright spot and the first dark fringe. The geometry of the situation can be used to find the diameter of the central bright spot on the moon.

SOLUTION   From Equation 27.5,

$$\theta = \sin^{-1}\left(\frac{1.22\lambda}{D}\right) = \sin^{-1}\left[\frac{1.22(694.3 \times 10^{-9} \text{ m})}{0.20 \text{ m}}\right] = 4.2 \times 10^{-6} \text{ rad}$$

From the following figure, we see that $2\theta = \dfrac{d}{R}$.

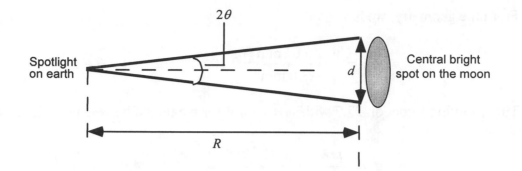

Therefore, the diameter of the spot on the moon is

$$d = 2\theta R = 2(4.2 \times 10^{-6} \text{ rad})(3.77 \times 10^{8} \text{ m}) = \boxed{3.2 \times 10^{3} \text{ m}}$$

---

37. ***REASONING AND SOLUTION***   According to Equation 27.7, the angles that correspond to the first-order ($m = 1$) maximum for the two wavelengths in question are:

a. for $\lambda = 660$ nm $= 660 \times 10^{-9}$ m,

$$\theta = \sin^{-1}\left(m\frac{\lambda}{d}\right) = \sin^{-1}\left[(1)\left(\frac{660 \times 10^{-9} \text{ m}}{1.1 \times 10^{-6} \text{ m}}\right)\right] = \boxed{37°}$$

b. for $\lambda = 410$ nm $= 410 \times 10^{-9}$ m,

$$\theta = \sin^{-1}\left(m\frac{\lambda}{d}\right) = \sin^{-1}\left[(1)\left(\frac{410 \times 10^{-9} \text{ m}}{1.1 \times 10^{-6} \text{ m}}\right)\right] = \boxed{22°}$$

---

41. ***REASONING AND SOLUTION***   The geometry of the situation is shown below.

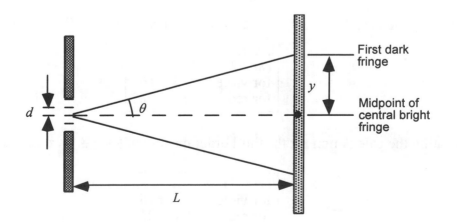

From the geometry, we have

$$\tan\theta = \frac{y}{L} = \frac{0.60 \text{ mm}}{3.0 \text{ mm}} = 0.20 \qquad \text{or} \qquad \theta = 11.3°$$

Then, solving Equation 27.7 with $m = 1$ for the separation d between the slits, we have

$$d = \frac{m\lambda}{\sin\theta} = \frac{(1)(780 \times 10^{-9} \text{ m})}{\sin 11.3°} = \boxed{4.0 \times 10^{-6} \text{ m}}$$

---

43. **REASONING** The angle θ that locates the first-order maximum produced by a grating with 3300 lines/cm is given by Equation 27.7, $\sin\theta = m\lambda/d$, with the order of the fringes given by $m = 0, 1, 2, 3, \ldots$ Any two of the diffraction patterns will overlap when their angular positions are the same.

SOLUTION Since the grating has 3300 lines/cm, we have

$$d = \frac{1}{3300 \text{ lines/cm}} = 3.0 \times 10^{-4} \text{ cm} = 3.0 \times 10^{-6} \text{ m}$$

a. In first order, $m = 1$; therefore, for violet light,

$$\theta = \sin^{-1}\left(m\frac{\lambda}{d}\right) = \sin^{-1}\left[(1)\left(\frac{410 \times 10^{-9} \text{ m}}{3.0 \times 10^{-6} \text{ m}}\right)\right] = \boxed{7.9°}$$

Similarly for red light,

$$\theta = \sin^{-1}\left(m\frac{\lambda}{d}\right) = \sin^{-1}\left[(1)\left(\frac{660 \times 10^{-9} \text{ m}}{3.0 \times 10^{-6} \text{ m}}\right)\right] = \boxed{13°}$$

b. Repeating the calculation for the second order maximum ($m = 2$), we find that

$(m = 2)$	
for violet	$\theta = 16°$
for red	$\theta = 26°$

c. Repeating the calculation for the third order maximum ($m = 3$), we find that

$(m = 3)$	
for violet	$\theta = 24°$
for red	$\theta = 41°$

d. Comparisons of the values for θ calculated in parts (a), (b) and (c) show that the
$\boxed{\text{second and third orders overlap}}$.

---

47. **REASONING**   The slit separation d is given by Equation 27.1 with $m = 1$; namely
$d = \lambda / \sin\theta$. As shown in Example 1 in the text, the angle θ is given by $\theta = \tan^{-1}(y/L)$.

SOLUTION The angle θ is

$$\theta = \tan^{-1}\left(\frac{0.037 \text{ m}}{4.5 \text{ m}}\right) = 0.47°$$

Therefore, the slit width d is

$$d = \frac{\lambda}{\sin\theta} = \frac{490 \times 10^{-9} \text{ m}}{\sin 0.47°} = \boxed{6.0 \times 10^{-5} \text{ m}}$$

---

51. **REASONING** According to Rayleigh's criterion, the two taillights must be separated by a
distance s sufficient to subtend an angle $\theta_{min} \approx 1.22\lambda / D$ at the pupil of the observer's eye.
Since this angle must be expressed in radians, we relate θ_{min} to the distances s and L.

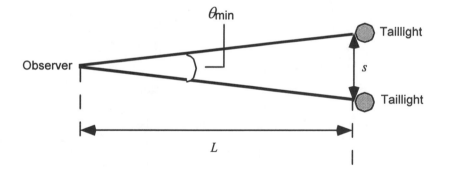

SOLUTION The wavelength λ is 660 nm. Therefore, we have from Equation 27.6

$$\theta_{min} \approx 1.22\frac{\lambda}{D} = 1.22\left(\frac{660 \times 10^{-9} \text{ m}}{7.0 \times 10^{-3} \text{ m}}\right) = 1.2 \times 10^{-4} \text{ rad}$$

According to Equation 8.1, the distance L between the observer and the taillights is

$$L = \frac{s}{\theta_{min}} = \frac{1.2 \text{ m}}{1.2 \times 10^{-4} \text{ rad}} = \boxed{1.0 \times 10^{4} \text{ m}}$$

55. **REASONING**   For a diffraction grating, the angular position θ of a principal maximum on the screen is given by Equation 27.7 as $\sin\theta = m\lambda / d$ with $m = 0, 1, 2, 3, \ldots$

SOLUTION   When the fourth-order principal maximum of light A exactly overlaps the third-order principal maximum of light B, we have

$$\sin\theta_A = \sin\theta_B$$

$$\frac{4\lambda_A}{d} = \frac{3\lambda_B}{d} \qquad \text{or} \qquad \frac{\lambda_A}{\lambda_B} = \boxed{\frac{3}{4}}$$

CHAPTER 28 │*SPECIAL RELATIVITY*

1. **REASONING** Since the "police car" is moving relative to the earth observer, the earth observer measures a greater time interval Δt between flashes. Since both the proper time Δt_0 (as observed by the officer) and the dilated time Δt (as observed by the person on earth) are known, the speed of the "police car" relative to the observer can be determined from the time dilation relation, Equation 28.1.

SOLUTION According to Equation 28.1, the dilated time interval between flashes is $\Delta t = \Delta t_0 / \sqrt{1 - (v^2/c^2)}$, where Δt_0 is the proper time. Solving for the speed v, we find

$$v = c\sqrt{1 - \left(\frac{\Delta t_0}{\Delta t}\right)^2} = (3.0 \times 10^8 \text{ m/s})\sqrt{1 - \left(\frac{1.5 \text{ s}}{2.5 \text{ s}}\right)^2} = \boxed{2.4 \times 10^8 \text{ m/s}}$$

3. **REASONING AND SOLUTION** The period Δt_0 of the rotating radar antenna is given by Equation 10.4 as

$$\Delta t_0 = \frac{2\pi}{\omega_0} = \frac{2\pi}{0.25 \text{ rad/s}} = 25 \text{ s}$$

This period, or time interval, is a proper time interval, because an earth-based observer measures the beginning and ending of each revolution at the same place. An observer moving relative to the earth measures a dilated time interval given by Equation 28.1 as

$$\Delta t = \frac{\Delta t_0}{\sqrt{1 - \frac{v^2}{c^2}}} = \frac{25 \text{ s}}{\sqrt{1 - \frac{(0.80 \, c)^2}{c^2}}} = 42 \text{ s}$$

The angular speed ω of the rotating antenna as measured by the moving observer is, therefore,

$$\omega = \frac{2\pi}{\Delta t} = \frac{2\pi}{42 \text{ s}} = \boxed{0.15 \text{ rad/s}}$$

9. ***REASONING AND SOLUTION*** The length L_0 that the person measures for the UFO when it lands is the proper length, since the UFO is at rest with respect to the person. Therefore, from Equation 28.2 we have

$$L_0 = \frac{L}{\sqrt{1 - \dfrac{v^2}{c^2}}} = \frac{230 \text{ m}}{\sqrt{1 - \dfrac{(0.90\,c)^2}{c^2}}} = \boxed{530 \text{ m}}$$

---

11. WWW ***REASONING AND SOLUTION*** The diameter D of the planet, as measured by a moving spacecraft, is given in terms of the proper diameter D_0 by Equation 28.2. Taking the ratio of the diameter D_A of the planet measured by spaceship A to the diameter D_B measured by spaceship B, we find

$$\frac{D_A}{D_B} = \frac{D_0\sqrt{1 - \dfrac{v_A{}^2}{c^2}}}{D_0\sqrt{1 - \dfrac{v_B{}^2}{c^2}}} = \frac{\sqrt{1 - \dfrac{(0.60\,c)^2}{c^2}}}{\sqrt{1 - \dfrac{(0.80\,c)^2}{c^2}}} = \boxed{1.3}$$

---

15. WWW ***REASONING*** Only the sides of the rectangle that lie in the direction of motion will experience length contraction. In order to make the rectangle look like a square, each side must have a length of $L = 2.0$ m. Thus, we move along the long side, taking the proper length to be $L_0 = 3.0$ m. We can solve for the speed using Equation 28.2. Then, with this speed, we can use the relation for length contraction to find L for the short side as we move along it.

SOLUTION From Equation 28.2, $L = L_0\sqrt{1 - (v^2/c^2)}$, we find that

$$v = c\sqrt{1 - \left(\frac{L}{L_0}\right)^2} = c\sqrt{1 - \left(\frac{2.0 \text{ m}}{3.0 \text{ m}}\right)^2} = 0.75\,c$$

Moving at this speed along the short side, we take $L_0 = 2.0$ m and find L:

$$L = L_0\sqrt{1 - \left(\frac{v}{c}\right)^2} = (2.0 \text{ m})\sqrt{1 - \left(\frac{0.75\,c}{c}\right)^2} = 1.3 \text{ m}$$

The observed dimensions of the rectangle are, therefore, $\boxed{3.0 \text{ m} \times 1.3 \text{ m}}$, since the long side is not contracted due to motion along the short side.

---

19. ***REASONING AND SOLUTION*** The magnitude of the relativistic momentum p of the rocket is given by Equation 28.3, $p = mv/\sqrt{1 - (v^2/c^2)}$. Solving this expression for the speed v of the rocket gives

$$v = \frac{p}{\sqrt{m^2 + \dfrac{p^2}{c^2}}} = \frac{3.15 \times 10^{13} \text{ kg} \cdot \text{m/s}}{\sqrt{(1.40 \times 10^5 \text{ kg})^2 + \dfrac{(3.15 \times 10^{13} \text{ kg} \cdot \text{m/s})^2}{(3.00 \times 10^8 \text{ m/s})^2}}} = \boxed{1.80 \times 10^8 \text{ m/s}}$$

---

23. ***REASONING*** According to the work-energy theorem, Equation 6.3, the work that must be done on the electron to accelerate it from rest to a speed of $0.990c$ is equal to the kinetic energy of the electron when it is moving at $0.990c$.

SOLUTION Using Equation 28.6, we find that

$$\text{KE} = mc^2 \left(\frac{1}{\sqrt{1 - (v^2/c^2)}} - 1 \right)$$

$$= (9.11 \times 10^{-31} \text{ kg})(3.00 \times 10^8 \text{ m/s})^2 \left(\frac{1}{\sqrt{1 - (0.990\,c)^2/c^2}} - 1 \right) = \boxed{5.0 \times 10^{-13} \text{ J}}$$

---

27. ***REASONING AND SOLUTION***
a. In Section 28.6 it is shown that when the speed of a particle is $0.01c$ (or less), the relativistic kinetic energy becomes nearly equal to the nonrelativistic kinetic energy. Since the speed of the particle here is $0.001c$, the ratio of the relativistic kinetic energy to the nonrelativistic kinetic energy is $\boxed{1.0}$.

b. Taking the ratio of the relativistic kinetic energy, Equation 28.6, to the nonrelativistic kinetic energy, $\frac{1}{2}mv^2$, we find that

$$\frac{mc^2\left(\dfrac{1}{\sqrt{1-(v^2/c^2)}}-1\right)}{\frac{1}{2}mv^2} = 2\left(\frac{c}{v}\right)^2\left(\frac{1}{\sqrt{1-(v^2/c^2)}}-1\right)$$

$$= 2\left(\frac{c}{0.970\,c}\right)^2\left(\frac{1}{\sqrt{1-(0.970\,c)^2/c^2}}-1\right) = \boxed{6.6}$$

---

31. **REASONING** Let's define the following relative velocities, assuming that the spaceship and exploration vehicle are moving in the positive direction.

   v_{ES} = velocity of **E**xploration vehicle relative to the **S**paceship.
   v_{EO} = velocity of **E**xploration vehicle relative to an **O**bserver on earth = +0.70c
   v_{SO} = velocity of **S**paceship relative to an **O**bserver on earth = +0.50c

The velocity v_{ES} can be determined from the velocity-addition formula, Equation 28.7:

$$v_{ES} = \frac{v_{EO} + v_{OS}}{1 + \dfrac{v_{EO}v_{OS}}{c^2}}$$

The velocity v_{OS} of the observer on earth relative to the spaceship is not given. However, we know that v_{OS} is the negative of v_{SO}, so $v_{OS} = -v_{SO} = -(+0.50c) = -0.50c$.

SOLUTION The velocity of the exploration vehicle relative to the spaceship is

$$v_{ES} = \frac{v_{EO} + v_{OS}}{1 + \dfrac{v_{EO}v_{OS}}{c^2}} = \frac{+0.70c + (-0.50c)}{1 + \dfrac{(+0.70c)(-0.50c)}{c^2}} = \boxed{+0.31c}$$

---

33. $\boxed{\text{WWW}}$ **REASONING** Since the crew is initially at rest relative to the escape pod, the length of 45 m is the proper length L_0 of the pod. The length of the escape pod as determined by an observer on earth can be obtained from the relation for length contraction given by Equation 28.2, $L = L_0\sqrt{1-\left(v_{PE}^2/c^2\right)}$. The quantity v_{PE} is the speed of the escape pod relative to the earth, which can be found from the velocity-addition formula, Equation 28.7. The following are the relative velocities, assuming that the direction away from the earth is the positive direction:

v_{PE} = velocity of the escape **P**od relative to **E**arth.

v_{PR} = velocity of escape **P**od relative to the **R**ocket = $-0.55c$. This velocity is negative because the rocket is moving away from the earth (in the positive direction), and the escape pod is moving in an opposite direction (the negative direction) relative to the rocket.

v_{RE} = velocity of **R**ocket relative to **E**arth = $+0.75c$

These velocities are related by the velocity-addition formula, Equation 28.7.

SOLUTION The relative velocity of the escape pod relative to the earth is

$$v_{PE} = \frac{v_{PR} + v_{RE}}{1 + \dfrac{v_{PR} v_{RE}}{c^2}} = \frac{-0.55c + 0.75c}{1 + \dfrac{(-0.55c)(+0.75c)}{c^2}} = +0.34c$$

The speed of the pod relative to the earth is $0.34c$. The length of the pod as determined by an observer on earth is

$$L = L_0 \sqrt{1 - \frac{v_{PE}^2}{c^2}} = (45 \text{ m}) \sqrt{1 - \frac{(0.34c)^2}{c^2}} = \boxed{42 \text{ m}}$$

---

37. $\boxed{\text{WWW}}$ **REASONING** All standard meter sticks at rest have a length of 1.00 m for observers who are at rest with respect to them. Thus, 1.00 m is the proper length L_0 of the meter stick. When the meter stick moves with speed v relative to an earth-observer, its length $L = 0.500$ m will be a contracted length. Since both L_0 and L are known, v can be found directly from Equation 28.2, $L = L_0 \sqrt{1 - \left(v^2/c^2\right)}$.

SOLUTION Solving Equation 28.2 for v, we find that

$$v = c \sqrt{1 - \left(\frac{L}{L_0}\right)^2} = (3.00 \times 10^8 \text{ m/s}) \sqrt{1 - \left(\frac{0.500 \text{ m}}{1.00 \text{ m}}\right)^2} = \boxed{2.60 \times 10^8 \text{ m/s}}$$

---

41. **REASONING** The total linear momentum of the system is conserved, since no net external force acts on the system. Therefore, the final total momentum $p_1 + p_2$ of the two fragments must equal the initial total momentum, which is zero since the particle is initially at rest. As a result, $p_1 = -p_2$, where Equation 28.3 must be used for the magnitudes of the momenta p_1 and p_2. Thus, we find

$$\frac{m_1 v_1}{\sqrt{1-(v_1^2/c^2)}} = \frac{-m_2 v_2}{\sqrt{1-(v_2^2/c^2)}}$$

SOLUTION Letting fragment 2 be the more-massive fragment, we have that $v_1 = 0.800c$, $m_1 = 1.67 \times 10^{-27}$ kg, and $m_2 = 5.01 \times 10^{-27}$ kg. Squaring both sides of the above equation, rearranging terms, substituting the known values for v_1, m_1, and m_2, we find that

$$\frac{v_2^2}{1-(v_2^2/c^2)} = \frac{m_1^2 v_1^2}{m_2^2 \left[1-(v_1^2/c^2)\right]} = \frac{(1.67 \times 10^{-27}\text{ kg})^2 (0.800\ c)^2}{(5.01 \times 10^{-27}\text{ kg})^2 \left[1-(0.800\ c/c)^2\right]} = 0.1975\ c^2$$

Therefore,

$$v_2^2 = 0.1975\ c^2 \left[1-(v_2^2/c^2)\right] = 0.1975\ c^2 - 0.1975\ v_2^2$$

Solving for v_2 gives

$$v_2 = \pm\sqrt{\frac{0.1975\ c^2}{1.1975}} = \pm 0.406\ c$$

We reject the positive root, since then both fragments would be moving in the same direction after the break-up and the system would have a non-zero momentum. According to the principle of conservation of momentum, the total momentum after the break-up must be zero, just as it was before the break-up. The momentum of the system will be zero only if the velocity v_2 is opposite to the velocity v_1. Hence, we chose the negative root and

$$\boxed{v_2 = -0.406\ c}.$$

CHAPTER 29 | *PARTICLES AND WAVES*

PROBLEMS

1. **REASONING** The energy of the photon is related to its frequency by Equation 29.2, $E = hf$. Equation 16.1, $v = f\lambda$, relates the frequency and the wavelength for any wave.

   SOLUTION Combining Equations 29.2 and 16.1, and noting that the speed of a photon is c, the speed of light in a vacuum, we have

   $$\lambda = \frac{c}{f} = \frac{c}{(E/h)} = \frac{hc}{E} = \frac{(6.63\times 10^{-34}\ \text{J}\cdot\text{s})(3.0\times 10^8\ \text{m/s})}{6.4\times 10^{-19}\ \text{J}} = 3.1\times 10^{-7}\ \text{m} = \boxed{310\ \text{nm}}$$

5. **REASONING** According to Equation 29.3, the work function W_0 is related to the photon energy hf and the maximum kinetic energy KE_{max} by $W_0 = hf - \text{KE}_{max}$. This expression can be used to find the work function of the metal.

   SOLUTION KE_{max} is 6.1 eV. The photon energy (in eV) is, according to Equation 29.2,

   $$hf = \left(6.63\times 10^{-34}\ \text{J}\cdot\text{s}\right)\left(3.00\times 10^{15}\ \text{Hz}\right)\left(\frac{1\ \text{eV}}{1.60\times 10^{-19}\ \text{J}}\right) = 12.4\ \text{eV}$$

   The work function is, therefore,

   $$W_0 = hf - \text{KE}_{max} = 12.4\ \text{eV} - 6.1\ \text{ev} = \boxed{6.3\ \text{eV}}$$

9. **REASONING AND SOLUTION** The number of photons per second, N, entering the owl's eye is $N = SA/E$, where S is the intensity of the beam, A is the area of the owl's pupil, and E is the energy of a single photon. Assuming that the owl's pupil is circular, $A = \pi r^2 = \pi\left(\frac{1}{2}d\right)^2$, where d is the diameter of the owl's pupil. Combining Equations 29.2 and 16.1, we have $E = hf = hc/\lambda$. Therefore,

   $$N = \frac{SA\lambda}{hc} = \frac{(5.0\times 10^{-13}\ \text{W/m}^2)\pi\left[\frac{1}{2}(8.5\times 10^{-3}\ \text{m})\right]^2 (510\times 10^{-9}\ \text{m})}{(6.63\times 10^{-34}\ \text{J}\cdot\text{s})(3.0\times 10^8\ \text{m/s})} = \boxed{73\ \text{photons/s}}$$

11. $\boxed{\text{WWW}}$ *REASONING* The heat required to melt the ice is given by $Q = mL_f$, where m is the mass of the ice and L_f is the latent heat of fusion for water (see Section 12.8). Since, according to Equation 29.2, each photon carries an energy of $E = hf$, the energy content of N photons is $E_{Total} = Nhf$. According to Equation 16.1, $f = c/\lambda$, so we have

$$E_{Total} = \frac{Nhc}{\lambda}$$

If we assume that all of the photon energy is used to melt the ice, then, $E_{Total} = Q$, so that

$$\underbrace{\frac{Nhc}{\lambda}}_{E_{Total}} = \underbrace{mL_f}_{Q}$$

This expression may be solved for N to determine the required number of photons.

SOLUTION
a. We find that

$$N = \frac{mL_f \lambda}{hc} = \frac{(2.0 \text{ kg})(33.5 \times 10^4 \text{ J/kg})(620 \times 10^{-9} \text{ m})}{(6.63 \times 10^{-34} \text{ J·s})(3.00 \times 10^8 \text{ m/s})} = \boxed{2.1 \times 10^{24} \text{ photons}}$$

b. The number N' of molecules in 2.0-kg of water is

$$N' = (2.0 \text{ kg}) \left(\frac{1 \text{ mol}}{18 \times 10^{-3} \text{ kg}} \right) \left(\frac{6.022 \times 10^{23} \text{ molecules}}{1 \text{ mol}} \right) = 6.7 \times 10^{25} \text{ molecules}$$

Therefore, on average, the number of water molecules that one photon converts from the ice phase to the liquid phase is

$$\frac{N'}{N} = \frac{6.7 \times 10^{25} \text{ molecules}}{2.1 \times 10^{24} \text{ photons}} = \boxed{32 \text{ molecules/photon}}$$

---

15. *REASONING AND SOLUTION*
a. According to Equation 29.6, the magnitude of the momentum of the incident photon is

$$p = \frac{h}{\lambda} = \frac{6.626 \times 10^{-34} \text{ J·s}}{0.3120 \times 10^{-9} \text{ m}} = \boxed{2.124 \times 10^{-24} \text{ kg·m/s}}$$

b. The wavelength of the scattered photon is, from Equation 29.7,

$$\lambda' = \lambda + \frac{h}{mc}(1 - \cos\theta)$$

where θ is the scattering angle. Combining this expression with Equation 29.6, we find that the magnitude of the momentum of the scattered photon is

$$p' = \frac{h}{\lambda'} = \frac{h}{\lambda + \left(\dfrac{h}{mc}\right)(1 - \cos\theta)}$$

$$= \frac{6.626 \times 10^{-34} \text{ J}\cdot\text{s}}{0.3120 \times 10^{-9} \text{ m} + \left[\dfrac{6.626 \times 10^{-34} \text{ J}\cdot\text{s}}{(9.109 \times 10^{-31} \text{ kg})(2.998 \times 10^8 \text{ m/s})}\right](1 - \cos 135.0°)}$$

$$= \boxed{2.096 \times 10^{-24} \text{ kg}\cdot\text{m/s}}$$

---

19. WWW **REASONING** The change in wavelength that occurs during Compton scattering is given by Equation 29.7:

$$\lambda' - \lambda = \frac{h}{mc}(1 - \cos\theta) \qquad \text{or} \qquad (\lambda' - \lambda)_{\text{max}} = \frac{h}{mc}(1 - \cos 180°) = \frac{2h}{mc}$$

$(\lambda' - \lambda)_{\text{max}}$ is the maximum change in the wavelength, and to calculate it we need a value for the mass m of a nitrogen molecule. This value can be obtained from the mass per mole M of nitrogen (N_2) and Avogadro's number N_A, according to $m = M / N_A$ (see Section 14.1).

SOLUTION Using a value of $M = 0.028 \text{ kg/mol}$, we obtain the following result for the maximum change in the wavelength:

$$(\lambda' - \lambda)_{\text{max}} = \frac{2h}{mc} = \frac{2h}{\left(\dfrac{M}{N_A}\right)c} = \frac{2(6.63 \times 10^{-34} \text{ J}\cdot\text{s})}{\left(\dfrac{0.028 \text{ kg/mol}}{6.02 \times 10^{23} \text{ mol}^{-1}}\right)(3.00 \times 10^8 \text{ m/s})}$$

$$= \boxed{9.50 \times 10^{-17} \text{ m}}$$

---

23. **REASONING** In order for the person to diffract to the same extent as the sound wave, the de Broglie wavelength of the person must be equal to the wavelength of the sound wave.

SOLUTION
a.   Therefore,

$$\lambda_{\text{sound}} = \lambda_{\text{person}}$$

$$\lambda_{sound} = \frac{h}{m_{person}v_{person}}$$

Solving for v_{person}, and using the relation $\lambda_{sound} = v_{sound}/f_{sound}$ (Equation 16.1), we have

$$v_{person} = \frac{h}{m_{person}(v_{sound}/f_{sound})} = \frac{hf_{sound}}{m_{person}v_{sound}}$$

$$= \frac{(6.63 \times 10^{-34} \text{ J} \cdot \text{s})(128 \text{ Hz})}{(55.0 \text{ kg})(343 \text{ m/s})} = \boxed{4.50 \times 10^{-36} \text{ m/s}}$$

b. At the speed calculated in part (a), the time required for the person to move a distance of one meter is

$$t = \frac{x}{v} = \frac{1.0 \text{ m}}{4.50 \times 10^{-36} \text{ m/s}} \underbrace{\left(\frac{1.0 \text{ h}}{3600 \text{ s}}\right)\left(\frac{1 \text{ day}}{24.0 \text{ h}}\right)\left(\frac{1 \text{ year}}{365.25 \text{ days}}\right)}_{\substack{\text{Factors to convert} \\ \text{seconds to years}}} = \boxed{7.05 \times 10^{27} \text{ years}}$$

---

27. ***REASONING AND SOLUTION*** The de Broglie wavelength λ of the woman is given by Equation 29.8 as $\lambda = h/p$, where p is the magnitude of her momentum. The magnitude of the momentum is $p = mv$, where m is the woman's mass and v is her speed. According to Equation 3.6b of the equations of kinematics, the speed v is given by $v = \sqrt{2a_y y}$, since the woman jumps from rest. In this expression, $a_y = -9.80 \text{ m/s}^2$ and $y = -9.5 \text{ m}$. With these considerations we find that

$$\lambda = \frac{h}{mv} = \frac{h}{m\sqrt{2a_y y}} = \frac{6.63 \times 10^{-34} \text{ J} \cdot \text{s}}{(41 \text{ kg})\sqrt{2(-9.80 \text{ m/s}^2)(-9.5 \text{ m})}} = \boxed{1.2 \times 10^{-36} \text{ m}}$$

---

31. $\boxed{\text{WWW}}$ ***REASONING AND SOLUTION*** According to the uncertainty principle, the minimum uncertainty in the momentum can be determined from $\Delta p_y \Delta y = h/(2\pi)$. Since $p_y = mv_y$, it follows that $\Delta p_y = m\Delta v_y$. Thus, the minimum uncertainty in the velocity of the oxygen molecule is given by

$$\Delta v_y = \frac{h}{2\pi m \Delta y} = \frac{6.63 \times 10^{-34} \text{ J} \cdot \text{s}}{2\pi(5.3 \times 10^{-26} \text{ kg})(0.25 \times 10^{-3} \text{ m})} = \boxed{8.0 \times 10^{-6} \text{ m/s}}$$

---

35. WWW *REASONING AND SOLUTION* If we assume that the number of particles that strikes the screen outside the bright fringe is negligible, the particles that pass through the slit will hit the screen somewhere in the central bright fringe. The angular width of the central bright fringe is equal to 2θ, where θ is shown in Figure 29.15. From this figure we see that

$$\theta = \tan^{-1}\left(\frac{\Delta p_y}{p_x}\right)$$

According to Equation 29.10, the minimum uncertainty in Δp_y occurs when

$$(\Delta p_y)(\Delta y) = \frac{h}{2\pi}$$

Since the electron can pass anywhere through the width W of the slit, the uncertainty in the y position of the electron is $\Delta y = W$. Thus, the minimum uncertainty can be written as

$$(\Delta p_y)W = \frac{h}{2\pi} \qquad \text{or} \qquad \Delta p_y = \frac{h}{2\pi W}$$

Using the fact that $p_x = h/\lambda$, we find

$$\theta = \tan^{-1}\left(\frac{h}{p_x \, 2\pi W}\right) = \tan^{-1}\left[\frac{h}{(h/\lambda)2\pi W}\right] = \tan^{-1}\left(\frac{\lambda}{2\pi W}\right)$$

$$= \tan^{-1}\left[\frac{633\times 10^{-9} \text{ m}}{2\pi(0.200\times 10^{-3} \text{ m})}\right] = 2.89\times 10^{-2} \text{ degrees}$$

Therefore, the minimum range of angles over which the particles spread is $\boxed{-0.0289° \le \theta \le +0.0289°}$.

---

39. *REASONING AND SOLUTION* The de Broglie wavelength λ is given by Equation 29.8 as $\lambda = h/p$, where p is the magnitude of the momentum of the particle. The magnitude of the momentum is $p = mv$, where m is the mass and v is the speed of the particle. Using this expression in Equation 29.8, we find that $\lambda = h/(mv)$, or

$$v = \frac{h}{m\lambda} = \frac{6.63\times 10^{-34} \text{ J}\cdot\text{s}}{(1.67\times 10^{-27} \text{ kg})(1.30\times 10^{-14} \text{ m})} = 3.05\times 10^{7} \text{ m/s}$$

The kinetic energy of the proton is

$$KE = \tfrac{1}{2}mv^2 = \tfrac{1}{2}(1.67 \times 10^{-27} \text{ kg})(3.05 \times 10^7 \text{ m/s})^2 = \boxed{7.77 \times 10^{-13} \text{ J}}$$

---

43. **REASONING** The width of the central bright fringe in the diffraction patterns will be identical when the electrons have the same de Broglie wavelength as the wavelength of the photons in the red light. The de Broglie wavelength of one electron in the beam is given by Equation 29.8, $\lambda_{\text{electron}} = h/p$, where $p = mv$.

SOLUTION Following the reasoning described above, we find

$$\lambda_{\text{red light}} = \lambda_{\text{electron}}$$

$$\lambda_{\text{red light}} = \frac{h}{m_{\text{electron}} v_{\text{electron}}}$$

Solving for the speed of the electron, we have

$$v_{\text{electron}} = \frac{h}{m_{\text{electron}} \lambda_{\text{red light}}} = \frac{6.63 \times 10^{-34} \text{ J} \cdot \text{s}}{(9.11 \times 10^{-31} \text{ kg})(661 \times 10^{-9} \text{ m})} = \boxed{1.10 \times 10^3 \text{ m/s}}$$

CHAPTER 30 | *THE NATURE OF THE ATOM*

1. **REASONING** Assuming that the hydrogen atom is a sphere of radius r_{atom}, its volume V_{atom} is given by $\frac{4}{3}\pi r_{atom}^3$. Similarly, if the radius of the nucleus is $r_{nucleus}$, the volume $V_{nucleus}$ is given by $\frac{4}{3}\pi r_{nucleus}^3$.

SOLUTION

a. According to the given data, the nuclear dimensions are much smaller than the orbital radius of the electron; therefore, we can treat the nucleus as a point about which the electron orbits. The electron is normally at a distance of about 5.3×10^{-11} m from the nucleus, so we can treat the atom as a sphere of radius $r_{atom} = 5.3 \times 10^{-11}$ m. The volume of the atom is

$$V_{atom} = \frac{4}{3}\pi r_{atom}^3 = \frac{4}{3}\pi(5.3 \times 10^{-11}\text{ m})^3 = \boxed{6.2 \times 10^{-31}\text{ m}^3}$$

b. Similarly, since the nucleus has a radius of approximately $r_{nucleus} = 1 \times 10^{-15}$ m, its volume is

$$V_{nucleus} = \frac{4}{3}\pi r_{nucleus}^3 = \frac{4}{3}\pi(1 \times 10^{-15}\text{ m})^3 = \boxed{4 \times 10^{-45}\text{ m}^3}$$

c. The percentage of the atomic volume occupied by the nucleus is

$$\frac{V_{nucleus}}{V_{atom}} \times 100\% = \frac{r_{nucleus}^3}{r_{atom}^3} \times 100\% = \left(\frac{1 \times 10^{-15}\text{ m}}{5.3 \times 10^{-11}\text{ m}}\right)^3 \times 100\% = \boxed{7 \times 10^{-13}\ \%}$$

---

5. **REASONING** The copper nucleus has a total charge $Ze = 29 \times 1.6 \times 10^{-19}$ C, and radius $r_{nucleus} = 4.8 \times 10^{-15}$ m. The required work can be found from Equation 19.1: $W_{AB} = \text{EPE}_A - \text{EPE}_B$. According to Equation 19.3, $\text{EPE} = Vq$, where $q = e$ for a proton. Therefore, $W_{AB} = V_A e - V_B e$. The proton is initially at infinity, so $V_A = 0$ V and $V_B = kZe/r_{nucleus}$, according to Equation 19.6. Thus, $W_{AB} = -kZe^2/r_{nucleus}$.

SOLUTION The work done by the electric force is, therefore,

$$W_{AB} = \frac{-kZe^2}{r_{nucleus}} = \frac{-(8.99 \times 10^9 \text{ N} \cdot \text{m}^2/\text{C}^2)(29)(1.6 \times 10^{-19} \text{ C})^2}{4.8 \times 10^{-15} \text{ m}} = -1.39 \times 10^{-12} \text{ J}$$

$$W_{AB} = \left(-1.39 \times 10^{-12} \text{ J}\right)\underbrace{\left(\frac{1.0 \text{ eV}}{1.6 \times 10^{-19} \text{ J}}\right)}_{\substack{\text{Converts from joules} \\ \text{to electron volts}}} = \boxed{-8.7 \times 10^6 \text{ eV}}$$

---

7. $\boxed{\text{WWW}}$ **REASONING AND SOLUTION**   The radii for Bohr orbits are given by Equation 30.10:

$$r_n = (5.29 \times 10^{-11} \text{ m})\frac{n^2}{Z}$$

The Bohr radii for a doubly ionized lithium atom Li^{2+} are given by Equation 30.10 with $Z = 3$; therefore, the radius of the $n = 5$ Bohr orbit in a doubly ionized lithium atom Li^{2+} is

$$r_5 = (5.29 \times 10^{-11} \text{ m})\frac{5^2}{3} = \boxed{4.41 \times 10^{-10} \text{ m}}$$

---

11.   **REASONING**   According to Equation 30.14, the wavelength λ emitted by the hydrogen atom when it makes a transition from the level with n_i to the level with n_f is given by

$$\frac{1}{\lambda} = \frac{2\pi^2 m k^2 e^4}{h^3 c} (Z^2)\left(\frac{1}{n_f^2} - \frac{1}{n_i^2}\right) \quad \text{with} \quad n_i, n_f = 1, 2, 3, \dots \quad \text{and} \quad n_i > n_f$$

where $2\pi^2 m k^2 e^4/(h^3 c) = 1.097 \times 10^7 \text{ m}^{-1}$ and $Z = 1$ for hydrogen. Once the wavelength for the particular transition in question is determined, Equation 29.2 ($E = hf = hc/\lambda$) can be used to find the energy of the emitted photon.

SOLUTION   In the Paschen series, $n_f = 3$. Using the above expression with $Z = 1$, $n_i = 7$ and $n_f = 3$, we find that

$$\frac{1}{\lambda} = (1.097 \times 10^7 \text{ m}^{-1}) (1^2) \left(\frac{1}{3^2} - \frac{1}{7^2}\right) \qquad \text{or} \qquad \lambda = 1.005 \times 10^{-6} \text{ m}$$

The photon energy is

$$E = \frac{hc}{\lambda} = \frac{(6.63 \times 10^{-34}\ \text{J} \cdot \text{s})(3.00 \times 10^8\ \text{m/s})}{1.005 \times 10^{-6}\ \text{m}} = \boxed{1.98 \times 10^{-19}\ \text{J}}$$

---

15. ***REASONING AND SOLUTION***   For the Paschen series, $n_f = 3$. The range of wavelengths occurs for values of $n_i = 4$ to $n_i = \infty$. Using Equation 30.14, we find that the shortest wavelength occurs for $n_i = \infty$ and is given by

$$\frac{1}{\lambda} = \left(1.097 \times 10^7\ \text{m}^{-1}\right)(1)^2 \left(\frac{1}{n_f^2} - \frac{1}{n_i^2}\right) = \left(1.097 \times 10^7\ \text{m}^{-1}\right)\left(\frac{1}{3^2}\right) \quad \text{or} \quad \underset{\substack{\text{Shortest wavelength in} \\ \text{Paschen series}}}{\underline{\lambda = 8.204 \times 10^{-7}\ \text{m}}}$$

The longest wavelength in the Paschen series occurs for $n_i = 4$ and is given by

$$\frac{1}{\lambda} = \left(1.097 \times 10^7\ \text{m}^{-1}\right)\left(\frac{1}{3^2} - \frac{1}{4^2}\right) \quad \text{or} \quad \underset{\substack{\text{Longest wavelength in} \\ \text{Paschen series}}}{\underline{\lambda = 1.875 \times 10^{-6}\ \text{m}}}$$

For the Brackett series, $n_f = 4$. The range of wavelengths occurs for values of $n_i = 5$ to $n_i = \infty$. Using Equation 30.14, we find that the shortest wavelength occurs for $n_i = \infty$ and is given by

$$\frac{1}{\lambda} = \left(1.097 \times 10^7\ \text{m}^{-1}\right)(1)^2 \left(\frac{1}{n_f^2} - \frac{1}{n_i^2}\right) = \left(1.097 \times 10^7\ \text{m}^{-1}\right)\left(\frac{1}{4^2}\right) \quad \text{or} \quad \underset{\substack{\text{Shortest wavelength in} \\ \text{Brackett series}}}{\underline{\lambda = 1.459 \times 10^{-6}\ \text{m}}}$$

The longest wavelength in the Brackett series occurs for $n_i = 5$ and is given by

$$\frac{1}{\lambda} = \left(1.097 \times 10^7\ \text{m}^{-1}\right)\left(\frac{1}{4^2} - \frac{1}{5^2}\right) \quad \text{or} \quad \underset{\substack{\text{Longest wavelength in} \\ \text{Brackett series}}}{\underline{\lambda = 4.051 \times 10^{-6}\ \text{m}}}$$

Since the longest wavelength in the Paschen series falls within the Brackett series, the wavelengths of the two series overlap.

---

17. ***REASONING***   Singly ionized helium, He$^+$, is a hydrogen-like species with $Z = 2$. The wavelengths of the series of lines produced when the electron makes a transition from higher energy levels into the $n_f = 4$ level are given by Equation 30.14 with $Z = 2$ and $n_f = 4$:

$$\frac{1}{\lambda} = \left(1.097 \times 10^7 \text{ m}^{-1}\right)(2^2)\left(\frac{1}{4^2} - \frac{1}{n_i^2}\right)$$

SOLUTION   Solving this expression for n_i gives

$$n_i = \left[\frac{1}{4^2} - \frac{1}{4\lambda(1.097 \times 10^7 \text{ m}^{-1})}\right]^{-1/2}$$

Evaluating this expression at the limits of the range for λ, we find that $n_i = 19.88$ for $\lambda = 380$ nm, and $n_i = 5.58$ for $\lambda = 750$ nm. Therefore, the values of n_i for energy levels from which the electron makes the transitions that yield wavelengths in the range between 380 nm and 750 nm are $\boxed{6 \le n_i \le 19}$.

---

23. ☐WWW☐ *REASONING*   The values that ℓ can have depend on the value of n, and only the following integers are allowed: $\ell = 0, 1, 2, \ldots (n-1)$. The values that m_ℓ can have depend on the value of ℓ, with only the following positive and negative integers being permitted: $m_\ell = -\ell, \ldots -2, -1, 0, +1, +2, \ldots +\ell$.

SOLUTION   Thus, when $n = 6$, the possible values of ℓ are 0, 1, 2, 3, 4, 5. Now when $m_\ell = 2$, the possible values of ℓ are 2, 3, 4, 5, $\ldots$ These two series of integers overlap for the integers 2, 3, 4, and 5. Therefore, the possible values for the orbital quantum number ℓ that this electron could have are $\boxed{\ell = 2, 3, 4, 5}$.

---

27. ☐WWW☐ *REASONING*   Let θ denote the angle between the angular momentum L and its z-component L_z. We can see from the figure at the right that $L_z = L \cos \theta$. Using Equation 30.16 for L_z and Equation 30.15 for L, we have

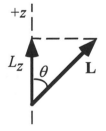

$$\cos \theta = \frac{L_z}{L} = \frac{m_\ell h / (2\pi)}{\sqrt{\ell(\ell+1)}\left[h/(2\pi)\right]} = \frac{m_\ell}{\sqrt{\ell(\ell+1)}}$$

The smallest value for θ corresponds to the largest value of $\cos \theta$. For a given value of ℓ, the largest value for $\cos \theta$ corresponds to the largest value for m_ℓ. But the largest possible value for m_ℓ is $m_\ell = \ell$. Therefore, we find that

$$\cos \theta = \frac{\ell}{\sqrt{\ell(\ell+1)}} = \sqrt{\frac{\ell}{\ell+1}}$$

SOLUTION   The smallest value for θ corresponds to the largest value for ℓ. When the electron is in the $n = 5$ state, the largest allowed value of ℓ is $\ell = 4$; therefore, we see that

$$\cos \theta = \sqrt{\frac{\ell}{\ell+1}} = \sqrt{\frac{4}{4+1}} \qquad \text{or} \qquad \theta = \cos^{-1}\left(\sqrt{4/5}\right) = \boxed{26.6°}$$

---

31. **REASONING**   In the theory of quantum mechanics, there is a selection rule that restricts the initial and final values of the orbital quantum number ℓ. The selection rule states that when an electron makes a transition between energy levels, the value of ℓ may not remain the same or increase or decrease by more than one. In other words, the rule requires that $\Delta \ell = \pm 1$.

    SOLUTION
    a. For the transition 2s $\rightarrow$ 1s, the electron makes a transition from the 2s state ($n = 2$, $\ell = 0$) to the 1s state ($n = 1$, $\ell = 0$). Since the value of ℓ is the same in both states, $\Delta \ell = 0$, and we can conclude that this energy level transition is $\boxed{\text{not permitted}}$.

    b. For the transition 2p $\rightarrow$ 1s, the electron makes a transition from the 2p state ($n = 2$, $\ell = 1$) to the 1s state ($n = 1$, $\ell = 0$). The value of ℓ changes so that $\Delta \ell = 0 - 1 = -1$, and we can conclude that this energy level transition is $\boxed{\text{permitted}}$.

    c. For the transition 4p $\rightarrow$ 2p, the electron makes a transition from the 4p state ($n = 4$, $\ell = 1$) to the 2p state ($n = 2$, $\ell = 1$). Since the value of ℓ is the same in both states, $\Delta \ell = 0$, and we can conclude that this energy level transition is $\boxed{\text{not permitted}}$.

    d. For the transition 4s $\rightarrow$ 2p, the electron makes a transition from the 4s state ($n = 4$, $\ell = 0$) to the 2p state ($n = 2$, $\ell = 1$). The value of ℓ changes so that $\Delta \ell = 1 - 0 = +1$, and we can conclude that this energy level transition is $\boxed{\text{permitted}}$.

    e. For the transition 3d $\rightarrow$ 3s, the electron makes a transition from the 3d state ($n = 3$, $\ell = 2$) to the 3s state ($n = 3$, $\ell = 0$). The value of ℓ changes so that $\Delta \ell = 0 - 2 = -2$, and we can conclude that this energy level transition is $\boxed{\text{not permitted}}$.

---

33. **REASONING**   This problem is similar to Example 10 in the text. We use Equation 30.14 with the initial value of n being $n_i = 2$, and the final value being $n_f = 1$. As in Example 10, we use a value of Z that is one less than the atomic number of the atom in question (in this

case, a value of $Z = 41$ rather than 42); this accounts approximately for the shielding effect of the single K-shell electron in canceling out the attraction of one nuclear proton.

SOLUTION Using Equation 30.14, we obtain

$$\frac{1}{\lambda} = (1.097 \times 10^7 \text{ m}^{-1})(41)^2 \left(\frac{1}{1^2} - \frac{1}{2^2}\right) \qquad \text{or} \qquad \boxed{\lambda = 7.230 \times 10^{-11} \text{ m}}$$

---

37. ***REASONING***   In the spectrum of X-rays produced by the tube, the cutoff wavelength λ_0 and the voltage V of the tube are related according to Equation 30.17, $V = hc/(e\lambda_0)$. Since the voltage is increased from zero until the K_α X-ray just appears in the spectrum, it follows that $\lambda_0 = \lambda_\alpha$ and $V = hc/(e\lambda_\alpha)$. Using Equation 30.14 for $1/\lambda_\alpha$, we find that

$$V = \frac{hc}{e\lambda_\alpha} = \frac{hcR(Z-1)^2}{e}\left(\frac{1}{1^2} - \frac{1}{2^2}\right)$$

In this expression we have replaced Z with $Z - 1$, in order to account for shielding, as explained in Example 10 in the text.

SOLUTION   The desired voltage is, then,

$$V = \frac{(6.63 \times 10^{-34} \text{ J} \cdot \text{s})(3.00 \times 10^8 \text{ m/s})(1.097 \times 10^7 \text{ m}^{-1})(47-1)^2}{(1.60 \times 10^{-19} \text{ C})}\left(\frac{1}{1^2} - \frac{1}{2^2}\right) = \boxed{21\ 600 \text{ V}}$$

---

41. WWW   ***REASONING***   The number of photons emitted by the laser will be equal to the total energy carried in the beam divided by the energy per photon.

SOLUTION   The total energy carried in the beam is, from the definition of power,

$$E_{\text{total}} = Pt = (1.5 \text{ W})(0.050 \text{ s}) = 0.075 \text{ J}$$

The energy of a single photon is given by Equations 29.2 and 16.1 as

$$E_{\text{photon}} = hf = \frac{hc}{\lambda} = \frac{(6.63 \times 10^{-34} \text{ J} \cdot \text{s})(3.00 \times 10^8 \text{ m/s})}{514 \times 10^{-9} \text{ m}} = 3.87 \times 10^{-19} \text{ J}$$

where we have used the fact that 514 nm $= 514 \times 10^{-9}$ m. Therefore, the number of photons emitted by the laser is

$$\frac{E_{\text{total}}}{E_{\text{photon}}} = \frac{0.075 \text{ J}}{3.87 \times 10^{-19} \text{ J/photon}} = \boxed{1.9 \times 10^{17} \text{ photons}}$$

45. **REASONING AND SOLUTION**   In the ground state, the electron in a hydrogen atom is in the $n = 1$ energy level. To ionize such a hydrogen atom, the electron must be taken from the $n = 1$ level to the level for which n is infinity. The electromagnetic radiation needed to accomplish this corresponds to a Lyman line in which the final state n is taken to be infinity. Equation 30.1 gives the reciprocal wavelength for the Lyman series.   With $n = \infty$, Equation 30.1 reveals that for the longest wavelength

$$\frac{1}{\lambda} = R\left(\frac{1}{1^2} - \frac{1}{n^2}\right) = (1.097 \times 10^7 \text{ m}^{-1})\left(\frac{1}{1^2} - \frac{1}{\infty^2}\right) \qquad \text{or} \qquad \boxed{\lambda = 91.2 \text{ nm}}$$

49. **REASONING**   According to the Bohr model, the energy E_n (in eV) of the electron in an orbit is given by Equation 30.13: $E_n = -13.6\left(Z^2/n^2\right)$.   In order to find the principal quantum number of the state in which the electron in a doubly ionized lithium atom Li^{2+} has the same total energy as a ground state electron in a hydrogen atom, we equate the right hand sides of Equation 30.13 for the hydrogen atom and the lithium ion.   This gives

$$-(13.6)\left(\frac{Z^2}{n^2}\right)_{\text{H}} = -(13.6)\left(\frac{Z^2}{n^2}\right)_{\text{Li}} \qquad \text{or} \qquad n_{\text{Li}}^2 = \left(\frac{n^2}{Z^2}\right)_{\text{H}} Z_{\text{Li}}^2$$

This expression can be evaluated to find the desired principal quantum number.

SOLUTION   For hydrogen, $Z = 1$, and $n = 1$ for the ground state.   For lithium Li^{2+}, $Z = 3$.   Therefore,

$$n_{\text{Li}}^2 = \left(\frac{n^2}{Z^2}\right)_{\text{H}} Z_{\text{Li}}^2 = \left(\frac{1^2}{1^2}\right)(3^2) \qquad \text{or} \qquad \boxed{n_{\text{Li}} = 3}$$

53. **REASONING**   A wavelength of 410.2 nm is emitted by the hydrogen atoms in a high-voltage discharge tube.   This transition lies in the visible region (380–750 nm) of the hydrogen spectrum. Thus, we can conclude that the transition is in the Balmer series and, therefore, that $n_{\text{f}} = 2$.   The value of n_{i} can be found using Equation 30.14, according to which the Balmer series transitions are given by

$$\frac{1}{\lambda} = R \left(1^2\right)\left(\frac{1}{2^2} - \frac{1}{n_{\text{i}}^2}\right) \qquad n = 3, 4, 5, \ldots$$

This expression may be solved for n_i for the energy transition that produces the given wavelength.

SOLUTION   Solving for n_i, we find that

$$n_i = \frac{1}{\sqrt{\dfrac{1}{2^2} - \dfrac{1}{R\lambda}}} = \frac{1}{\sqrt{\dfrac{1}{2^2} - \dfrac{1}{(1.097 \times 10^7 \text{ m}^{-1})(410.2 \times 10^{-9} \text{ m})}}} = 6$$

Therefore, the initial and final states are identified by $\boxed{n_i = 6 \text{ and } n_f = 2}$.

CHAPTER 31 | *NUCLEAR PHYSICS AND RADIOACTIVITY*

PROBLEMS

1. **REASONING** For an element whose chemical symbol is X, the symbol for the nucleus is $^A_Z X$ where A represents the number of protons and neutrons (the nucleon number) and Z represents the number of protons (the atomic number) in the nucleus.

SOLUTION

a. The symbol $^{195}_{78} X$ indicates that the nucleus in question contains $Z = 78$ protons, and $N = A - Z = 195 - 78 = \boxed{117 \text{ neutrons}}$. From the periodic table, we see that $Z = 78$ corresponds to $\boxed{\text{platinum, Pt}}$.

b. Similar reasoning indicates that the nucleus in question is $\boxed{\text{sulfur, S}}$, and the nucleus contains $N = A - Z = 32 - 16 = \boxed{16 \text{ neutrons}}$.

c. Similar reasoning indicates that the nucleus in question is $\boxed{\text{copper, Cu}}$, and the nucleus contains $N = A - Z = 63 - 29 = \boxed{34 \text{ neutrons}}$.

d. Similar reasoning indicates that the nucleus in question is $\boxed{\text{boron, B}}$, and the nucleus contains $N = A - Z = 11 - 5 = \boxed{6 \text{ neutrons}}$.

e. Similar reasoning indicates that the nucleus in question is $\boxed{\text{plutonium, Pu}}$, and the nucleus contains $N = A - Z = 239 - 94 = \boxed{145 \text{ neutrons}}$.

---

5. $\boxed{\text{WWW}}$ **REASONING AND SOLUTION** The surface area of a sphere is $\text{Area} = 4\pi r^2$. But according to Equation 31.2, the radius of a nucleus in meters is $r = (1.2 \times 10^{-15} \text{ m}) A^{1/3}$, where A is the nucleon number. With this expression for r, the surface area becomes $\text{Area} = 4\pi (1.2 \times 10^{-15} \text{ m})^2 A^{2/3}$. The ratio of the largest to the smallest surface area is, then,

$$\frac{\text{Largest area}}{\text{Smallest area}} = \frac{4\pi (1.2 \times 10^{-15} \text{ m})^2 A^{2/3}_{\text{largest}}}{4\pi (1.2 \times 10^{-15} \text{ m})^2 A^{2/3}_{\text{smallest}}} = \frac{209^{2/3}}{1^{2/3}} = \boxed{35.2}$$

9. **WWW** *REASONING* According to Equation 31.2, the radius of a nucleus in meters is $r = (1.2 \times 10^{-15} \text{ m}) A^{1/3}$, where A is the nucleon number. If we treat the neutron star as a uniform sphere, its density (Equation 11.1) can be written as

$$\rho = \frac{M}{V} = \frac{M}{\frac{4}{3}\pi r^3}$$

Solving for the radius r, we obtain,

$$r = \sqrt[3]{\frac{M}{\frac{4}{3}\pi \rho}}$$

This expression can be used to find the radius of a neutron star of mass M and density ρ.

SOLUTION As discussed in Conceptual Example 1, nuclear densities have the same approximate value in all atoms. If we consider a uniform spherical nucleus, then the density of nuclear matter is approximately given by

$$\rho = \frac{M}{V} \approx \frac{A \times (\text{mass of a nucleon})}{\frac{4}{3}\pi r^3} = \frac{A \times (\text{mass of a nucleon})}{\frac{4}{3}\pi \left[(1.2 \times 10^{-15} \text{ m}) A^{1/3}\right]^3}$$

$$= \frac{1.67 \times 10^{-27} \text{ kg}}{\frac{4}{3}\pi (1.2 \times 10^{-15} \text{ m})^3} = 2.3 \times 10^{17} \text{ kg} / \text{m}^3$$

Substituting values into the expression for r determined above, we have

$$r = \sqrt[3]{\frac{(0.40)(1.99 \times 10^{30} \text{ kg})}{\frac{4}{3}\pi (2.3 \times 10^{17} \text{ kg} / \text{m}^3)}} = \boxed{9.4 \times 10^3 \text{ m}}$$

---

11. *REASONING AND SOLUTION* The symbol $^{16}_{8}\text{O}$ indicates that the oxygen nucleus contains $A = 16$ nucleons. According to the binding energy per nucleon curve in Figure 31.6, the corresponding binding energy per nucleon (for $A = 16$) is 8.00 MeV/nucleon. Therefore, the total binding energy of $^{16}_{8}\text{O}$ is

$$(16 \text{ nucleons}) \left(\frac{8.00 \text{ MeV}}{\text{nucleon}}\right) = \boxed{128 \text{ MeV}}$$

15. **REASONING** Since we know the difference in binding energies for the two isotopes, we can determine the corresponding mass defect. Also knowing that the isotope with the larger binding energy contains one more neutron than the other isotope gives us enough information to calculate the atomic mass difference between the two isotopes.

SOLUTION The mass defect corresponding to a binding energy difference of 5.03 MeV is

$$(5.03 \text{ MeV}) \left(\frac{1 \text{ u}}{931.5 \text{ MeV}} \right) = 0.005 \, 40 \text{ u}$$

Since the isotope with the larger binding energy has one more neutron ($m = 1.008 \, 665 \text{ u}$) than the other isotope, the difference in atomic mass between the two isotopes is

$$1.008 \, 665 \text{ u} - 0.005 \, 40 \text{ u} = \boxed{1.003 \, 27 \text{ u}}$$

---

21. **REASONING AND SOLUTION** The general form for β^- decay is

$$\underset{\substack{\text{Parent} \\ \text{nucleus}}}{{}^{A}_{Z}\text{P}} \rightarrow \underset{\substack{\text{Daughter} \\ \text{nucleus}}}{{}^{A}_{Z+1}\text{D}} + \underset{\substack{\beta^-\ \text{particle} \\ \text{(electron)}}}{{}^{0}_{-1}\text{e}}$$

Therefore, the β^- decay process for ${}^{35}_{16}\text{S}$ is $\boxed{{}^{35}_{16}\text{S} \rightarrow {}^{35}_{17}\text{Cl} + {}^{0}_{-1}\text{e}}$.

---

23. $\boxed{\text{WWW}}$ **REASONING AND SOLUTION** The general form for β^+ decay is

$$\underset{\substack{\text{Parent} \\ \text{nucleus}}}{{}^{A}_{Z}\text{P}} \rightarrow \underset{\substack{\text{Daughter} \\ \text{nucleus}}}{{}^{A}_{Z-1}\text{D}} + \underset{\substack{\beta^+\ \text{particle} \\ \text{(positron)}}}{{}^{0}_{+1}\text{e}}$$

a. Therefore, the β^+ decay process for ${}^{18}_{9}\text{F}$ is $\boxed{{}^{18}_{9}\text{F} \rightarrow {}^{18}_{8}\text{O} + {}^{0}_{+1}\text{e}}$.

b. Similarly, the β^+ decay process for ${}^{15}_{8}\text{O}$ is $\boxed{{}^{15}_{8}\text{O} \rightarrow {}^{15}_{7}\text{N} + {}^{0}_{+1}\text{e}}$.

---

27. **REASONING** Energy is released during the β decay. To find the energy released, we determine how much the mass has decreased because of the decay and then calculate the equivalent energy. The reaction and masses are shown below:

$$\underset{21.994\,434\ \text{u}}{{}^{22}_{11}\text{Na}} \rightarrow \underset{21.991\,383\ \text{u}}{{}^{22}_{10}\text{Ne}} + \underset{5.485\,799 \times 10^{-4}\ \text{u}}{{}^{0}_{+1}\text{e}}$$

SOLUTION The decrease in mass is

$$21.994\ 434\ \text{u} - (21.991\ 383\ \text{u} + 5.485\ 799 \times 10^{-4}\ \text{u} + 5.485\ 799 \times 10^{-4}\ \text{u}) = 0.001\ 954\ \text{u}$$

where the extra electron mass takes into account the fact that the atomic mass for sodium includes the mass of 11 electrons, whereas the atomic mass for neon includes the mass of only 10 electrons.

Since 1 u is equivalent to 931.5 MeV, the released energy is

$$(0.001\ 954\ \text{u})\left(\frac{931.5\ \text{MeV}}{1\ \text{u}}\right) = \boxed{1.82\ \text{MeV}}$$

---

31. ***REASONING AND SOLUTION*** According to Equation 31.5, $N = N_0 e^{-\lambda t}$, the decay constant is

$$\lambda = -\frac{1}{t}\ln\left(\frac{N}{N_0}\right) = -\frac{1}{20\ \text{days}}\ln\left(\frac{8.14 \times 10^{14}}{4.60 \times 10^{15}}\right) = 0.0866\ \text{days}^{-1}$$

The half-life is, from Equation 31.6,

$$T_{1/2} = \frac{0.693}{\lambda} = \frac{0.693}{0.0866\ \text{days}^{-1}} = \boxed{8.00\ \text{days}}$$

---

35. ***REASONING*** We can find the decay constant from Equation 31.5, $N = N_0 e^{-\lambda t}$. If we multiply both sides by the decay constant λ, we have

$$\lambda N = \lambda N_0 e^{-\lambda t} \qquad \text{or} \qquad A = A_0 e^{-\lambda t}$$

where A_0 is the initial activity and A is the activity after a time t. Once the decay constant is known, we can use the same expression to determine the activity after a total of six days.

SOLUTION Solving the expression above for the decay constant λ, we have

$$\lambda = -\frac{1}{t}\ln\left(\frac{A}{A_0}\right) = -\frac{1}{2\ \text{days}}\ln\left(\frac{285\ \text{disintegrations}/\min}{398\ \text{disintegrations}/\min}\right) = 0.167\ \text{days}^{-1}$$

Then the activity four days after the second day is

$$A = (285 \text{ disintegrations}/\text{min})e^{-(0.167 \text{ days}^{-1})(4.00 \text{ days})} = \boxed{146 \text{ disintegrations}/\text{min}}$$

---

39. **REASONING** According to Equation 31.5, the number of nuclei remaining after a time t is $N = N_0 e^{-\lambda t}$. Using this expression, we find the ratio N_A / N_B as follows:

$$\frac{N_A}{N_B} = \frac{N_{0A} e^{-\lambda_A t}}{N_{0B} e^{-\lambda_B t}} = e^{-(\lambda_A - \lambda_B)t}$$

where we have used the fact that initially the numbers of the two types of nuclei are equal $\left(N_{0A} = N_{0B}\right)$. Taking the natural logarithm of both sides of the equation above shows that

$$\ln\left(N_A / N_B\right) = -\left(\lambda_A - \lambda_B\right)t \quad \text{or} \quad \lambda_A - \lambda_B = \frac{-\ln\left(N_A / N_B\right)}{t}$$

SOLUTION Since $N_A / N_B = 3.00$ when $t = 3.00$ days, it follows that

$$\lambda_A - \lambda_B = \frac{-\ln(3.00)}{3.00 \text{ days}} = -0.366 \text{ days}^{-1}$$

But we need to find the half-life of species B, so we use Equation 31.6, which indicates that $\lambda = 0.693 / T_{1/2}$. With this expression for λ, the result for $\lambda_A - \lambda_B$ becomes

$$0.693\left(\frac{1}{T_{1/2}^A} - \frac{1}{T_{1/2}^B}\right) = -0.366$$

Since $T_{1/2}^B = 1.50 \text{ days}$, the result above can be solved to show that $\boxed{T_{1/2}^A = 7.23 \text{ days}}$.

---

41. **REASONING AND SOLUTION** The answer can be obtained directly from Equation 31.5, combined with Equation 31.6:

$$\frac{N}{N_0} = e^{-\lambda t} = e^{-(0.693)t/T_{1/2}} = e^{-(0.693)(41\,000 \text{ yr})/(5730 \text{ yr})} = 0.0070$$

The percent of atoms remaining is $\boxed{0.70\%}$.

45. **REASONING** According to Equation 31.5, $N = N_0 e^{-\lambda t}$. If we multiply both sides by the decay constant λ, we have

$$\lambda N = \lambda N_0 e^{-\lambda t} \qquad \text{or} \qquad A = A_0 e^{-\lambda t}$$

where A_0 is the initial activity and A is the activity after a time t. The decay constant λ is related to the half-life through Equation 31.6: $\lambda = 0.693 / T_{1/2}$. We can find the age of the fossils by solving for the time t. The maximum error can be found by evaluating the limits of the accuracy as given in the problem statement.

SOLUTION The age of the fossils is

$$t = -\frac{T_{1/2}}{0.693} \ln\left(\frac{A}{A_0}\right) = -\frac{5730 \text{ yr}}{0.693} \ln\left(\frac{0.10 \text{ Bq}}{0.23 \text{ Bq}}\right) = \boxed{6900 \text{ yr}}$$

The maximum error can be found as follows:

When there is an error of +10 %, $A = 0.10$ Bq $+ 0.010$ Bq $= 0.11$ Bq , and we have

$$t = -\frac{5730 \text{ yr}}{0.693} \ln\left(\frac{0.11 \text{ Bq}}{0.23 \text{ Bq}}\right) = 6100 \text{ yr}$$

Similarly, when there is an error of –10 %, $A = 0.10$ Bq $- 0.010$ Bq $= 0.090$ Bq , and we have

$$t = -\frac{5730 \text{ yr}}{0.693} \ln\left(\frac{0.090 \text{ Bq}}{0.23 \text{ Bq}}\right) = 7800 \text{ yr}$$

The maximum error in the age of the fossils is 7800 yr – 6900 yr = $\boxed{900 \text{ yr}}$.

---

49. **REASONING AND SOLUTION** The number of radioactive nuclei that remains in a sample after a time t is given by Equation 31.5, $N = N_0 e^{-\lambda t}$, where λ is the decay constant. From Equation 31.6, we know that the decay constant is related to the half-life by $T_{1/2} = 0.693 / \lambda$; therefore, $\lambda = 0.693 / T_{1/2}$ and we can write

$$\frac{N}{N_0} = e^{-(0.693/T_{1/2})t} \qquad \text{or} \qquad \frac{t}{T_{1/2}} = -\frac{1}{0.693} \ln\left(\frac{N}{N_0}\right)$$

When the number of radioactive nuclei decreases to one-millionth of the initial number, $N / N_0 = 1.00 \times 10^{-6}$; therefore, the number of half-lives is

$$\frac{t}{T_{1/2}} = -\frac{1}{0.693} \ln (1.00 \times 10^{-6}) = \boxed{19.9}$$

---

53. $\boxed{\text{WWW}}$ *REASONING AND SOLUTION* As shown in Figure 31.18, if the first dynode produces 3 electrons, the second produces 9 electrons (3^2), the third produces 27 electrons (3^3), so the Nth produces 3^N electrons. The number of electrons that leaves the 14th dynode and strikes the 15th dynode is

$$3^{14} = \boxed{4\ 782\ 969 \text{ electrons}}$$

CHAPTER 32 | *IONIZING RADIATION, NUCLEAR ENERGY, AND ELEMENTARY PARTICLES*

1. **REASONING AND SOLUTION** The exposure in roentgens is given by Equation 32.1:

$$\text{Exposure (in roentgens)} = \left(\frac{1}{2.58 \times 10^{-4}}\right)\frac{q}{m}$$

The total charge q in the beam is equal to the number of ions in the beam multiplied by the charge on each ion, namely $e = 1.6 \times 10^{-19}$ C. Therefore, the exposure is

$$\text{Exposure} = \left(\frac{1}{2.58 \times 10^{-4}}\right)\frac{(1.7 \times 10^{12})(1.6 \times 10^{-19} \text{ C})}{4.0 \times 10^{-3} \text{ kg}} = \boxed{0.26 \text{ roentgens}}$$

5. **WWW** **REASONING AND SOLUTION** The energy absorbed is given by the absorbed dose in grays times the mass of the person in kilograms, according to Equation 32.2. To find the absorbed dose from the biologically equivalent dose (BED), we turn to Equation 32.4:

$$\text{Absorbed dose in rad} = \frac{\text{BED}}{\text{RBE}} = \frac{45 \times 10^{-3} \text{ rem}}{12}$$

Using the fact that 0.01 Gy = 1 rad, we find that

$$\text{Absorbed dose in grays} = \left(\frac{45 \times 10^{-3}}{12} \text{ rad}\right)\left(\frac{0.01 \text{ Gy}}{1 \text{ rad}}\right)$$

$$\text{Energy} = (\text{Absorbed dose in grays})(\text{Mass}) = \left(\frac{45 \times 10^{-5}}{12} \text{ Gy}\right)(75 \text{ kg}) = \boxed{2.8 \times 10^{-3} \text{ J}}$$

7. **REASONING** According to Equation 32.2, the absorbed dose is the energy absorbed divided by the mass of absorbing material:

$$\text{Absorbed dose} = \frac{\text{Energy absorbed}}{\text{Mass of absorbing material}}$$

The energy absorbed in this case is the sum of three terms: (1) the heat needed to melt a mass m of ice at 0.0 °C into liquid water at 0.0 °C, which is mL_f, according to the definition of the latent heat of fusion L_f (see Section 12.8); (2) the heat needed to raise the temperature of liquid water by an amount ΔT, which is $cm\Delta T$, where c is the specific heat capacity and ΔT is the change in temperature from 0.0 to 100.0 °C, according to Equation 12.4; (3) the heat needed to vaporize liquid water at 100.0 °C into steam at 100.0 °C, which is mL_v, according to the definition of the latent heat of vaporization L_v (see Section 12.8). Once the energy absorbed is determined, the absorbed dose can be determined using Equation 32.2.

SOLUTION   Using the value of $c = 4186$ J/(kg·C°) for liquid water from Table 12.2 and the values of $L_f = 33.5 \times 10^4$ J/kg and $L_v = 22.6 \times 10^5$ J/kg from Table 12.3, we find that

$$\frac{\text{Absorbed dose}}{\text{in grays}} = \frac{\text{Energy}}{\text{mass}} = \frac{mL_f + cm\Delta T + mL_v}{m} = L_f + c\Delta T + L_v$$

$$= 33.5 \times 10^4 \text{ J/kg} + \left[4186 \text{ J/}\left(\text{kg} \cdot \text{C°}\right)\right]\left(100.0 \text{ C°}\right) + 22.6 \times 10^5 \text{ J/kg}$$

$$= 3.01 \times 10^6 \text{ J/kg}$$

Using the fact that 0.01 Gy = 1 rd, we find that

$$\text{Absorbed dose} = \left(3.01 \times 10^6 \text{ Gy}\right)\left(\frac{1 \text{ rd}}{0.01 \text{ Gy}}\right) = \boxed{3.01 \times 10^8 \text{ rd}}$$

---

11. *REASONING AND SOLUTION*   When a nitrogen $^{14}_{7}$N nucleus absorbs a deuterium $^{2}_{1}$H nucleus during a nuclear reaction, the compound nucleus formed will have $A = 14 + 2 = 16$, and $Z = 7 + 1 = 8$. Reference to the periodic table reveals that the compound nucleus must, therefore, be

$$\boxed{\text{oxygen } ^{16}_{8}\text{O}}$$

---

15. *REASONING*   Energy is released from this reaction. Consequently, the combined mass of the daughter nucleus $^{12}_{6}$C and the α particle $^{4}_{2}$He is less than the combined mass of the parent nucleus $^{14}_{7}$N and $^{2}_{1}$H. The mass defect is equivalent to the energy released. We proceed by determining the difference in mass in atomic mass units and then use the fact that 1 u is equivalent to 931.5 MeV (see Section 31.3).

SOLUTION The reaction and the atomic masses are as follows:

$$\underbrace{{}^{2}_{1}\text{H}}_{2.014\,102\,\text{u}} + \underbrace{{}^{14}_{7}\text{N}}_{14.003\,074\,\text{u}} \rightarrow \underbrace{{}^{12}_{6}\text{C}}_{12.000\,000\,\text{u}} + \underbrace{{}^{4}_{2}\text{He}}_{4.002\,603\,\text{u}}$$

The mass defect Δm for this reaction is

$$\Delta m = 2.014\,102\text{ u} + 14.003\,074\text{ u} - 12.000\,000\text{ u} - 4.002\,603\text{ u} = 0.014\,573\text{ u}$$

Since 1 u = 931.5 MeV, the energy released is

$$(0.014\,573\text{ u})\left(\frac{931.5\text{ MeV}}{1\text{ u}}\right) = \boxed{13.6\text{ MeV}}$$

---

19. *REASONING* The rest energy of the uranium nucleus can be found by taking the atomic mass of the ${}^{235}_{92}\text{U}$ atom, subtracting the mass of the 92 electrons, and then using the fact that 1 u is equivalent to 931.5 MeV (see Section 31.3). According to Table 31.1, the mass of an electron is $5.485\,799 \times 10^{-4}$ u. Once the rest energy of the uranium nucleus is found, the desired ratio can be calculated.

SOLUTION The mass of ${}^{235}_{92}\text{U}$ is 235.043 924 u. Therefore, subtracting the mass of the 92 electrons, we have

$$\text{Mass of }{}^{235}_{92}\text{U nucleus} = 235.043\,924\text{ u} - 92(5.485\,799\times10^{-4}\text{u}) = 234.993\,455\text{ u}$$

The energy equivalent of this mass is

$$(234.993\,455\text{ u})\left(\frac{931.5\text{ MeV}}{1\text{ u}}\right) = 2.189\times10^{5}\text{ MeV}$$

Therefore, the ratio is

$$\frac{200\text{ MeV}}{2.189\times10^{5}\text{ MeV}} = \boxed{9.0\times10^{-4}}$$

---

23. *REASONING* We first determine the energy released by 1.0 kg of ${}^{235}_{92}\text{U}$. Using the data given in the problem statement, we can then determine the number of kilograms of coal that must be burned to produce the same energy.

SOLUTION The energy equivalent of one atomic mass unit is given in the text (see Section 31.3) as

$$1\text{ u} = 1.4924\times10^{-10}\text{ J} = 931.5\text{ MeV}$$

Therefore, the energy released in the fission of 1.0 kg of $^{235}_{92}$U is

$$(1.0 \text{ kg of } ^{235}_{92}\text{U}) \left(\frac{1.0 \times 10^3 \text{ g/kg}}{235 \text{ g/mol}} \right) \left(\frac{6.022 \times 10^{23} \text{ nuclei}}{1.0 \text{ mol}} \right)$$

$$\times \left(\frac{2.0 \times 10^2 \text{ MeV}}{\text{nuclei}} \right) \left(\frac{1.4924 \times 10^{-10} \text{ J}}{931.5 \text{ MeV}} \right) = 8.2 \times 10^{13} \text{ J}$$

When 1.0 kg of coal is burned, about 3.0×10^7 J is released; therefore the number of kilograms of coal that must be burned to produce an energy of 8.2×10^{13} J is

$$m_{\text{coal}} = \left(8.2 \times 10^{13} \text{ J} \right) \left(\frac{1.0 \text{ kg}}{3.0 \times 10^7 \text{ J}} \right) = \boxed{2.7 \times 10^6 \text{ kg}}$$

---

27. $\boxed{\text{WWW}}$ **REASONING**   We first determine the total power generated (used and wasted) by the plant. Energy is power times the time, according to Equation 6.10, and given the energy, we can determine how many kilograms of $^{235}_{92}$U are fissioned to produce this energy.

SOLUTION   Since the power plant produces energy at a rate of 8.0×10^8 W when operating at 25 % efficiency, the total power produced by the power plant is

$$\left(8.0 \times 10^8 \text{ W} \right) 4 = 3.2 \times 10^9 \text{ W}$$

The energy equivalent of one atomic mass unit is given in the text (see Section 31.3) as

$$1 \text{ u} = 1.4924 \times 10^{-10} \text{ J} = 931.5 \text{ MeV}$$

Since each fission produces 2.0×10^2 MeV of energy, the total mass of $^{235}_{92}$U required to generate 3.2×10^9 W for a year (3.156×10^7 s) is

$$\overbrace{\left(3.2 \times 10^9 \text{ J/s} \right) \left(3.156 \times 10^7 \text{ s} \right)}^{\text{Power times time gives energy in joules}}$$

$$\times \underbrace{\left(\frac{931.5 \text{ MeV}}{1.4924 \times 10^{-10} \text{ J}} \right)}_{\substack{\text{Converts joules to} \\ \text{MeV}}} \underbrace{\left(\frac{1.0 \ ^{235}_{92}\text{U nucleus}}{2.0 \times 10^2 \text{ MeV}} \right)}_{\substack{\text{Converts MeV to} \\ \text{number of nuclei}}} \underbrace{\left(\frac{0.235 \text{ kg}}{6.022 \times 10^{23} \ ^{235}_{92}\text{U nuclei}} \right)}_{\substack{\text{Converts number of nuclei to} \\ \text{kilograms}}} = \boxed{1200 \text{ kg}}$$

31. **_REASONING_**   To find the energy released per reaction, we follow the usual procedure of determining how much the mass has decreased because of the fusion process. Once the energy released per reaction is determined, we can determine the mass of lithium $_3^6\text{Li}$ needed to produce 3.8×10^{10} J.

SOLUTION   The reaction and the masses are shown below:

$$\underbrace{_1^2\text{H}}_{2.014\text{ u}} \;+\; \underbrace{_3^6\text{Li}}_{6.015\text{ u}} \;\rightarrow\; \underbrace{2\,_2^4\text{He}}_{2(4.003\text{ u})}$$

The mass defect is, therefore, 2.014 u $+ 6.015$ u $- 2(4.003$ u$) = 0.023$ u. Since 1 u is equivalent to 931.5 MeV, the released energy is 21 MeV, or since the energy equivalent of one atomic mass unit is given in Section 31.3 as $1\text{ u} = 1.4924 \times 10^{-10}$ J $= 931.5$ MeV ,

$$(21\text{ MeV})\left(\frac{1.4924 \times 10^{-10}\text{ J}}{931.5\text{ MeV}} \right) = 3.4 \times 10^{-12}\text{ J}$$

In 1.0 kg of lithium $_3^6\text{Li}$, there are

$$(1.0\text{ kg of }_3^6\text{Li})\left(\frac{1.0 \times 10^3\text{ g}}{1.0\text{ kg}} \right)\left(\frac{6.022 \times 10^{23}\text{ nuclei/mol}}{6.015\text{ g/mol}} \right) = 1.0 \times 10^{26}\text{ nuclei}$$

Therefore, 1.0 kg of lithium $_3^6\text{Li}$ would produce an energy of

$$(3.4 \times 10^{-12}\text{ J/nuclei})(1.0 \times 10^{26}\text{ nuclei}) = 3.4 \times 10^{14}\text{ J}$$

If the energy needs of one household for a year is estimated to be 3.8×10^{10} J, then the amount of lithium required is

$$\frac{3.8 \times 10^{10}\text{ J}}{3.4 \times 10^{14}\text{ J/kg}} = \boxed{1.1 \times 10^{-4}\text{ kg}}$$

35. **_REASONING AND SOLUTION_**   The lambda particle contains three different quarks, one of which is the up quark u, and contains no antiquarks. Therefore, the remaining two quarks must be selected from the down quark d, the strange quark s, the charmed quark c, the top quark t, and the bottom quark b. Since the lambda particle has an electric charge of zero and since u has a charge of $+2e/3$, the charges of the remaining two quarks must add up to a total charge of $-2e/3$. This eliminates the quarks c and t as choices, because they each have a

charge of +2e/3. We are left, then, with d, s, and b as choices for the remaining two quarks in the lambda particle. The three possibilities are as follows:

$$\boxed{(1) \; u,d,s \qquad (2) \; u,d,b \qquad (3) \; u,s,b}$$

---

39. $\boxed{\text{WWW}}$ **REASONING** The momentum of a photon is given in the text as $p = E/c$ (see the discussion leading to Equation 29.6). This expression applies to any massless particle that travels at the speed of light. In particular, if the neutrino has no mass and travels at the speed of light, it applies to the neutrino. Once the momentum of the neutrino is determined, the de Broglie wavelength can be calculated from Equation 29.6 ($p = h/\lambda$).

SOLUTION
a. The momentum of the neutrino is, therefore,

$$p = \frac{E}{c} = \left(\frac{35 \text{ MeV}}{3.00 \times 10^8 \text{ m/s}} \right)\left(\frac{1.4924 \times 10^{-10} \text{ J}}{931.5 \text{ MeV}} \right) = \boxed{1.9 \times 10^{-20} \text{ kg} \cdot \text{m/s}}$$

where we have used the fact that 1.4924×10^{-10} J = 931.5 MeV (see Section 31.3).

b. According to Equation 29.6, the de Broglie wavelength of the neutrino is

$$\lambda = \frac{h}{p} = \frac{6.63 \times 10^{-34} \text{ J} \times \text{s}}{1.9 \times 10^{-20} \text{ kg} \cdot \text{m/s}} = \boxed{3.5 \times 10^{-14} \text{ m}}$$

---

43. $\boxed{\text{WWW}}$ **REASONING** The reaction given in the problem statement is written in the shorthand form: $^{17}_{8}\text{O} \, (\gamma, \alpha n) \, ^{12}_{6}\text{C}$. The first and last symbols represent the initial and final nuclei, respectively. The symbols inside the parentheses denote the incident particles or rays (left side of the comma) and the emitted particles or rays (right side of the comma).

SOLUTION Using the reasoning above and noting that an α particle is a helium nucleus, $^{4}_{2}\text{He}$, we have

$$\boxed{\gamma + \, ^{17}_{8}\text{O} \; \rightarrow \; ^{12}_{6}\text{C} + \, ^{4}_{2}\text{He} + \, ^{1}_{0}\text{n}}$$

---

45. **REASONING** To find the energy released per reaction, we follow the usual procedure of determining how much the mass has decreased because of the fusion process. Once the energy released per reaction is determined, we can determine the amount of gasoline that must be burned to produce the same amount of energy.

SOLUTION   The reaction and the masses are shown below:

$$\underbrace{3\ _{1}^{2}\text{H}}_{3(\,2.0141\ u\,)} \rightarrow \underbrace{_{2}^{4}\text{He}}_{4.0026\ u} + \underbrace{_{1}^{1}\text{H}}_{1.0078\ u} + \underbrace{_{0}^{1}\text{n}}_{1.0087\ u}$$

The mass defect is, therefore, $3(2.0141\ u) - 4.0026\ u - 1.0078\ u - 1.0087\ u = 0.0232\ u$. Since 1 u is equivalent to 931.5 MeV, the released energy is 21.6 MeV, or since it is shown in Section 31.3 that $931.5\ \text{MeV} = 1.4924 \times 10^{-10}$ J, the energy released per reaction is

$$(21.6\ \text{MeV})\left(\frac{1.4924 \times 10^{-10}\ \text{J}}{931.5\ \text{MeV}}\right) = 3.46 \times 10^{-12}\ \text{J}$$

To find the total energy released by all the deuterium fuel, we need to know the number of deuterium nuclei present. The number of deuterium nuclei is

$$(6.1 \times 10^{-3}\ \text{g})\left(\frac{6.022 \times 10^{23}\ \text{nuclei/mol}}{2.0141\ \text{g/mol}}\right) = 1.8 \times 10^{21}\ \text{nuclei}$$

Since each reaction consumes three deuterium nuclei, the total energy released by the deuterium fuel is

$$\frac{1}{3}\ (3.46 \times 10^{-12}\ \text{J/nuclei})(1.8 \times 10^{21}\ \text{nuclei}) = 2.1 \times 10^{9}\ \text{J}$$

If one gallon of gasoline produces 2.1×10^{9} J of energy, then the number of gallons of gasoline that would have to be burned to equal the energy released by all the deuterium fuel is

$$(2.1 \times 10^{9}\ \text{J})\left(\frac{1.0\ \text{gal}}{2.1 \times 10^{9}\ \text{J}}\right) = \boxed{1.0\ \text{gal}}$$

Notes

Notes

Notes

Notes